The Satellite TV Handbook

Anthony T. Easton is president of the Easton Corporation, a San Francisco engineering and market research firm specializing in developing high-technology products and services for US and international markets. Mr. Easton is also founding chairman of the Graduate Department of Telecommunications Management at Golden Gate University, the nation's first MBA program in the field.

Prior to joining Golden Gate University, Mr. Easton was a member of the faculties of Strayer College and The American University in Washington, DC. Previous assignments have included positions with the NASA Apollo project, Bendix Radio Corporation, Western Electric, General Electric Corporation, and National Entertainment Television, Inc.

Mr. Easton received his bachelor's degree in Engineering Sciences from Johns Hopkins University, and holds a dual-concentrate MS Degree in Management of Technology and International Business from The American University. Mr. Easton is a member of the Institute of Electrical and Electronic Engineers, the World Future Society, and the Explorers Club, among others. He is a current or past director of several organizations including Island Television Corporation, a low power tv operator, and Bank One International, a new federally chartered commercial bank. He is listed in *Who's Who in America*.

Mr. Easton has written numerous books and articles in the field of telecommunications management and futures technologies, and has conducted over 100 college-level courses in telecommunications, computer sciences, and applied management fields. He has spoken at symposiums and conferences throughout the world in the fields of high-technology and telecommunications.

Mr. Easton can be reached through The Satellite Center, P.O. Box 330045, San Francisco, CA 94133 (415/673-7000).

The Satellite TV Handbook

by
Anthony T. Easton

Howard W. Sams & Co., Inc.
4300 WEST 62ND ST. INDIANAPOLIS, INDIANA 46268 USA

International Standard Book Number: 0-672-22055-5
Library of Congress Catalog Card Number: 83-60155

Edited by *Welborn Associates*
Illustrated by *R. E. Lund and James Lentz*

Printed in the United States of America.

Preface

First run movies, championship boxing, live news when it happens, "secret" network television feeds—beamed direct from space to you. To your own personal backyard TVRO satellite dish antenna. . . . It is perfectly legal and downright cheap to install. Be the first in your neighborhood to own your own earth station. . . . Satellite television is here today, and it's red-hot! Over 150,000 homes throughout the USA and Canada now have their own small rooftop or backyard dishes which can pick up more than 90 television channels with razor-sharp clarity. Within the next few years, a million homes will be tuned in to television pictures from the satellites.

Literally hundreds of communications satellites circle this tiny planet of ours, relaying millions of telephone calls daily, and beaming television pictures from the four corners of the earth. And what television! Everything is up there, full-length feature films, sports 24 hours-per-day, round-the-clock news channels, rock music extravaganzas, religious progams, children's shows, cultural, political, and educational channels. Uncut and uncensored network television feeds and "prefeeds," international and foreign-language channels. And much, much more. . . . Satellite television is here today!

Moreover, the cost of buying or building your own backyard antenna has collapsed over the past few years. As recently as 1978, $15,000 (and up) was the going price for a satellite TVRO dish antenna. Today, the "do-it-yourself" satellite tv enthusiast can put together his whole TVRO satellite antenna for under $500! The average video hobbyist can walk out of the satellite TVRO dealers showroom with a good backyard package for $1900 to $2900. And, the prices are continuing to tumble downward.

This book covers the whole world of satellite television from top to bottom. How to buy your own system wholesale, what's on the birds, how to even go into business by starting your own "mini-CATV" company to share your backyard dish with your neighbors—and make some money doing it!

A detailed Apple®* II computer program is included to automatically find the birds anywhere in the world. So is a complete listing of satellite TVRO manufacturers, distributors, and dealers. A description of the equipment, how it all works, and how to put it together, along with over a half dozen appendices and hundreds of pictures and drawings to round out the story are all given in these pages.

The adventure of satellite television continues to grow. New secret audio and radio channels, hidden news services, and private videoconference feeds pop up every day. Dozens of news organizations shuttle their stories back and forth from coast-to-coast and continent-to-continent via the television "birds." Over 13 US and Canadian satellites are now carrying television feeds, another 10 or so operate internationally. This number will double within the next few years. There is a world of satellite television at your fingertips. It is as near as your own tv set and your rooftop dish. Welcome to the adventure!

ANTHONY TERRY EASTON

To "Muffin," who taught me the meaning of Happiness

*Apple is a registered trademark of Apple Computer, Inc.

Contents

CHAPTER 1

The Satellite and Arthur C. Clarke: How It All Began

INTRODUCTORY HISTORY OF SATELLITE TECHNOLOGY

It all began in ancient China some 2000 years before the birth of Christ. By then, the Chinese artisans were already launching the first rudimentary rockets, putting their festive fireworks displays into earth suborbits of several hundred feet or so. This "state-of-the-art" in satellite launching technology was to remain a constant until the basis of modern celestial mechanics was developed by Sir Issac Newton at the end of the 17th century. With the creation of calculus and advances in the science of astronomy, the necessary theoretical tools needed to launch an artificial satellite into earth orbit were available. But, another 250 years would pass before rockets were developed with enough power to test these theories. In 1903, the Russian physicist, K. E. Tsiolkovsky, published a paper on the use of high energy liquid-fuel rockets. In 1926 the American scientist, Robert H. Goddard, launched the first liquid propellant rocket, and the space race was on.

The Germans recognized the military significance of the rocket, and built thousands of "V-2s" which rained death and destruction on a wartime London. As the 50s rolled around, both the Americans and Russians were using the talents of the German rocket scientists to develop and construct true intercontinental missiles.

The Russians were first in space with the Sputnik I, a soccer-ball sized satellite which was orbited in 1957. Carrying only a tiny radio transmitter which lasted but a few days, Sputnik's impact far exceeded its meager capabilities. Within months, the US launched the Explorer I satellite, a far more sophisticated device carrying instruments which detected the previously unknown Van Allen radiation belt. The space race was on in earnest.

SATELLITES AND THE SIXTIES

By the early 60s, manned flights were being prepared in both the United States and Russia. Astronaut John Glenn flew the first orbital

Mercury mission; this series of flights lead rapidly to the Gemini and Apollo programs, and the construction of the world's largest liquid fuel rocket: the Saturn V-1B. The Russians provided a parallel course of development through their Vostok, Salyut, and Soyuz missions, although they were never able to build sufficiently large rockets and life support systems to land a man on the moon.

The extraordinary advantages of satellite technology were apparent on both sides of the iron curtain. The military sought surveillance devices which could fly higher and faster than a supersonic airplane. Cartographers hoped to accurately map the globe since the precise shape of the earth was then unknown. The Navy wanted to establish a sophisticated global navigational system for use by ships at sea, and weather scientists dreamed of seeing a "bird's eye view" of hemispheric and global weather patterns. Geologists were interested in mapping oil and mineral deposits through the use of on-board satellite cameras and detectors. Agricultural experts hoped to analyze plant growth, and develop methods of predetermining crop yields through the use of infrared photography.

The world telecommunication authorities and International Record Carriers looked to a communications satellite revolution, as did governments on both national and international levels.

All of these dreams were quickly fulfilled. By the mid-1960s, hundreds of satellites had been launched, and a half dozen countries, including the United States, the USSR, France, China, India, and Italy were operating their own satellite systems. National pride, the initial motivation, soon gave way to economic necessity as both the developed and "third world" looked upon satellite technology as a cornucopia of new wealth and riches. Three of the many uses of satellites in communications are shown in Fig. 1—1.

HOW A SATELLITE IS LAUNCHED

How does a one ton box of transistors, fuel cells, stabilization motors, and sensors get lifted into space? How is it maneuvered into a precise earth orbit under the command of ground-based controllers?

The answer lies in the physical laws of gravity and orbital mechanics, as described by Newton in 1687. The attraction of two bodies toward each other is called the gravitational force, and is measured in units of weight such as pounds or kilograms. To put 2000 pounds into earth orbit, rocket motors must provide a continuous thrust to lift the object several hundred or more miles straight up, providing an acceleration sufficient to counter the constant 32 feet-per-second per second downward pull of the earth.

The rocket motors must accelerate to a speed of several thousand miles per hour to accomplish this task. At a velocity of about seven miles

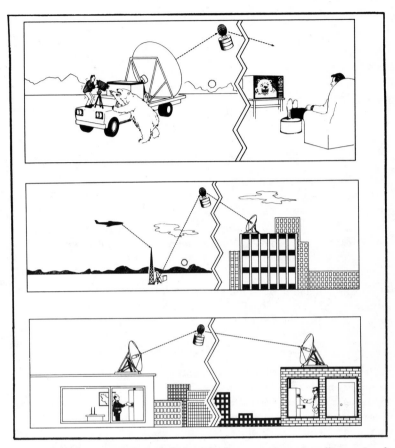

Fig. 1-1. Communications satellites can be used for (from top to bottom) satellite tv, aircraft communications, and the transmission of electronic mail.
(*Courtesy Telsat Canada*)

per second (roughly 25,000 miles per hour), a rocket launched from the surface of the earth will successfully escape the gravitational pull of the earth, and sail off into space on an eternal cosmic journey (Fig. 1—2).

Low earth-orbit satellites, traveling many thousands of miles per hour, circle the planet dozens of times per day speeding far faster than the earth turns on its daily rotation. A military surveillance satellite 100 miles up takes about 90 minutes to complete each orbit. At altitudes greater than 23,000 miles above the earth a satellite takes more than 24 hours to completely circle the earth. The orbit which coincides with the daily rotation of the earth is called the *geosynchronous orbit* or *Clarke orbit*, in honor of the science fiction writer Arthur C. Clarke. This circular orbit,

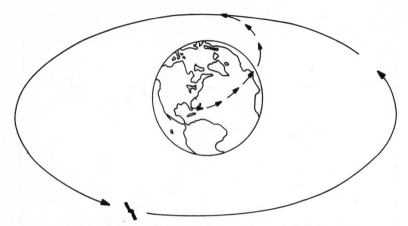

Fig. 1-2. An earth-orbit when the initial launch velocity of the satellite is below 25,000 mph. Above 25,000 mph the spacecraft will fly off into deep space, breaking free of the earth's gravitational field.

which cuts a plane through the Equator, is about 22,300 miles above the surface of the earth. Launching of such a satellite is a complicated procedure. Fig. 1–3 shows the launch procedure.

CLARKE'S GEOSYNCHRONOUS ORBIT

Clarke first described this very special orbit in an article he wrote for the British publication *Wireless World*, published in 1945. In it, he outlined the possibilities for establishing a global communications system using three satellites placed in geosynchronous orbits at equal distances from each other. Clarke observed that these three satellites would appear to remain entirely motionless in space to observers viewing them from the planet's surface (Fig. 1–4). As the earth rotates beneath them, the satellites would also fly around the planet, completing one daily orbital revolution in precise synchronization with the 24-hour rotation of the earth.

Since Clarke was a science fiction writer, most of the scientific community at the time chalked up this idea as but an interesting fantasy. After all, long distance telephone circuits were quite crude in those days. Calls were relayed from Europe to North America using unreliable and static-prone high-frequency radio signals bounced off the ionosphere. The first transatlantic telephone cable was not laid until 1956, well over a decade after Clarke's article appeared.

A number of scientists from the Bell Telephone Laboratories in New Jersey were inspired by the idea of using satellites to communicate. This group set out to develop a nongeosynchronous satellite communications

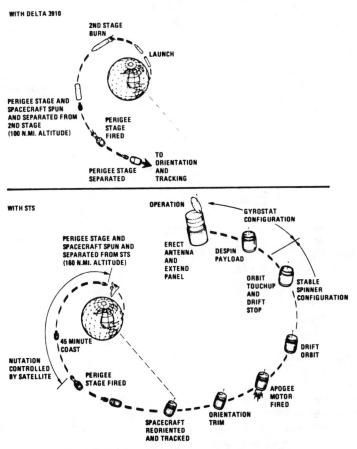

Fig. 1-3. The launching of a communications satellite.
(Courtesy Hughes Corp.)

system. On July 10, 1962, the tiny Telstar I satellite was launched from Cape Canaveral, Florida. Weighing only a few dozen pounds, the satellite was carried aloft by a modified Thor-Delta rocket. Its orbit was quite low, and massive ground-station antennas were needed to track the satellite as it whizzed by overhead, but Telstar brought the first live television pictures from Great Britain to the United States. Since Telstar was in a polar orbit, it was only in the proper position over the Atlantic for a few minutes at a time so the period of time pictures could be sent was very limited.

The Telstar satellite inspired the imagination of John F. Kennedy, and captured the attention of Congress, which proceeded to adopt the

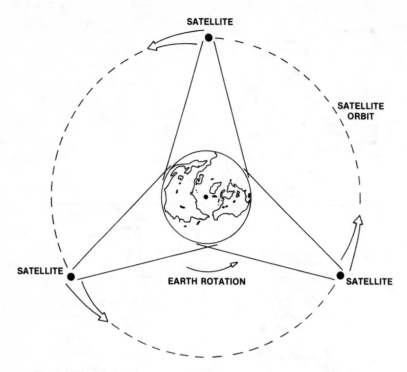

Fig. 1-4. The Clarke orbit as seen from a vantage point above the North Pole.

Communications Satellite Act immediately following the Telstar success. The Act created Comsat, a private corporation to develop satellite systems. The Intelsat series of satellites are placed into geosynchronous orbits using American-developed rockets launched by NASA at its Cape Kennedy facility. The international satellites are placed in key positions over the Atlantic, Pacific, and Indian Oceans, under control from Intelsat's operations center at L'Efant Plaza in Washington, DC. At present, over a dozen Intelsat satellites provide telephone, telex, and television services for public and private communications to all points on the face of earth.

RADIO WAVES AND TELEVISION SATELLITES

The communications satellite is an electronic "mirror" aimed earthward from a specific point in space. The satellites utilize radio waves operating in the microwave band to relay signals which are "uplinked" from earth, and "downlinks" them back to the planet.

All physical matter vibrates at some periodic frequency. The law of

Quantum Mechanics points out that even the smallest subatomic particles are constantly in motion. Oscillating waves form the basis for all human and electronic communications. Sine waves, the most common waveform, are named for the repeating geometric pattern which describes their motion. A musical instrument or human vocal cord produces a series of air pressure waves and troughs in a sine-wave pattern (Fig. 1–5A). In the human hearing range, these waves range in frequency from 20 hertz* for the lowest tones to about 20,000 hertz for the highest musical pitches. Radio waves (Fig. 1–5B), on the other hand, extend to millions of hertz.

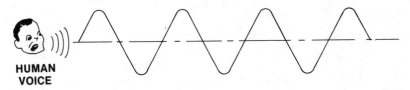

HUMAN VOICE

(A) Audio wave.

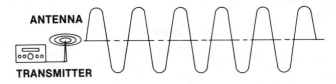

ANTENNA

TRANSMITTER

(B) Radio wave.

Fig. 1-5. Sine waves.

Sine waves can be described by their properties as shown in Fig. 1–6. The frequency is the number of complete oscillations or cycles that occur per second (called the "pitch" of an audio wave). The amplitude describes the power level of a sine wave (known as the "loudness" of an audio sine wave). The phase describes the relationship over time that two sine waves of identical amplitude and frequency can have with each other, and can vary from 0 to 360 degrees. By systematically manipulating these three properties, useful information can be transmitted over long distances. This process is called "modulation."

Radio waves are sine waves produced electromagnetically through the interaction of perpendicular magnetic and electrical fields. Radio waves are generated by an oscillator/amplifier known as a radio

*Hertz is the term used to describe cycles per second. Thus 1 hertz equals 1 cycle per second; 20,000 hertz equals 20,000 cycles per second. Higher frequencies are usually designated by kilohertz (kHz) or megahertz (MHz). 1 kHz equals 1,000 Hz; 1 MHz equals 1,000,000 Hz.

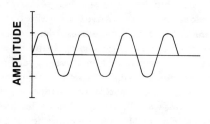

(A) Low-amplitude signal.

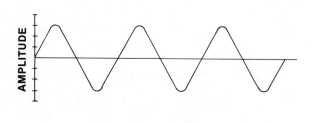

(B) Low-frequency signal.

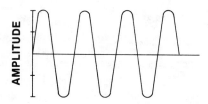

(C) High-amplitude signal.

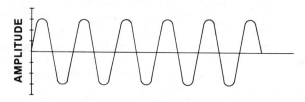

(D) High-frequency signal.

Fig. 1-6. Frequency and amplitude of sine waves.

transmitter, and are radiated into the atmosphere via an antenna. Low-frequency radio waves which have very long wavelengths are often used to communicate with submarines under water, and can penetrate hundreds of feet below the surface of the oceans. Medium-frequency radio waves carry the standard am (amplitude modulation) broadcast stations whose transmitters vary the output amplitude or power of the signal in direct relation to the announcers voice or a musical instrument. Very-high-frequency radio waves are used by the fm (or frequency modulation) broadcasting stations, where the frequency of the radio transmission is shifted up and down in step with the audio program coming from the studio.

Vhf (very high frequency) and uhf (ultrahigh frequency) radio waves are used to carry the North American television Channels 2 through 13, and 14 through 70 respectively. Terrestrial tv broadcast stations commonly vary both the amplitude and frequency of the transmitted signal. Each tv station operates two separate transmitters. The video transmitter is amplitude modulated with the visual picture information. The aural transmitter is frequency modulated with the corresponding audio soundtrack. At these megahertz frequencies, most random, natural, and man-made noise is of the impulse kind, which can affect am transmission schemes. Fm systems, however, are relatively immune to atmospheric and man-made noise. Thus, during an electrical storm, the television fm sound will remain static free, while the am picture jumps and breaks up with each lightning flash.

Am radio operates in a frequency band from 535,000 Hz (or 535 kilohertz) to 1600 kHz and the fm radio band ranges from 88 to 108 megahertz (MHz). Vhf television, Channel 2, begins at a frequency of 54 MHz and extends upwards to 60 MHz where vhf Channel 3 is located. Each television channel consumes a bandwidth of six megahertz, the amount of spectrum needed to transmit the rapidly changing color picture information. The upward progression of vhf tv channels is interrupted by the fm radio band which occupies the space between Channels 6 and 7 on a North American television set. Uhf television channels begin at 470 MHz (Channel 14) and run up to 806 MHz (Channel 70). The older Channels 71 through 83, which are still present on many television receivers, have been reassigned for use by mobile radiotelephone and other nontelevision services. In North America, the specifications of the commercial television system were developed years ago by the National Television Standards Commission; thus the system is known as the NTSC standard. This system is incompatible with the European PAL and SECAM systems in both channel spacing and modulation techniques.

Microwave frequencies begin at about 1,000 MHz (1 gigahertz) and rise upwards in frequency to the light-wave portion of the electromagnetic spectrum. As wavelengths become shorter and shorter, they begin to take on more and more of the properties of visible light waves;

they can be reflected by mirrors, and are attenuated by rain. They travel by line of sight. and do not readily scatter, bounce, or bend in their travels over the surface of the earth. Communication satellites operate at microwave frequencies ranging from 3½ to 6 gigahertz (known as the "C-Band"), and from 12 to 14 gigahertz (known as the "Ku-Band"). The communications satellite (Fig. 1—7) is simply an electronic repeater located rather far (22,300 miles) away. It can be seen visibly by telescope, and its microwave signals can be picked up by a backyard parabolic dish antenna.

Medical diathermy equipment, microwave ovens, and terrestrial radar systems also use frequencies operating in the microwave portion of the electromagnetic spectrum. Telephone companies also route the vast majority of continental telephone calls over microwave circuits. The Bell toll network uses terrestrial microwave transmission repeaters scattered at 40 mile intervals across the country. Because of this enormous use of microwave systems, the average backyard is awash in a veritable soup of microwave radiation. These spurious signals can interfere with each other and the home satellite tv viewer. Luckily, it is usually

Fig. 1-7. The Westar IV communications satellite launched June 8, 1982. It has a 24-transponder capacity, a design life of 10 years, and is located at 99° west latitude.
(Courtesy Western Union)

easy to eliminate such interference by simply moving the dish antenna a few feet or so in any direction. Table 1–1 gives a summary of the electromagnetic spectrum.

NETWORK TELEVISION AND
THE RISE OF SATELLITE COMMUNICATIONS

For over 30 years the US television networks have used the Bell Telephone backbone microwave network to carry programming from their New York City control centers to their affiliate tv stations nationwide. Over the decades, these terrestrial communication networks have become very sophisticated. Automatic switching centers can instantly connect or disconnect any tv station from any network feed in a matter of seconds. By pressing a few buttons on computerized consoles, managers at AT&T's network control center in midtown New York can instantly reconfigure the Bell national television microwave network. A single NFL Monday night football game can be delivered to hundreds of ABC affiliates, or dozens of simultaneous Sunday games can be fed to multiregional NBC and CBS "mininetworks."

Big time television networking is sophisticated and expensive. One television channel consumes spectrum that could otherwise carry 2000 telephone calls simultaneously. The cost to operate the telecommunications facilities for each of the three US television networks runs upwards of $100 million per year. And, until the launch of the domestic communications satellites by Comsat in the mid-1970s, no other options were available. The Canadians were the first to recognize the cost savings of a satellite television system. By 1972, Telsat Canada had built and launched the first Anik satellite system. CBC broadcasts could reach the most distant outback in the Northwest Territories through relatively inexpensive television receive-only (TVRO) earth stations. Each TVRO terminal cost about $50,000 at that time.

Both the RCA Corporation and Western Union quickly capitalized the idea by constructing their own Satcom and Westar satellite systems to

Table 1-1. The Electromagnetic Spectrum

Band	Approximate Frequencies	Usage
Very Low Frequencies (VLF)	50,000–100,000 Hz	Submarine Communications
Low Frequencies (LF)	1000,000 HZ–535 kHz	Navigational Aids
Medium Frequencies (MF)	535 kHz–1600 kHz	Am Radio
High Frequencies (HF)	1.6–50 MHz	Shortwave Radio
Very High Frequencies (VHF)	60–216 MHZ	VHF TV (Channels 2–13)
Ultrahigh Frequencies (UHF)	470–806 MHz	UHF TV (Channels 14–70)
Microwaves (MW)	1–20 GHz	Satellite Communications
Light	100–10,000 GHz	Infrared and Visible Light

Charting The Satellites

from Space . . .

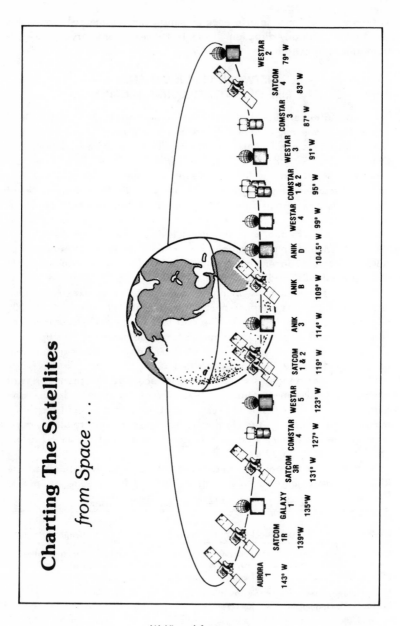

(A) Viewed from space.

Fig. 1-8. Location of
(Courtesy Satellite Channel

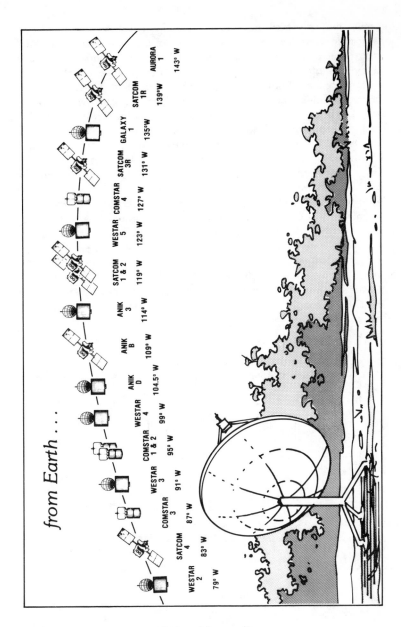

(B) Viewed from earth.

communication satellites.
Chart, Westsat Communications)

augment the terrestrial microwave networks. Each satellite could carry tens of thousands of data and telephone calls, and multiple tv pictures. Comsat which had previously focused upon international service only, had by the late 70s constructed and launched a series of domestic Comstar satellites. Comsat leased these exclusively to AT&T and GTE for national toll telephone use. As the new decade rolled around, over a dozen US and Canadian communication satellites were operating, beaming dozens of national television network feeds, "ad hoc" broadcasts, teleconferences, and cable television programs back to earth from space. Fig. 1–8 shows the location of 15 satellites.

The original communications satellites carried 12 separate signal repeaters or "transponders" on board, each capable of relaying 2000 telephone conversations or one full color television picture. By careful reuse of these scarce microwave frequencies through the technique of vertical and horizontal polarization, the newer communication satellites now operate with 24 transponders, providing a simultaneous capacity of 24 television channels. Fig. 1–9 shows the frequencies and polarization of the popular Hughes-built series of satellites. Figs. 1–10 and 1–11 show the component locations on two satellites. These satellites operate in the C-Band range. A separate series of Ku-Band satellites designed to carry from three to 12 television channels is being readied for launch in the mid-1980s.

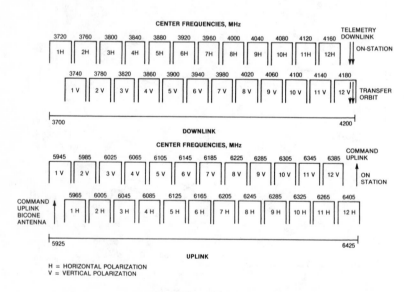

FREQUENCY AND POLARIZATION PLAN OF HUGHES COMMUNICATIONS SATELLITES

Fig. 1-9. Transponder frequency vs polarization for Hughes-built satellites.
(Courtesy Hughes Corp.)

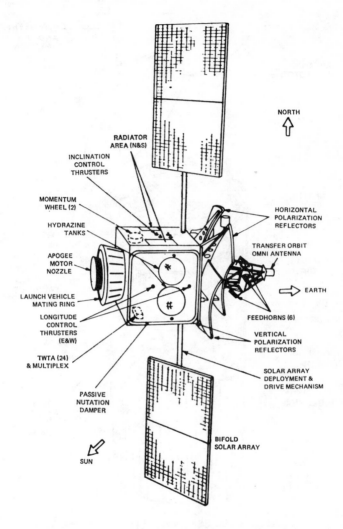

Fig. 1-10. Component locations on RCA Satcom 3 Satellite.
(Courtesy RCA America Communications, Inc.)

The major television networks were cautious, at first, about using the new satellite technologies. The terrestrial Bell network was expensive, but dependable. The US cable tv organizations were free to monopolize the capacity of the television satellites. One programmer in particular, Home Box Office (HBO) a subsidiary of Time-Life began in 1975 to use the RCA Satcom system to deliver its HBO pay tv movies to cable com-

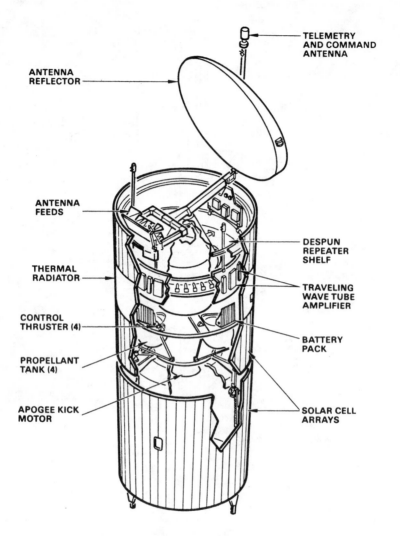

TELEMETRY AND COMMAND ANTENNA

ANTENNA REFLECTOR

ANTENNA FEEDS

DESPUN REPEATER SHELF

THERMAL RADIATOR

TRAVELING WAVE TUBE AMPLIFIER

CONTROL THRUSTER (4)

BATTERY PACK

PROPELLANT TANK (4)

APOGEE KICK MOTOR

SOLAR CELL ARRAYS

Fig. 1-11. Component locations on ANIK D Satellite.
(Courtesy Telsat Canada)

panies scattered throughout the country. Cable television had been around since the early 50s, but had been slow to grow. By the mid-70s, however, enough cities and towns had been wired by the cable companies to make it economically feasible for pay-tv programming to be uplinked from the New York control centers via the satellites to the cable systems TVRO, instead of being delivered on magnetic tape through the mail. The impact of the new satellite-based networks was felt immediately. New franchises sprung up overnight, and cable-tv stocks shot through the roof.

Distribution by satellite of national television programming via the satellites was proven to be far cheaper. A single satellite transponder was capable of delivering a television channel to any point on the continent, and it could be rented for $1 million a year, less than 1% of the "big three" television networks costs. Other satellite television services quickly followed, and by 1982 the old guard television networks finally announced that henceforth they would use a new AT&T "network bird" (the Comstar/Telstar series) to deliver their programming to their local affiliates. Satellite television had come of age.

CHAPTER 2

Cable Television Wires America

Dr. Lee de Forest invented the vacuum tube amplifier in 1906. By the 1920s several others had begun to experiment with these devices to transmit pictures electronically. In the 1930s a number of experimental television stations were operating in the United States and, when the San Francisco Worlds Fair opened in 1936, television was the hit of the show. Newspapers predicted that tv would sweep the US in months, but as the war clouds of World War II darkened the horizon, the industrial capacity of the nation was transferred from the civilian technologies to military projects. The full-scale introduction of commercial television would be delayed until the conclusion of the war.

By the end of the war, the old radio networks, CBS and NBC, were ready to capitalize upon the newer cathode-ray picture tubes and orthicon television cameras which had been developed in the military electronics boom. During the late 40s, experimental television stations were started along the eastern seaboard of the US, and pioneer tv sets were introduced by Dumont, Philco, and Bendix Radio. Post-war America took to television at once. Sears, the giant retailer, began to sell its own tv receivers and the radio networks formed national television counterparts to feed live tv programs from their New York City studios over AT&T's budding microwave network.

As the new decade rolled around, major US cities saw two and three stations start up in their markets, and the tv set became the new status symbol for that famous family—the Jones. But smaller towns, those that were down the road apiece from the big cities, also wanted tv. Nestled in the valleys and blocked by adjacent mountains, these suburban and semirural areas received spotty pictures at best. Television signals travel by line-of-sight and are easily reflected by any large flat object such as buildings, mountains, water towers, even statues as shown in Fig. 2–1. Ghosts result when the television receiver picks up both the original line-of-sight transmission and one or more of the undesirable reflected signals.

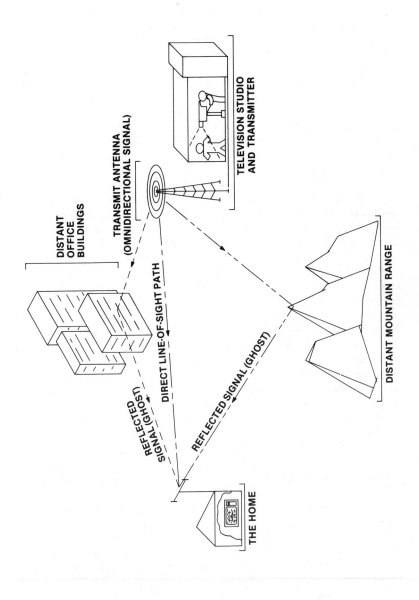

Fig. 2-1. Television line-of-sight and reflected signals.

Local engineers and business people took up the challenge. One such man was John Walson, who owned an appliance store in Mahanoy City, Pennsylvania. In 1947, when he began selling television sets, it was almost impossible to receive the three Philadelphia network stations located some 86 miles away from Mahanoy City because the town is surrounded by mountains. Even though Mahanoy City had a population of about 10,000, most of the early television customers came from communities on top of the mountains, such as Frackville, Vulcan, and Hazleton. Because of this situation, Mr. Walson erected an antenna tower on top of a nearby mountain so he that could take his customers to that location and demonstrate his television receivers. In searching for ways to increase his sales, in 1947, Mr. Walson decided to purchase some heavy-duty, twin-lead Army surplus wire. This wire was run on trees from a nearby mountain to Mr. Walson's appliance warehouse on the south side of town.

It was not until the late Spring of 1948, however, that a larger antenna tower on top of Boston Mountain was constructed and the Electro-Voice boosters (amplifiers) were added. The Electro-Voice television set boosters were modified by Mr. Walson and positioned every 500 feet to give 6 dB gain on 12 channels, although only three were then available. In June, Mr. Robert Gray, District Manager for the Pennsylvania Power & Light Company, gave Mr. Walson verbal permission to string his wire on PP&L poles in Mahanoy City. The line was then extended to the Walson appliance store at Main and Pine Streets. Along the way, the homes of Robert D. Gray, Dr. Aaron Liachowitz, Frank Boyle, George Barlow, and several others were connected to this community antenna system. Television sets were tuned to three television stations, Channels 3, 6, and 10. Many people congregated in front of the appliance store window to view the television pictures. Subscribers to the community antenna television system were initially charged an installation fee of $100 which included the first year rental. Thereafter, the monthly service rental was $2.00. In the process of selling his tv sets, Mr. Walson had started the nation's first community antenna television (CATV) system—the Service Electric Company. Today, Service Electric is the largest individually owned cable company and the 18th largest cable operator in the nation, running franchises in 154 communities with 150,000 prime subscribers.

Fig. 2—2 shows a typical system of the 50s. The CATV companies constructed hilltop "antenna farms" of television receiving antennas to pick up distant signals. These signals were then retransmitted on a different tv channel (to eliminate interference with the original broadcast signal). The new signal is amplified and sent along a coaxial cable track circuit from house to house. Such a system strengthens weak signals and eliminates ghosts.

As new cable companies sprang up, new manufacturers appeared to

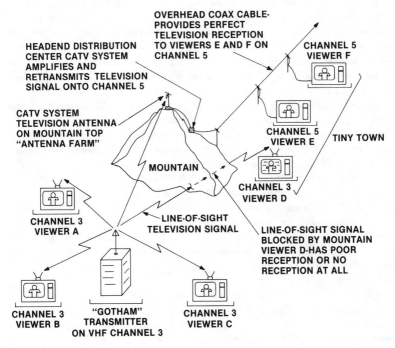

Fig. 2-2. A CATV system in the 1950s.

supply the fledgling industry with CATV* equipment. On May 18, 1950, Benjamin Tongue and Ike Blonder formed their new company with an initial capitalization of $5,000. Blonder-Tongue's initial product was the first commercially successful broadband tv booster amplifier. By amplifying the weak tv signals picked up from the distant cities, the Blonder-Tongue booster amplifiers improved tv reception in the outlying areas. Newly cabled towns could then enjoy television pictures which were better than the ghost-ridden images the city dweller obtained from his set-top rabbit-ear antennas.

CABLE TELEVISION AND THE REGULATORS

The Federal Communications Commission (FCC) in Washington has statutory control over all telecommunications and broadcasting facilities in the United States. As established by the Communications Act of 1934, over-the-air broadcasts and their reception are exempt from state and

---

*Originally the term CATV meant community antenna television; today the same CATV designation is used for cable television.

local regulation. Receive-only antenna systems are not regulated or licensed by the Federal Government. Unlike the United Kingdom, US viewers do not have to pay any annual tv receiver licensing fees. Private ownership of television stations and commercially supported programs eliminated the need to levy taxes for development of the system. The result is that the United States has over 1,000 television stations; Great Britain has only a few dozen.

When community antenna television systems were originally constructed, there were few federal regulations to hinder their expansion. However, local permission was required to operate in most jurisdictions, since the CATV company required the use of publicly owned streets and utility right-of-ways to wire up its franchise area. Cable rapidly became a sensitive local political issue. In 1965, the FCC established rules regarding CATV operations, particularly concerning the "importation" of distant television station signals by microwave. Additional rules established the next year required cable television companies to carry all local tv stations, and restricted the duplication of programs from distant cities. Although the cable companies fought desperately against these regulations, in June, 1968, the Supreme Court ruled that the FCC had the right to regulate cable television.

Today, cable television systems are treated much like utilities. Public hearings on every aspect of the cable business are common, and profits are tightly monitored, using the old utility concept of return on investment. Most of the tiny "mon and pop" cable companies have long since been acquired by Fortune 500 firms, whose giant legal staffs have the ability to follow the Byzantine cable regulations established by city and county governments. Little CATV, the stepchild of broadcast television has grown up to become today's cable television system—the multimillion dollar favored son of the investment banking community.

INSIDE CATV: THE TECHNOLOGY OF THE CABLE SYSTEM

The original NTSC (National Television Standards Commission) plan called for the creation of 13 television channels, ranging from roughly 50 MHz upwards in frequency. Problems in partitioning the post-war spectrum caused changes to be made almost immediately. First, the space reserved for Channel 1 was reassigned to other services, leaving Channels 2 through 13 remaining on the tv dial. This band was further divided into three segments: Channels 2 through 6, known as the "low band" vhf channels occupy the space from 54 MHz to 88 MHz with a 4-MHz gap between Channels 4 and 5. The "high band" vhf channels run from 174 to 216 MHz. The two vhf television bands are separated by the fm radio broadcasting band (88 to 108 MHz), and commercial and amateur radiocommunication services.

By the early 50s, it was evident that 12 channels would prove to be insufficient for national television growth, and 70 additional channels, designated 14 through 83, were set aside in a new ultrahigh frequency (uhf) band spaced contiguously from 470 MHz to 890 MHz. The FCC extended its mandate to cover the manufacturer of tv sets, requiring these companies to provide uhf as well as vhf tuners in each of the sets that they built. After this "chicken and egg" problem was solved, uhf growth took off. The last 14 uhf channels (70 through 83) were lightly used and were reallocated by the FCC for mobile telephone service in the early 70s. Although they no longer have a television function, the upper uhf channels still appear on many of the tv sets sold today.

The old limited-capacity twin-lead cable systems could only carry the vhf signals. To expand their capacity, seven extra "sub-band" channels were squeezed into frequencies located below Channel 2. An additional group of "mid-band" channels, labeled A through I, were inserted between Channels 6 and 7, beginning just above the fm broadcasting band at 120 MHz and running through 174 MHz. Few of the old community antenna television systems carried these additional channels, however. Most continued to feed the "standard 12" only.

The introduction of new coaxial cable technology saw the creation of an additional 17 "super band" channels, labeled J through Z, running from 216 through 318 MHz (Fig. 2—3). New super-high frequency coaxial cable has allowed for the recent creation of yet 14 more channels running from 318 MHz to 402 MHz. With installation of this new cable, 59 different tv programs and all fm radio channels can now be delivered to a subscribers home over a cable less than one-fourth inch in diameter. The conventional (broadcast) tv signals for uhf Channels 14 through 70 cannot be directly carried over cable systems. They are always converted to other channels by the cable system.

Since most of the current television sets are not able to tune all of these new channels, a converter is installed on the tv set to replace the internal tuner of the tv receiver. The signals fed through the cable are then converted by the convertor to an unused television channel— typically Channel 3. The output cable from the convertor is permanently connected to the antenna terminals of the tv set and the television is left permanently tuned to Channel 3. Fig. 2—4 shows four typical systems. A modern convertor used with the Qube system which also has facilities for the user to "talk back" to obtain instant viewer response is pictured in Fig. 2—5.

CATV TODAY

Subscriber demand for additional channels seemed to increase as fast as CATV operators could add new channels. Larger and taller antenna arrays were constructed. Independent common carriers were created to bring distant television station signals to the cable com-

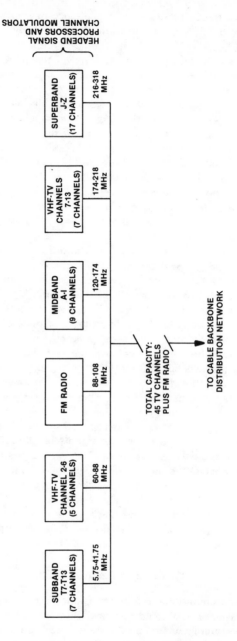

Fig. 2-3. A 45-channel cable system.

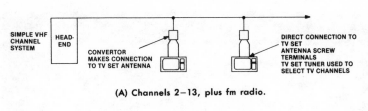

(A) Channels 2—13, plus fm radio.

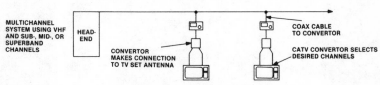

(B) Channels 2—13, T7—T13, A—I, and J—Z.

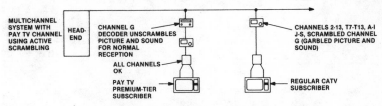

(C) Channels 2—13, T7—T13, A—I, and J—Z with pay tv on channel G.

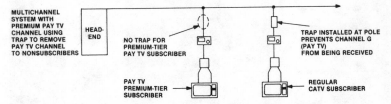

(D) Channels 2—13, T7—T13, A—I, and J—Z with scrambled
pay tv on channel G

Fig. 2-4. Four types of cable-tv systems.

panies. Master distribution centers transmitted signals via private microwave circuits. These networks reached 100 miles or more in length, as regional networks were formed, and distant cities began to exchange their signals. As the FCC relaxed its cable regulations, intrastate viewing and multistate consortiums became popular.

It was inevitable that cable television and the broadcasting community would eventually unite. CATV could deliver a perfect picture directly from the tv studio to the viewers homes. Broadcasting stations saw the

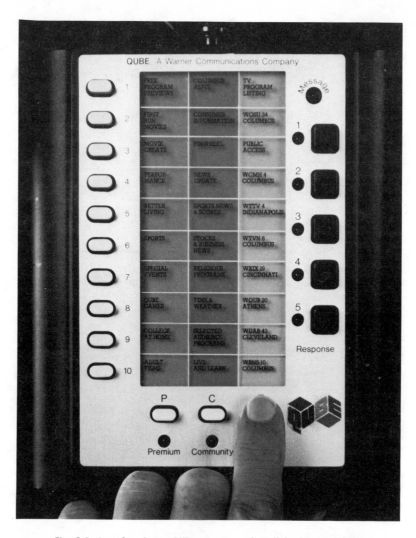

Fig. 2-5. A modern 2-way CATV converter with "talk-back" capabilities.
(Courtesy Warner/Amex Corp.).

CATV industry as a delivery vehicle of the highest quality and their old nemisis, the ghost, could be eliminated. Local CATV operators built community access studio facilities, and began installing video playback machines to run old movies. By the early 70s both Time-Life and Viacom International were providing first run movies to CATV subscribers by rolling tapes on videotape machines at the cable company headends.

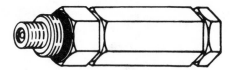

Fig. 2-6. Pay-tv signal trap

To prevent the "basic-cable" subscriber from viewing the premium movie service, a series of scrambling systems were developed. The least expensive method consists of placing the premium movie service on a sub-, mid-, or super-band channel which could not be received by a conventional tv set. The pay-movie subscriber needs a special decoder to extract the premium programming feed, by converting its frequency to a standard vhf channel tunable by the tv set.

Other scrambling techniques were developed. The premium service can be sent through the cable on a normal vhf channel. The signal is then "trapped" at the customer's house by using a filter or signal trap (Fig. 2–6). The trap is inserted in series with the drop wire at the "tap" point of connection to the main trunk line. The filter completely attenuates the pay-tv channel, and is removed when the customer subscribes to the additional service.

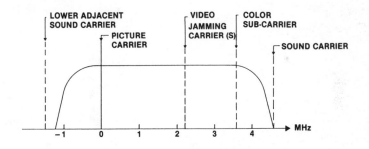

(A) Jamming carrier(s) in video channel.

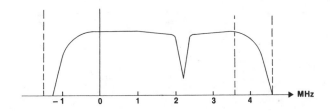

(B) Effect of trap to remove the jamming carrier(s).

Fig. 2-7. Adding a jamming signal to pay-tv channel.

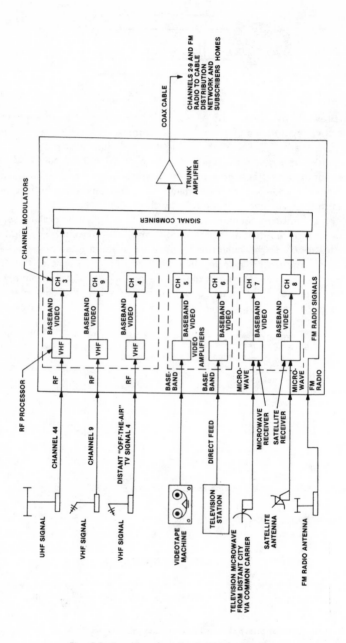

Fig. 2-8. A cable system headend installation.

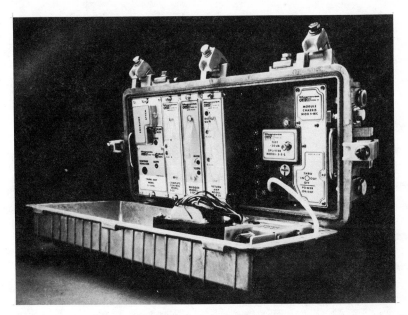

Fig. 2-9. A trunk amplifier
(Courtesy Magnavox CATV Systems, Inc.)

Another technique uses an interference signal simultaneously sent over the pay-movie channel along with the tv programming. When the customer subscribes to the pay-movie service, a "notch" filter is then installed at the customer's home to remove this obliterating signal as shown in Fig. 2—7.

The system illustrated in Fig. 2—8 utilizes all of the inputs discussed. Here off-the-air signals, video tape signals, a direct feed, microwave signals from an over land system, signals from a satellite receiver, plus fm radio signals are processed at the headend, or CATV offices. The signals are amplified and sent out over the system. Waterproof, line-mounted trunk amplifiers (Fig. 2—9) are inserted at various points in the system to amplify the signal to provide the proper signal to the various subscriber locations (Fig. 2—10).

CABLE GETS THE BIRD

By the time the first man landed on the moon, at the end of the 60s, the development of communication satellites was well under way. RCA and Western Union launched domestic satellites to relay telephone and data communications traffic nationwide. Western Union was first to loft its three-satellite system, designated as Westar-1, 2 and 3, each of

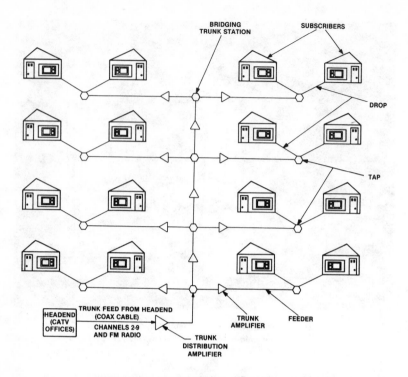

Fig. 2-10. A typical cable distribution network.

which carried 12 satellite transponders. RCA was quick to follow with two 24-transponder satellites known as Satcom 1 and 2. Through the early 70s, the technology was under used; a recession curtailed demand and transponders went begging for customers.

By 1975, Time-Life Films realized that the process of duplicating and "bicycling" hundreds of tapes nationwide for pay-cable use could be eliminated by simply beaming their HBO movie service to all of the cable companies simultaneously via the RCA Satcom 1 satellite. Viacom was quick to follow with its Showtime movie service. Satellite television had been born!

The first cable TVRO satellite earth stations were very expensive. FCC restrictions required the use of unnecessarily large antennas, 25 feet or more in diameter. Manufacturers of this type of equipment were limited; most systems were designed originally for use by the military. It was not uncommon to spend $50,000 to install a satellite antenna system. But the cost advantages were significant. One facility could simultaneously receive 24 television channels from space, and thousands of videotape recorders could be tossed out. Equipment maintenance budgets could be slashed. Earth stations were soon appearing at cable headends everywhere, and the cost to buy a TVRO system began to drop rapidly in price.

By 1980, dozens of other entrepreneurs had joined the HBO bandwagon and were using the television satellites to create new cable television networks. As the number of TVRO system installations increased and antenna costs fell, the number of-cable subscribers increased. The cabling of America, which had been sluggishly plodding along in the 60s, experienced a rebirth of activity. In the late 70s, the FCC modified the TVRO antenna regulations to allow for the use of 15-foot diameter dishes. A few years later the licensing requirement was eliminated altogether. The backyard satellite antenna business boomed as earth station costs descended from their stratospheric prices. Today, a home TVRO earth station costs well under $3000 as a complete package, and the technically skilled hobbyist can easily assemble a personal system for under $1000!

CHAPTER **3**

Cable TV Goes Into Space: What's on the Birds

Cable television has been the driving force behind the television satellite explosion. As the new 35, 50, and 100-channel CATV systems have appeared, dozens of program suppliers and television networks have sprung up to fill transponders of a half-dozen birds, beaming their signals earthward to cable television systems throughout the US and Canada.

At present, over 200 simultaneous channels of television programming can be carried by satellite in the United States; the Canadian Anik series can handle another 20 or so. By 1985, this capacity will have increased to over 600 transponders capable of carrying tv programming. Multiple picture-per-transponder, diplexing techniques, direct-to-home DBS broadcasting satellites, and new higher frequency satellites will expand this capacity to over 1000 tv pictures. Not all of these transponders will be allocated for television use; telephone, telex, and data communications users will account for much of the spare capacity. Also, corporate video teleconferencing will grow by a tenfold over the next few years.

Similar trends are occurring in Canada and overseas. Canadians have phased out the use of two older 12-transponder satellites, originally carrying 15 television channels of programming. Their new generation of Anik D satellites provides for a dozen CBC channels, pay-movie services, and four television "superstations" downlinked to cable companies throughout the nation.

Intelsat transponders are used to relay dozens of global television news feeds from network bureaus in London, New York, South America, and Asia. Visnet, the world's largest television newswire service, headquartered in London, feeds hours of programming each day both across the Atlantic and between Europe and Africa. Other Intelsat transponders are leased on a full-time basis for regional and domestic feeds throughout the world.

The Russians and Indonesians operate their own internal television networks with stations interconnected by a satellite. Arabsat, the Middle

East consortium, will feed Arabic television throughout the Middle East and to and from Europe by 1984. Several other countries plan to lease Intelsat regional spot transponder beams. Still others plan to launch their own domestic satellite systems within the next several years.

The major satellite television usage, however, is centered in North America, with over 80 different satellite television channels presently beaming their signals direct from space (Fig. 3–1).

TELEVISION SATELLITE PROGRAMMING

There are eleven different types of satellite programming which can be segmented by type: pay-tv channels, regional and national superstations, sports networks, news networks, religious channels, educational networks, public affairs feeds, commercially supported channels, the "Big 4" television networks, video conferences, and private communications networks (such as NASA's space shuttle video).

Fig. 3-1. Six of the many cable networks.

PAY-TELEVISION NETWORKS

Perhaps the best known of the pay-television networks is the Home Box Office (HBO) service with five active transponders on the RCA Satcom 3 satellite. HBO has over 9 million subscribers who pay between $8 and $20 per month to watch their movie service (Fig. 3–2). Time-Life, the owners of HBO, maintains an additional pool of six other transponders available for expansion as required. HBO provides a basic 24-hour-a-day first-run movie service with separate satellite feeds to the East and West coasts. A complementary "fill in" general family audience

TUESDAY, FEB. 1

6:00 am	CLOWN WHITE p.20
7:00	THE MAGIC OF LASSIE (G) p.19
9:00	STAR WARS (PG-2:01)
11:00	SUPERMAN II (PG) p.5
1:30 pm	VIDEO JUKEBOX
2:00	OLIVIA NEWTON-JOHN IN CONCERT p.16
3:30	THE MAGIC OF LASSIE (G) p.19
5:15	SUPERMAN II (PG) p.5
7:30	STAR WARS (PG-2:01)
9:30	NOT NECESSARILY THE NEWS p.6
10:00	GEORGE CARLIN AT CARNEGIE HALL p.6
11:00	NEIGHBORS (R) p.10
12:40 am	PENNIES FROM HEAVEN (R) p.9
2:30	STAR WARS (PG-2:01)
4:35	GEORGE CARLIN AT CARNEGIE HALL p.6

WEDNESDAY, FEB. 2

6:00 am	GULLIVER'S TRAVELS (G) p.20
7:30	FRAGGLE ROCK "You Can't Do That Without a Hat" p.12
8:00	CONVOY (PG) p.14
10:00	TAKE THIS JOB AND SHOVE IT (PG-1:40)
12:00 pm	MAKING LOVE BETTER p.15
12:30	HISTORY OF PRO FOOTBALL p.15
2:00	HBO MAGAZINE p.7
3:00	CONVOY (PG) p.14
5:00	FRAGGLE ROCK "You Can't Do That Without a Hat" p.12
5:30	GULLIVER'S TRAVELS (G) p.20
7:00	HBO MAGAZINE p.7
8:00	SHARKEY'S MACHINE (R-1:59)
10:00	FOUR FRIENDS (R) p.9
12:00 am	SLEEPING DOGS p.11
1:50	ROLLOVER (R) p.5
3:50	SHARKEY'S MACHINE (R-1:59)

THURSDAY, FEB. 3

6:00 am	DOT AND KANGAROO p.20
7:30	ALL SUMMER IN A DAY p.20
8:00	LIAR'S MOON (PG) p.14
10:00	STAR WARS (PG-2:01)
12:00 pm	GOLDEN RENDEZVOUS (1:42)
2:00	LIAR'S MOON (PG) p.14
4:00	AIR SUPPLY IN HAWAII p.16
5:00	THE BIG CATS p.20
6:00	DOT AND THE KANGAROO p.20
7:30	INSIDE THE NFL p.15
8:30	STAR WARS (PG-2:01)
10:30	GEORGE CARLIN AT CARNEGIE HALL p.6
11:30	LIAR'S MOON (PG) p.14
1:20 am	NOT NECESSARILY THE NEWS p.6
1:50	INSIDE THE NFL p.15
2:50	STAR WARS (PG-2:01)
4:55	GEORGE CARLIN AT CARNEGIE HALL p.6

FRIDAY, FEB. 4

6:00 am	GULLIVER'S TRAVELS (G) p.20
7:30	INSIDE THE NFL p.15
8:30	OLIVIA NEWTON-JOHN IN CONCERT p.16
10:00	THE BAD NEWS BEARS (PG-1:42)
12:00 pm	HBO MAGAZINE p.7
1:00	ACROSS THE GREAT DIVIDE (G) p.17
3:00	TAKE THIS JOB AND SHOVE IT (PG-1:40)
5:00	OLIVIA NEWTON-JOHN IN CONCERT p.16
6:30	FRAGGLE ROCK "You Can't Do That Without a Hat" p.12
7:00	INSIDE THE NFL p.15
8:00	MAKING LOVE (R-1:52)
10:00	NEIGHBORS (R) p.10
11:40	VICE SQUAD (R) p.16
1:20 am	OLIVIA NEWTON-JOHN IN CONCERT p.16
2:50	THE BAD NEWS BEARS (PG-1:42)
4:35	MAKING LOVE (R-1:52)

Fig. 3-2. HBO movie chanel schedule.

service, Cinemax is also uplinked to two separate transponders to provide for coverage across the various time zones.

Viacom Cablevision, the nation's largest cable multiple-system television operator (MSO), owns the Showtime pay movie channel. Showtime (Fig. 3–3) provides a program service similar to HBO, with simultaneous

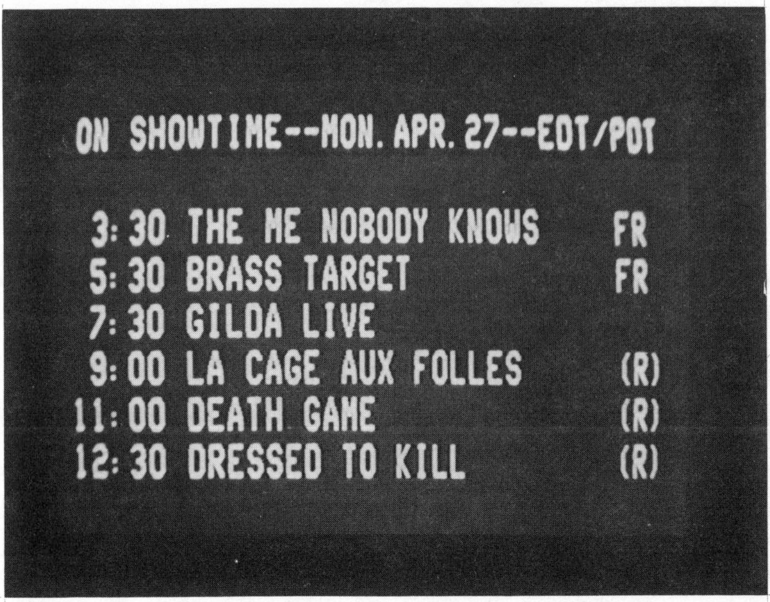

ON SHOWTIME--MON. APR. 27--EOT/POT

3:30 THE ME NOBODY KNOWS   FR
5:30 BRASS TARGET   FR
7:30 GILDA LIVE
9:00 LA CAGE AUX FOLLES   (R)
11:00 DEATH GAME   (R)
12:30 DRESSED TO KILL   (R)

Fig. 3-3. Showtime television schedule.

East and West transponder feeds on the Satcom 3 satellite, and like HBO, fills in its daily movie clock with special events and other made-for-cable programs. Warner-Amex, the 6th largest MSO, operates The Movie Channel (TMC), a 24-hour-per-day service which exclusively features first-run, genel-release movies in stereo. TMC (Fig. 3–4) is downlinked via a single Satcom 3 transponder for simultaneous carriage by the cable companies from coast to coast. The Home Theatre Network, a subsidiary of Group W, Westinghouse Broadcasting, operates a limited-service evening-hour G and PG movie service located on Satcom 3. Group W is America's third largest MSO cable company.

SelecTV, the successful Los Angeles subscription television (STV) station feeds its first-run pay-movie service to other over-the-air STV affiliates throughout the country. SelecTv may be found on the Westar 5 satellite.

A group of smaller pay tv movie services which deliver their programs to hospitals, MDS (multipoint distribution system) operators, and satellite master antenna television (SMATV) systems may also be found on the Satcom 4 and Comstar 4 satellites. Spotlight delivers first-run movies, concerts, and entertainment specials in stereo over Satcom 3.

Other pay television programmers specialize in "narrowcasting" to target audiences. GalaVision, owned by the Spanish International Network (SIN), feeds Spanish language movies and special events from the

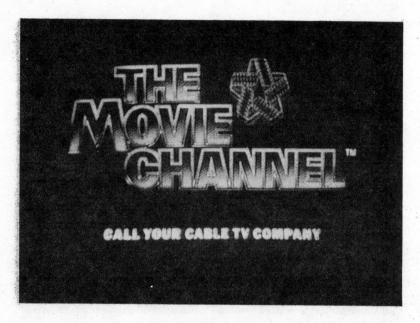

Fig. 3-4. The Movie Channel.

US, Mexico, South America, and Spain over Westar 4. The Black Entertainment Television Network (BET) provides a variety of programming featuring black actors and athletes, including sports events, music specials, and films. BET is delivered to dozens of cable companies via Westar 5.

Several adult programming services have carved out a nice market by providing sexually oriented entertainment via satellite. The Playboy Channel delivers its programs late nights via the Satcom 4 satellite. No X rated or explicit sexual programming is allowed to be fed via satellite, however. Federal Communications Commission regulations are somewhat ambiguous, but the Justice Department rules concerning pornography do apply to the use of Federally licensed facilities.

TV SUPERSTATIONS

A number of regional television stations have "gone national" by having their signal carried to cable television companies via satellite transmission. The cost to carry these popular old-movie stations is very low for the cable company (on the order of from $.10 to $.15 per subscriber per month. The superstations make their profits by charging higher commercial rates for their advertising due to their expanded audience base.

Ted Turner, who created the nation's first superstation, Channel 17 in

Fig. 3-5. WTBS-TV Atlanta superstation.

Atlanta, has seen WTBS-TV (Fig. 3—5) become one of the top five most-watched satellite-delivered cable channels. Ted's winning combination of classic movies purchased years ago at bargain basement prices plus exclusive coverage of the Turner-owned sports teams, has enabled him to finance another television satellite channel, The Cable News Network (CNN). WTBS operates 24 hours a day on Satcom 3.

WOR-TV, Channel 9 in New York City, is Gotham's independent station, delivering a variety of news and entertainment programming via Westar 5. WGN-TV, the independent Channel 9 superstation in Chicago, also operates on a 24-hour schedule, featuring classic movies and variety programming including the Phil Donahue Show, which originates in its studios. WGN-TV downlinks on Satcom 3.

XEW-TV, Mexico City, provides Spanish language programming on the Westar 4 satellite under a special Western Union tariff (Fig. 3—6). KUSK-TV, a country and western station in Prescott, Arizona will be uplinking to low power television (LPTV) stations in the future, as the FCC begins to issue LPTV licenses in the thousands.

THE SPORTS NETWORKS

Three major sports networks are fed over US satellites. The Entertainment and Sports Programming Network (ESPN) operates 24 hours a day,

Fig. 3-6. XEW-TV Mexico City.

bringing over 600 professional and collegiate games per year to sports enthusiasts as well as two hours of early morning business news. ESPN (Fig. 3—7), which is owned by Getty Oil, operates on the Satcom 3 satellite. Hughes Television Network delivers big name events and sports extravaganzas such as nationally known golf tournaments, to independent television stations. Hughes usually operates on both Westar 3 and Westar 4 satellites. The USA network, affiliated with Time-Life, is a commercially supported operation which carries the Thursday Night Baseball, Madison Square Garden events, and the NHL, NBA, and NASL ballgames. It is located on Satcom 3. During championship boxing matches, the Don King Sports and Entertainment Network may be found on Satcom 3. In addition to these specialized sports networks, each football weekend during the fall, the "Big 3" television networks also uplink most NFL football games via various satellites (Fig. 3—8). You will also find sports feeds from the major independent stations (Fig. 3—9).

THE NEWS NETWORKS

Cable News Network (CNN), Ted Turner's second brainchild (after Superstation WTBS), operates 24 hours a day from its Atlanta headquarters. CNN (Fig. 3—10) provides news, weather, and sports to over 14 million people and 1,000 cable networks, operating bureaus in Washington DC, New York, and Los Angeles, with live feeds hourly. A second

Fig. 3-7. ESPN sports network.

service, CNN Headline News (CNN-2), summarizes the key late-breaking news items also on a 24-hour clock providing "instant news capsules" throughout the day. CNN operates two transponders on the Satcom 3 satellite.

The Satellite News Channel (SNC), a Westinghouse-ABC joint programming venture, feeds five different transponders on the Westar 5 satellite with 24-hour per day regional news coverage, SNC operates as a headline service, summarizing the news several times each hour.

The Independent Network News (INN), operated by a consortium of independent vhf and uhf tv stations, uplinks an evening news program and special summaries via its New York City flagship station and Washington DC desk. INN Programming (Fig. 3—11) can be found on the Westar 3 satellite.

NBC, ABC, and CBS also lease satellite transponders for internal news coordination and for feeding both nightly news and prefeeding other programs, including soap operas (Fig. 3—12) and late night shows. These same transponders are often used for news bureau coordination, and interesting tidbits of conversations between the bureau personnel can often be heard on satellite. One to two hours prior to East Coast network news airtime, the private news video circuits are often humming with activity. The Comstar 3 satellite, AT&T's new "network bird," is handling more and more of the traffic, with other feeds occasionally appearing on Westar and Satacom birds.

ALL TIMES EASTERN

SUNDAY, SEPTEMBER 12 (First Weekend)

Atlanta at New York Giants (CBS) 1:00PM - good chance of being carried on
COMSTAR 3, TR-17 and on WESTAR 4

Chicago at Detroit (CBS) 1:00PM - not likely

Houston at Cincinnati (NBC) 1:00PM - very likely - should be found on
WESTAR 4, SATCOM 1, TR-20, SATCOM 2, TR23, and also partially on COMSTAR 3, TR- 1

Kansas City at Buffalo (NBC) 1:00PM - not likely

Los Angeles at Green Bay (CBS) 1:00PM - very likely - should be found on
SATCOM 1, TR-1 or 3

St. Louis at New Orleans (CBS) 1:00 - not likely

Tampa Bay at Minnesota (CBS) 1:00 - Chance at being on
WESTAR 1/2, WESTAR 3, WESTAR 4 or WESTAR 5

Washington at Philadelphia (CBS) 1:00 - not likely

New England at Baltimore (NBC) 2:00 - not likely

Cleveland at Seattle (NBC) 4:00 - very likely - Look for it on
WESTAR 4, COMSTAR 1/2, WESTAR 5, SATCOM 4, TR-24, and WESTAR 3

Miami at New York Jets (NBC) 4:00 - big Eastern regional game. Good chance of being
on WESTAR 4 and SATCOM 4; TR-24

LA Raiders at San Francisco (NBC) 4:00 - definitely on - big one for NBC doubleheader
Sunday. Should be on COMSTAR 3, TR-1, SATCOM 2, TR-23, WESTAR 4,
also possibbly SATCOM 1, TR-20.

San Diego at Denver (NBC) 4:00 - possibility of being on COMSTAR 1/2,
also SATCOM 4, TR-24

MONDAY, SEPTEMBER 13

Pittsburgh at Dallas (ABC) 9:00 - You'll find this one like all other Monday, Thursday
and Sunday night football games on COMSTAR 3, TR-13, WESTAR 4,
TR-5, SATCOM 2, TR-23 and SATCOM 1, TR-20.

Fig. 3-8. Sunday network football schedule
(Courtesy Westsat Communications)

THE RELIGIOUS CHANNELS

The granddaddy of all religious programming is the Christian Broadcasting Network, now known as the Continental Broadcasting Net
(CBN). CBN features 24 hours a day of Christian programming including
the popular 700 Club (Fig. 3—13), as well as general entertainment and
special events. CBN is carried via Satcom 3 free of charge to any cable
television system, and over 17 million people can tune into CBN at any
time. Individuals and organizations may also pickup CBN without restrictions, and the Virginia-based channel encourages people to contact
it directly.

People That Love Television Network (PTL) operates another 24 hour

Fig. 3-9. Baseball feed from WPIX New York.

Fig. 3-10. The Cable News Network (CNN).

Fig. 3-11. The Independent News Network (INN).

Fig. 3-12. ABC television network feed.

per day Christian talk and variety channel. PTL may be found on the Satcom 3 satellite. Trinity Broadcasting Network (TBN) uplinks the feeds of a Los Angeles-based uhf television station, providing 24-hour per day of Christian and inspirational programming carried via the Satcom 4 satellite. The National Christian Network (NCN) is also the Satcom 4 satellite. NCN is a nondenominational programming service which has also filed a request with the FCC to operate a direct broadcasting satellite service of its own.

Although the nondenominational Christian services are the oldest religious programmers found on the satellites, both the National Jewish Television Network and the Catholic Eternal Word Television Network sublease air time on Satcom 3 to distribute their faiths' inspirational and religious programming.

PUBLIC AFFAIRS AND EDUCATIONAL CHANNELS

Perhaps the most popular public affairs programming is that of Cable Satellite Public Affairs Network (C-SPAN) headquartered in Washington DC. C-SPAN (Fig. 3—14) provides gavel-to-gavel coverage of the House of Representatives when it is in session, and operates via the Satcom 3 satellite. During those hours of the day when the Congress is not in

Fig. 3-13. The CBN network.

session, C-SPAN provides other special-interest public affairs programming, including interviews with key public figures, politicians, and statesmen.

The Appalachian Community Service Network (ACSN) now known as The Learning Channel offers educational programming for adults at the secondary, post-secondary, and college levels (Fig. 3—15). It also provides for a variety of continuing education course. ACSN programming is sponsored by a number of higher educational institutions and authorities. It is delivered via Satcom 3. Full educational credit is available for a number of courses and ACSN programming is available to the general public both through their cable system (when carried) and local school districts.

Several other organizations, including the American Educational Television Network (AETN), provide specialty seminars and continuing education instruction for professional groups and individuals. These seminars are often sponsored by professional associations in the medical, legal, and accounting fields on a close-group arrangement. HSC-TV, produced by the Academy of Health Science for the medical profession is located on Westar 3. The American Network, providing both first-run movies and medical programming to hospitals, may be found on Westar 5.

Fig. 3-14. C-SPAN coverage of Washington DC.

Fig. 3-15. The ACSN network.

COMMERCIALLY SUPPORTED TELEVISION NETWORKS

A number of programming organizations specialize in creating services catered to specific but widespread audiences. These are usually advertising-supported operations. The oldest is the Spanish International Network (SIN) which feeds Spanish speaking variety shows, dramas, news, etc. from Puerto Rico, Mexico, Spain, and South America to several dozen independent stations and large cable systems which have a total viewership capacity of 25 million households. SIN (Fig. 3–16) concentrates in those cities which have large Spanish-speaking populations with particular strength on the US coasts. Its programming is delivered via the Satcom 4 satellite.

Modern Satellite Network (MSN), a division of Modern Talking Pictures, provides entertainment and information programs to daytime audiences. Its shows, typically 30 minutes in length, are sponsored by major corporations, and often produced at these corporations' in-house television studio facilities. Howard Ruff's "Ruff House" is a particularly successful show which airs weekly, presenting that economist's financial views and investment services. MSN is delivered to cable systems via the Satcom 3 satellite.

Southern Satellite Systems operates the Satellite Programming Network (SPN) similarly to MSM, feeding its cable viewers via the Satcom 4

Fig. 3-16. The Spanish International Network (SIN).

satellite. SPN's unique programming (Fig. 3—19) includes a number of hours per week of German, Italian, and French cultural shows, talk shows, corporate public relations and syndicated shows, and a classic movie service. SPN operates 24 hours per day.

The USA (Fig. 3—18) network originated by running Madison Square Garden sports events. The network has been significantly broadened to include a variety of commercially sponsored children's programming known as Calliope, the English Channel (British movies) and recent tv series. Sports events are still heavily carried by the network which is now affiliated with Home Box Office. It may be found on Satcom 3.

The ABC Alpha Repertory Television Service (ARTS) Channel presents operas, artistic productions, and cultural events, primarily originating from the Lincoln Center. Fig. 3—19 shows a typical offering. It is carried on the Satcom 3 satellite. Bravo, another performing and cultural arts channel is downlinked on Satcom 4.

The Cable Health Network (CHN), a creation of Viacom Cablevision, is a sister service of Showtime. Featuring Dr. Art Ulene, CHN provides a wide range of programs, news items, exercise shows, and information of physical fitness and general health. CHN uses a transponder on the Satcom 3 satellite.

The Weather Channel, the creation of Good Morning America's meteorologist, John Coleman, provides just that: 24-hours per day of local and national weather presented by a variety of "weather hosts." The Weather Channel is also carried over Satcom 3.

JULY 1983
PROGRAM GRID

8252 SOUTH HARVARD   TULSA, OK 74136   (918) 481-0881   TELEX 796322

EASTERN TIME	PACIFIC TIME	MONDAY	TUESDAY	WEDNESDAY	THURSDAY	FRIDAY	SATURDAY	SUNDAY
6:30	3:30						MOVIE (Cont d)	MOVIE (Cont d)
7:00	4:00	SPN MOVIE					POST TIME	INTERNATIONAL BYLINE
7:30	4:30						SCUBA WORLD	TRAVELLER'S WORLD
8:00	5:00	INTERNATIONAL BYLINE					MATCH BASS FISHING	HOME BASED BUSINESS
8:30	5:30	INVESTOR'S ACTION LINE	MONEY TALKS	MONEYWORKS	HOME BASED BUSINESS	REAL ESTATE ACTION LINE	JIMMY HOUSTON OUTDOORS	THE BIBLE ANSWERS
9:00	6:00	FRAN CARLTON					THE GOOD EARTH JOURNAL	THE HYDE PARK HOUR
9:30	6:30	JANET SLOANE AEROBIC DANCE/EXERCISE					SEWING WITH NANCY	INSIGHT
10:00	7:00	MEDICINE MAN	THE AMERICAN BABY	MEDICINE MAN	THE GOOD EARTH JOURNAL	THE AMERICAN BABY	THE NAME OF THE GAME IS GOLF	KENNETH COPELAND
10:30	7:30	THE PICTURE OF HEALTH					THE GOOD LIFE	
11:00	8:00	THE BODY BUDDIES						ORAL ROBERTS
11:30	8:30	HOME BASED BUSINESS	CONNIE MARTINSON	MOVIEWEEK	SEWING WITH NANCY	THE GOOD EARTH JOURNAL	CONNIE MARTINSON	JIMMY SWAGGART
NOON	9:00	THE PERSONAL COMPUTER SHOW	THE GOOD EARTH JOURNAL	PET ACTION LINE	CAREER WOMAN	NEW ANTIQUES	TELEPHONE AUCTION	
12:30	9:30	THE GOURMET	NEW ANTIQUES	MICROWAVES ARE FOR COOKING	MOVIEWEEK	THE GOURMET		REAL ESTATE ACTION LINE
1:00	10:00	MURIEL STEVENS					FINANCIAL INQUIRY	MATCH BASS FISHING
1:30	10:30	CAREER WOMAN	THE GOOD LIFE	THE PERSONAL COMPUTER SHOW	PHOTOGRAPHER'S EYE	TELEPHONE AUCTION	PHOTOGRAPHER'S EYE	THE NAME OF THE GAME IS GOLF
2:00	11:00	NEW ANTIQUES		SEWING WITH NANCY	THE AMERICAN BABY		POST TIME	REAL ESTATE ACTION LINE
2:30	11:30			SCANDINAVIAN WEEKLY		HOLLAND ON SATELLITE	HOME BASED BUSINESS	CHRISTIAN CHILDREN'S FUND
3:00	NOON	MEDITERRANEAN ECHOES	JAPAN 120		JAPAN 120		SCANDINAVIAN WEEKLY	PHOTOGRAPHER'S EYE
3:30	12:30			HELLO JERUSALEM		TRAVELLER'S WORLD		THE AMERICAN INVESTOR
4:00	1:00					INTERNATIONAL BYLINE		MONEY, MONEY, MONEY
4:30	1:30	PAUL RYAN					MEDITERRANEAN ECHOES	ONE IN THE SPIRIT
5:00	2:00	NOSTALGIA			TELEPHONE AUCTION	LOOKING EAST		HELLO JERUSALEM
5:30	2:30							
6:00	3:00	TELEPHONE AUCTION	MICROWAVES ARE FOR COOKING	CONNIE MARTINSON	MICROWAVES ARE FOR COOKING	TELEPHONE AUCTION	HOLLAND ON SATELLITE	
6:30	3:30		THE AMERICAN INVESTOR	LOFTON-ST. JOHN IN WASHINGTON	THE FIRST NIGHTER			JAPAN 120
7:00	4:00	MOVIEWEEK	MATCH BASS FISHING	MEDICINE MAN	PET ACTION LINE	THE PERSONAL COMPUTER SHOW	JAPAN 120	
7:30	4:30	PET ACTION LINE	JIMMY HOUSTON OUTDOORS	MONEY TALKS	THE PERSONAL COMPUTER SHOW	INVESTOR'S ACTION LINE		
8:00	5:00	PHOTOGRAPHER'S EYE	POST TIME	THE AMERICAN BABY	TRAVELLER'S WORLD	HOME BASED BUSINESS		SCANDINAVIAN WEEKLY
8:30	5:30	MONEYWORKS	SCUBA WORLD	NIKKI HASKELL	THE SHARPER IMAGE	THE FIRST NIGHTER		
9:00	6:00	TELEFRANCE — U.S.A.						
10:00	7:00							
11:00	8:00							
12:00	9:00						LOOKING EAST	MOVIEWEEK
12:30	9:30							
1:00	10:00	TRAVELLER'S WORLD	PAUL RYAN	THE FIRST NIGHTER	THE SHARPER IMAGE	NIKKI HASKELL	JOE BURTON JAZZ	MEDITERRANEAN ECHOES
1:30	10:30	ALL NIGHT AT THE MOVIES						
2:00	11:00							
6:30	3:30							

*Local advertising availability for cable systems during network breaks (average 2 minutes per hour)

Fig. 3-17. SPN's programming schedule.

Children's programming is well represented on the satellites through the operation of the Emmy-winning Nickelodeon Channel, a commercially supported service of the Warner-Amex Corporation. Nickelodeon (Fig. 3-20) is fun, providing old time adventure serials, new and old cartoons, and specials just for children. Nickelodeon is located on the

Fig. 3-18. USA network.

Satcom 3 satellite. The USA network, although not usually considered a children's channel, provides an outstanding service called Calliope, and both of these networks are carried by many cable companies as part of their basic monthly service. Finally, the newest of the "children's networks," the Disney Channel, provides premium family entertainment from the Disney Studio and EPCOT Disney World via East and West Coast feeds on Westar 5.

The Music Channel (MTV), another 24-hour per day service of Warner-Amex delivers "video records" in full stereophonic sound to the cable viewer via the Satcom 3 satellite. The Music Channel's 10 to 20 minute video pieces are often supplied by the record producers, and feature the latest pop groups performing their current hit singles. With a stereo decoder option built into the satellite TVRO receiver, the home backyard viewer can enjoy the wide range of hi-fi sounds through his or her own stereo system.

Although not yet offering its own cable channel, the Robert Wold Corporation, an affiliate of Cox Broadcasting, owns a number of satellite transponders on several satellite systems. Wold utilizes these channels to uplink special events and sports feeds for the Big 3 television networks as well as to deliver a series of live television programs, including Entertainment Tonight from Hollywood, and the Merv Griffin Show. Wold also provides the satellite services for video conferences and special

SUNDAY, APRIL 11
9:00 MOZART IN JERUSALEM (54 min.) Jean-Pierre Rampal performs Mozart flute concertos at the Jerusalem Music Centre with the Israel Philharmonic Orchestra. Conducted by Isaac Stern.
10:00 CARAVAGGIO (75 min.) A documentary about the controversial Italian artist of the late 16th–early 17th centuries.
11:25 NIGHTCAP: CONVERSATIONS ON THE ARTS AND LETTERS (27 min.) Hosted by Studs Terkel and Calvin Trillin. This program centers on the topic of the images of women with guests Colette Dowling, Erica Jong and Nora Ephron.

MONDAY, APRIL 12
9:00 PIERROT LUNAIRE (60 min.) One of the classics of modern dance is performed by the Ballet Rambert. Arnold Schoenberg's composition is set to dance by the brilliant American choreographer, Glen Tetley.
10:00 SOFT SELF-PORTRAIT (52 min.) An award-winning documentary on the familiar Spanish artist, Salvador Dali, best known as one of the leaders of the surrealist movement which arose in the 1920's.
11:00 ARTS AT SOTHEBY'S: AMERICANA (15 min.) This program, from Sotheby Parke Bernet, examines six major periods of Early American furniture. Includes auction of a rare piece of American porcelain, an 18th century Bonnin & Morris candy dish which sold for $60,000. Hosted by Gene Klavan.
11:25 WOMEN IN JAZZ: THE INNER VOICE (30 min.) This program takes a look at the "inner voice" that each jazz musician tries to express through music.

TUESDAY, APRIL 13
9:00 SHADES (62 min.) A trio of plays by Samuel Beckett: "Ghost Trio" and "...but the clouds..." (written especially for this program) and "Not I." Also included is documentary footage on the playwright's life.
10:10 JOSEPH PAPP PRESENTS: "WEDDING BAND" (100 min.) A drama written by Alice Childress set in 1918 in North Carolina about the love shared between a white man and a black woman and their struggle for acceptance in a very racial and cruel time. Starring Ruby Dee, J.D. Cannon, Eileen Heckart and Clarice Taylor.

SALVADOR DALI

WEDNESDAY, APRIL 14
9:00 MOBIL SHOWCASE: "TURNER—THE SUN IS GOD" (58 min.) A dramatized documentary on the English painter J.M.W. Turner. Leo McKern stars as Turner who gains enormous popularity and some wealth as a painter of renown, but who loses his sanity in the process.
10:00 WOMEN IN JAZZ: THE INNER VOICE (30 min.) This program takes a look at the "inner voice" that each jazz musician tries to express through music, with Melba Liston and Marian McPartland.
10:40 ARTS AT SOTHEBY'S: AMERICANA (15 min.) This program, from Sotheby Parke Bernet, examines six major periods of Early American furniture. Includes auction of a rare piece of American porcelain, an 18th century Bonnin & Morris candy dish which sold for $60,000. Hosted by Gene Klavan.
11:00 MOZART IN JERUSALEM (54 min.) Jean-Pierre Rampal performs Mozart flute concertos at the Jerusalem Music Centre with the Israel Philharmonic Orchestra. Conducted by Isaac Stern.

THURSDAY, APRIL 15
Repeat of Sunday, April 11

FRIDAY, APRIL 16
Repeat of Monday, April 12

SATURDAY, APRIL 17
Repeat of Tuesday, April 13

SUNDAY, APRIL 18
9:00 LUISA MILLER (135 min.) The opera by Giuseppe Verdi. In three acts. With Katia Ricciarelli as Luisa and Placido Domingo as Rodolfo. From the Royal Opera House at Covent Garden. Orchestra under the direction of Lorin Maazel.

MONDAY, APRIL 19
9:00 THE ROMANTIC AGE (88 min.) Four legendary prima ballerinas of our time—Alicia Alonso, Carla Fracci, Ghislaine Thesmar and Eva Evdokimova—perform famous ballets from the mid-19th century. Includes all four dancers performing the 1845 masterpiece "Le Grand Pas de Quatre."
10:50 THE METROPOLITAN MUSEUM: CURATORS CHOICES (25 min.) A film about the museum pieces personally chosen by the Met's curators.
11:20 WOMEN IN JAZZ: A MATTER OF STYLE (31 min.) This program examines some of the different jazz styles that have developed over the years.

TUESDAY, APRIL 20
9:00 TENNESSEE WILLIAMS (81 min.) A documentary on the life and work of a man widely considered to be among the best American playwrights of the 20th century. Included are excerpts from two of Williams' best known plays "A Streetcar Named Desire" and "The Glass Menagerie."
10:30 GREAT POETS, GREAT WRITERS: WILLIAM BUTLER YEATS "FIVE POEMS" (11 min.)
10:45 DESIGN FINNISH STYLE (53 min.) An award-winning documentary about Tapio Wirkkala, the Finnish designer, sculptor and craftsman especially associated with original dishes and bowls made of laminated veneers.

WEDNESDAY, APRIL 21
9:00 MOBIL SHOWCASE: "MEETING OF THE SPIRITS" (60 min.) A classical/jazz guitar concert recorded at Royal Albert Hall, featuring John McLoughlin, Paco Da Lucia and Larry Coryell.
10:00 THE METROPOLITAN MUSEUM: CURATORS CHOICES (25 min.) A film about the museum pieces personally chosen by the Met's curators.

ALPHA REPERTORY TELEVISION SERVICE

10:30 DESIGN FINNISH STYLE (53 min.) See April 20, 10:45.
11:30 GREAT POETS, GREAT WRITERS: WILLIAM BUTLER YEATS "FIVE POEMS" (11 min.)

THURSDAY, APRIL 22
Repeat of Sunday, April 18

FRIDAY, APRIL 23
Repeat of Monday, April 19

SATURDAY, APRIL 24
Repeat of Tuesday, April 20

SUNDAY, APRIL 25
9:00 AILEY DANCES (82 min.) The Alvin Ailey American Dance Theater performs the most exciting highlights from its repertoire. Intermission includes an interview with lead dancer Donna Wood.

ALVIN AILEY

10:30 ARTS VISITS WITH ROBERT ALTMAN (12 min.) The director of "Rattlesnake in a Cooler" talks about his work.
10:45 "RATTLESNAKE IN A COOLER" (57 min.) A dramatic monologue about a Kentucky doctor whose aimless drifting winds up in tragedy when he murders a state trooper. Directed by Robert Altman and written by Frank South.

MONDAY, APRIL 26
9:00 MAURICE BEJART: THE LOVE FOR DANCE (62 min.) Performance documentary of the famed choreographer.
10:10 LOTTE LENYA (50 min.) A documentary portrait of the late musical star and originator of the many roles in the most famous musicals by her husband Kurt Weill.
11:10 L.A. JAZZ (31 min.) Live jazz from the Lighthouse Cafe, hosted by Leonard Feather, with vocalist Jimmy Witherspoon and the Ahmad Jamal Trio.
11:45 GREAT PAINTERS: DEGAS (8 min.)

TUESDAY, APRIL 27
9:00 REUNION AND DARK PONY (59 min.) Two works written by David Mamet, directed by Lamont Johnson and starring Lindsay Crouse and Michael Higgins. An interview with David Mamet follows.
10:00 NIGHTCAP: CONVERSATIONS ON THE ARTS AND LETTERS (26 min.) Hosted by Studs Terkel and Calvin Trillin. This program examines the topic of censorship with guests Kurt Vonnegut, Ring Lardner, Jr., and Norma Klein.
10:40 JOSEPH PAPP PRESENTS: "THE DANCE AND THE RAILROAD" (60 min.) David Henry Hwang's play starring John Lone and Tzi Ma.

WEDNESDAY, APRIL 28
9:00 MOBIL SHOWCASE: "NOW SHE LIES THERE" (55 min.) A period drama from an anthology series set between the two world wars. This episode deals with an heiress with too much money and a taste for the fast life which ends in tragedy.
10:00 L.A. JAZZ (31 min.) See April 26, 11:10.
10:35 GREAT PAINTERS: DEGAS (8 min.)
10:45 "RATTLESNAKE IN A COOLER" (57 min.) See April 25, 10:45.

DAVID MAMET LINDSAY CROUSE

THURSDAY, APRIL 29
Repeat of Sunday, April 25

FRIDAY, APRIL 30
Repeat of Monday, April 26

ARTS is sponsored by General Motors Corp., American Telephone and Telegraph Co., Mobil Oil Corp., Polaroid Corp.

Fig. 3-19. ARTS program schedule.

PROGRAM SCHEDULE: OCTOBER 1-17 1982

NICKELODEON™ — The first channel for kids

MON.-FRI.

	TODAY'S SPECIAL (MON.-FRI.)	DUSTY'S TREEHOUSE	PIN-WHEEL	MATT AND JENNY (MON. & THURS.)	ADVENTURES IN RAINBOW COUNTRY (TUES. & FRI.)	YOU CAN'T DO THAT ON TELEVISION (WEDS.)	WHAT WILL THEY THINK OF NEXT? (MON. & THURS.)	WHAT WILL THEY THINK OF NEXT? (TUES. & FRI.)	WHAT WILL THEY THINK OF NEXT? (WED.)	STUDIO SEE (MON., WED. & THURS.)	THE TOMORROW PEOPLE	SPREAD YOUR WINGS (TUES. & FRI.)	BLACK BEAUTY (MON.-FRI.)	LIVEWIRE (MON.-FRI.)
E	8:00am 2:00pm	8:30am	9:00am	2:30pm 6:00pm	2:30pm 6:00pm	2:30pm 6:00pm	3:00pm 6:30pm	3:00pm	3:00pm 6:00pm	3:30pm	4:00pm 7:00pm	3:30pm 6:30pm	4:30pm 7:30pm	5:00pm 8:00pm
C	7:00am 1:00pm	7:30am	8:00am	1:30pm 5:00pm	1:30pm 5:00pm	1:30pm 5:00pm	2:00pm 5:30pm	2:00pm	2:00pm 5:00pm	2:30pm	3:00pm 6:00pm	2:30pm 5:30pm	3:30pm 6:30pm	4:00pm 7:00pm
M	6:00am 12:00n	6:30am	7:00am	12:30pm 4:00pm	12:30pm 4:00pm	12:30pm 4:00pm	1:00pm 4:30pm	1:00pm	1:00pm 4:00pm	1:30pm	2:00pm 5:00pm	1:30pm 4:30pm	2:30pm 5:30pm	3:00pm 6:00pm
P	5:00am 11:00am	5:30am	6:00am	11:30am 3:00pm	11:30am 3:00pm	11:30am 3:00pm	12:00n 3:30pm	12:00n	12:00n 3:00pm	12:30pm	1:00pm 4:00pm	12:30pm 3:30pm	1:30pm 4:30pm	2:00pm 5:00pm

SATURDAYS

	PINWHEEL	MATT AND JENNY	ADVENTURES IN RAINBOW COUNTRY	SPREAD YOUR WINGS	WHAT WILL THEY THINK OF NEXT?	THE TOMORROW PEOPLE
E	8:00am	1:00pm	1:30pm	2:00pm	2:30pm	12:30pm 4:30pm 7:00pm
C	7:00am	12:00n	12:30pm	1:00pm	1:30pm	11:30am 3:30pm 6:00pm
M	6:00am	11:00am	11:30am	12:00n	12:30pm	10:30am 2:30pm 5:00pm
P	5:00am	10:00am	10:30am	11:00am	11:30am	9:30am 1:30pm 4:00pm

	YOU CAN'T DO THAT ON TELEVISION	WHAT WILL THEY THINK OF NEXT?	BLACK BEAUTY	REGGIE JACKSON'S WORLD OF SPORTS	SPECIAL DELIVERY	LIVEWIRE	SPREAD YOUR WINGS
E	12:00n 4:00pm	3:00pm 7:00pm	3:00pm 7:30pm	4:00pm	5:00pm	5:30pm 8:00pm	6:30pm
C	11:00am 3:00pm	2:00pm 6:00pm	2:00pm 6:30pm	3:00pm	4:00pm	4:30pm 7:00pm	5:30pm
M	10:00am 2:00pm	1:30pm 5:00pm	1:00pm 5:30pm	2:00pm	3:00pm	3:30pm 6:00pm	4:30pm
P	9:00am 1:00pm	12:30pm 4:00pm	12:00n 4:30pm	1:00pm	2:00pm	2:30pm 5:00pm	3:30pm

SUNDAYS

	PINWHEEL	THE TOMORROW PEOPLE	YOU CAN'T DO THAT ON TELEVISION	REGGIE JACKSON'S WORLD OF SPORTS	BLACK BEAUTY	LIVEWIRE	WHAT WILL THEY THINK OF NEXT?
E	8:00am	12:30pm 4:30pm 7:00pm	12:00n 4:00pm	1:00pm 5:30pm	3:30pm 7:30pm	2:00pm 8:00pm	3:00pm
C	7:00am	11:30am 3:30pm 6:00pm	11:00am 3:00pm	12:00n 4:30pm	2:30pm 6:30pm	1:00pm 7:00pm	2:00pm
M	6:00am	10:30am 2:30pm 5:00pm	10:00am 2:30pm	11:00am 3:30pm	1:30pm 5:30pm	12:00n 6:00pm	1:00pm
P	5:00am	9:30am 1:30pm 4:00pm	10:00am 2:30pm	10:00am 2:30pm	12:30pm 4:30pm	11:00am 5:00pm	12:00n

	WHAT WILL THEY THINK OF NEXT?	SPECIAL DELIVERY	SPREAD YOUR WINGS
E	3:00pm	5:00pm	6:30pm
C	2:00pm	4:00pm	5:30pm
M	1:00pm	3:00pm	4:30pm
P	12:00n	2:00pm	3:30pm

SPECIAL DELIVERY SCHEDULE: OCTOBER 1982

(preempts regular scheduled programming)

	KID'S IN PERFORMANCE — SAT., OCT. 2	ROGER DALTREY — SUN., OCT. 3	HIS MAJESTY, THE SCARECROW OF OZ — SAT., OCT. 9	THE WORLD ACCORDING TO NICHOLAS, PART III, PRAIDY CATS — SUN., OCT. 10	WATER BABIES — MON., OCT. 11 (COLUMBUS DAY)	ELO, IN CONCERT — FRI., OCT. 15
E	5:00-5:30pm	5:00-5:30pm	5:00-5:30pm	5:00-5:30pm	1:00-2:30pm	8:00-9:00pm
C	4:00-4:30pm	4:00-4:30pm	4:00-4:30pm	4:00-4:30pm	12:00-1:30pm	7:00-8:00pm
M	3:00-3:30pm	3:00-3:30pm	3:00-3:30pm	3:00-3:30pm	11:00-12:30pm	6:00-7:00pm
P	2:00-2:30pm	2:00-2:30pm	2:00-2:30pm	2:00-2:30pm	10:00-11:30am	5:00-6:00pm

Fig. 3-20. Nickelodeon program schedule.

events such as corporate and stockholders meetings, and fundraising telethons. Fig. 3—21 shows the setup for a stereo simulcast by Robert Wold at the Grammy Awards from New York's Radio City Music Hall to 60 fm stations across the country.

Table 3—1 summarizes satellite-fed cable programming services.

THE BIG 4 NATIONAL TELEVISION NETWORKS

The three commercial networks, CBS, ABC, and NBC, are active users of satellite communications facilities for both television programming feeds and audio services. News shows (Fig. 3—22) and prefeeds from the coasts, as well as television programming like the Tonight Show, are relayed to and from the New York network control centers via satellite. Although the Westar 4 satellite has been frequently used for these services, the Comstar 3 bird has been designated the new "network satellite" by AT&T, with over 20 transponders dedicated for exclusive use by the three commercial networks and Robert Wold. As the network begins to systematically convert from terrestrial microwave circuits to satellite feeds, activity on this bird will rapidly intensify.

A selection of the three network's programming is also carried to Alaskan television stations several hours or days before appearing in the "lower-48." This special "Alaska Channel" can be found on the Alascom Aurora satellite.

In Canada, the national government network, CBC (Fig. 3—23), delivers its entire programming schedule via the Anik series of satellites. Anik B carries CBC-North on separate East and West Coast transponders; CBC-French is on a third transponder, and occasionally news and other transmissions are on a fourth transponder. Anik D delivers the CBC Parliamentary Network (daily live coverage of the Canadian House of Commons) from Ottawa (similar to the C-SPAN feed of the House of Representatives in the US). The second national Canadian network, CTV, also carries occasional programming via the Anik satellites.

Four new Canadian "mini-networks" are now feeding programs nationwide via Anik D. Telemedia Communications Television (TCTV) presents French programming from CHLT television, Sherbrooke, and CFTM, Montreal. British Columbia Television presents the leading CTV network station in the West, BCTV. CITV television is Alberta's leading independent station from Edmonton, while CHCH is Ontario's leading independent station from Hamilton. Together, these four services are known as Cancom and are uplinked in a fully encrypted mode for delivery to cable companies nationwide.

CBC and CTV buy popular network adventure shows and situation comedies from the same Hollywood producers who sell their products to US television. Thus, it is often possible to watch US shows several days

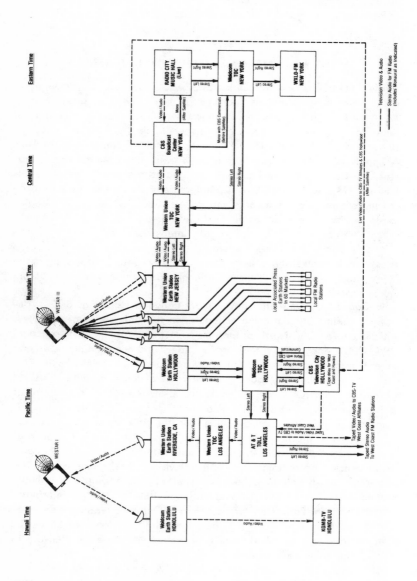

Fig. 3-21. Anatomy of a simulcast.

Table 3–1. Satellite-Fed Cable Networks*

Service	Affilliates	Subscribers
Television		
ACSN The Learning Channel	396	3,057,642
AP Cable News	445	4,700,000
ARIS (Hearst/ABC)	1,625	9,500,000
Black Entertainment Television (BET)	180	3,700,000
Cable Health Network (CHN)	1,028	10,282,777
Cable News Network (CNN)	3,369	19,159,000
CNN Headline News (CNN2)	476	3,516,617
C-SPAN America's Network	1,000	13,000,000
CBN Cable Network	3,830	20,770,000
Daytime (Hearst/ABC)	671	9,200,000
Dow Jones Cable News	110	1,115,000
Electronic Program Guide (EPG)	75	1,109,464
Entertainment and Sports Programming Network (ESPN)	5,733	22,238,479
Eternal Word Television Network (EWTN)	72	1,124,000
Financial News Network (FNN)	758	7,704,527
Modern Satellite Network (MSN)	459	7,805,020
Music Television (MTV)	1,500	12,000,000
The Nashville Network (TNN)	725	7,500,000
National Christian Network (NCN)	92	1,421,647
National Jewish Television (NJTV)	114	2,446,735
Nickelodeon	2,450	11,400,000
PTL Satellite Network	725	7,500,000
Reuters Monitor Service	15	5,000
Reuters News View	375	3,500,000
Satellite News Channel (SNC)	557	5,700,000
Satellite Programming Network (SPN)	375	5,600,000
SIN Television Network	199	25,531,400
Trinity Broadcasting Network	245	2,874,000
UPI News Cable	485	NA
USA Network	3,400	17,000,000
The Weather Channel (TWC)	840	8,300,000
WGN	3,976	11,084,783
WOR	796	4,703,757
WTBS	5,214	24,823,000
Audio		
KKGO-FM	3	100,000
Lifestyle	127	1.066,400
Moody Bible	23	316,723
Satellite Radio Network	40	NA
SCAN	8	100,000
Sunshine Entertainment Network	3	21,000
WFMT	168	912,997

(cont. on next page)

*As of May 1, 1983

Table 3—1. Cont. Satellite-Fed Cable Networks

Service	Affiliates	Subscribers
Pay Services	70	100,000
Bravo	1,600	2,000,000
Cinemax	425	100,000
The Disney Channel	22	200,000
EROS	160	125,000
Galavision	4,500	11,500,000
Home Box Office (HBO)	325	170,000
Home Theater Network Plus (HTN)	2,350	2,350,000
The Movie Channel (TMC)	3	452,995
ON TV	240	420,000
The Playboy Channel	35	215,000
SelectTv™	2,400	4,000,000
Showtime	231	750,000
Spotlight		

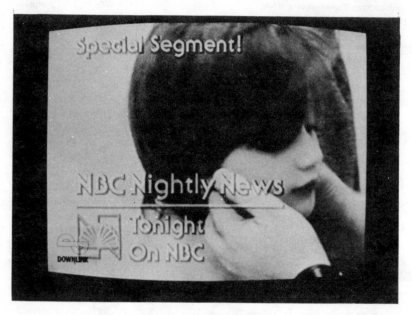

Fig. 3-22. NBC news feed.

earlier on the Anik satellite, providing one the answer to "who shot J.R.?" a day or two before local tv stations in the US air the program.

The Public Broadcasting Service (PBS) operates three full-time transponders and one occasional-use transponder on Westar 4 (Fig. 3—24). Four schedules, A, B, C, and D, are transmitted allowing West Coast

Fig. 3-23. Canadian Broadcasting Corp. (CBC) news.

viewers to watch Wall Street Week live at 5:30 Pacific time. It is actually uplinked from the Maryland Center for Public Broadcasting at 8:30 pm on the East Coast.

THE SECRET "TELEVISION NETWORKS"

Some fascinating occasional use, *ad hoc* and special television programs can be found on just about every satellite. Some of these feeds, such as Reuter's "Monitor" high speed business data service found on Satcom 3, are fed to desktop display terminals nationwide. This system requires a special data decoder to unscramble the signal, although with the appropriate computer program, an Apple® II or TRS-80® home computer should work nicely. (Reuter's will rent their special terminal to satellite tv users who wish to access the Monitor service via their TVRO satellite dishes.) UPI and AP news services transmit several channels of teletypewriter wirefeeds via satellite, as do a number of high-fidelity radio networks. These audio channels can be received by connecting a conventional shortwave multiband receiver to the video output of the TVRO satellite receiver, and then tuning the programming in via the shortwave set. (See Chapter 11 for detailed information on how to pick up these fascinating special audio and data transmissions.) Many of the

Fig. 3-24. Public Broadcasting Service (PBS).

audio networks are also located on one of the multiple audio subcarriers that several transponders provide when operating in a normal-video mode. These can be tuned in by the TVRO receiver directly, using the audio-subcarrier fine-tuning control located on the front panel. They include sports, Spanish-speaking, Black, and national news networks, National Public Radio, religious channels, and independent radio chain feeds.

NASA's Space Shuttle Channel is located on Satcom 1 (located at 139° W latitude). The space agency provides video coordination between Houston, Cape Kennedy, and Washington, as well as occasional shuttle video over this private contract channel. Robert Wold, Bonneville Satellite Corporation, Hughes Television, and others use the Western Union and RCA satellites to carry video conferences and political fundraising events on a daily basis. The US Chamber of Commerce in Washington delivers its Biznet, Business Video Conferencing Service over Satcom 4, and Main Events' "Superfights" uses the same satellite to feed sporting events and championship boxing matches to private-admission theaters nationwide.

Fig. 3–25 illustrates some of the many users of satellites.

WHO OWNS THE SATELLITES?

Five major common carrier organizations operate the majority of the domestic US and Canadian satellites. The first US satellite carrier, RCA

Americom, Inc. owns the Satcom series. Western Union Telegraph Corporation, the nation's first common carrier, founded in the midnineteenth century, owns the Westar series of satellites. The Communications Satellite Corporation (Comsat) owns the Comstar series of satellites which are leased to AT&T, and whose transponders are made available by Bell Telephone under FCC Tariff 260. AT&T is constructing its own new series of Telesat satellites to replace the aging Comstar satellites. North of the border, Telesat Canada, a crown corporation, owns and operates the Anik domestic Canadian satellites. (In addition, Hughes Satellite, the aerospace manufacturer, has launched an unusual "Condominium" Satellite, Galaxy I, which is jointly owned by Hughes and the Satellite's major users.)

Other companies have been authorized by the Federal Communications Commission to launch domestic US satellite systems, including Hughes Aircraft, GTE, SP Communications, and American Satellite Corporation. Most of these firms will have launched their new generation of satellites by the mid-1980s, providing over 600 new transponders available for carrying television services.

Common carriers usually own the actual satellite systems. But the birds themselves are constructed by two major manufacturers: RCA Astrocom (a sister company to RCA Americom, Inc.) and Hughes Satellite. Satellites are typically launched by NASA's Space Shuttle at their Cape Kennedy launch facility, or by the European Space Agency's Ariane rockets launched from the French Equitorial space facility. The common carriers control their own communications satellites through the use of a

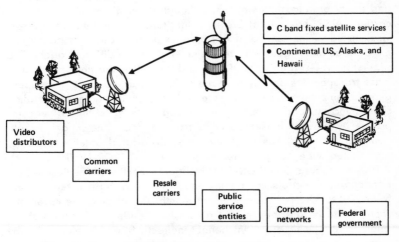

- C band fixed satellite services
- Continental U.S., Alaska, and Hawaii

Video distributors

Common carriers

Resale carriers

Public service entities

Corporate networks

Federal government

Fig. 3-25. Who uses the satellites.
(Courtesy Hughes Corp.)

67

central command ground station handling telemetry, tracking, and satellite control instructions from the ground.

The common carriers lease their satellite transponders to television programmers and networks under FCC tariffs for between $100,000 and $170,000 a month. In some instances, common carriers have substituted a flat one-time payment in lieu of a monthly charge, and one firm (Hughes Satellite) has sold actual "condominium" ownership rights in their physical satellites to the satellite programmers. The cost to purchase a transponder runs between $10 and $20 million, so one must have deep pockets indeed to play in the satellite television programming game.

THE FCC RULES THE HEAVENS

FCC rules and regulations treat satellite transponder channels provided by the common carriers as if they were ordinary terrestrial video circuits. Under the Communications Act of 1934, common carriers are prohibited from owning their own network programming services. Therefore, common carrier satellite owners must rent transponders to other organizations who provide for the program origination, usually through 15- to 30-foot uplink earth stations located on the premises of the programmer. The common carrier can also provide the programmer with uplinking capability at the common carrier's site. This allows the programmer to simply deliver his live video feeds or video tapes to the common carrier's earth station. The satellite downlink receive only earth stations are usually independently owned by cable television companies and the tv broadcasting stations, although at least one of the common carriers, AT&T, has a tariff provision for "turn key" and end-to-end service, including the renting of AT&T dishes on the customers premises.

FCC regulations governing television stations are different again. The superstations—WTBS-TV, WOR-TV, and WGN-TV—are not permitted to lease transponders directly from the common carriers. Several independent organizations have been created to pick up these superstation signals locally, and retransmit their programming to the cable companies by way of leased satellite transponders. Known as "Resale Common Carriers," Eastern Microwave, Syndicated Satellite Systems, and United Video deliver the programming of WOR-TV (New York), WTBS-TV (Atlanta) and WGN-TV (Chicago) respectively. The Resale Carriers charge the cable companies under their own tariff a fee ranging from $.10 to $.15 per subscriber per month. This nominal fee covers the cost of satellite distribution and allows the Resale Carriers to make a small profit. The superstations benefit indirectly from this relationship by obtaining significantly larger audiences, thus allowing them to increase their advertising rates.

THE INTERNATIONAL SATELLITES

The Canadian domestic satellite system can be seen throughout the United States. Anik satellites are owned and operated by Telesat Canada, an independent corporation formed by the Government and owned by the Canadian Treasury. Telesat Canada provides Canadian telephone companies with satellite circuits and interconnects the national telex networks of CN and CP Telecommunications. Telesat Canada leases transponders to the Canadian Broadcasting Company and other television programmers who deliver their multiprogramming services to Canadian cable television companies and Northwest Territories villages. Four new superstations, located in Toronto, Vancouver, Hamilton, and Edmonton, have banded together to form the Cancom programming service which feeds scrambled programming to CATV and SMATV systems throughout the country. The French Quebec independent television network also carries its shows throughout the country via Anik. All of these programs can be seen anywhere in the continental United States, although a viewer in the deep South may require a slightly larger TVRO antenna for optimum reception.

The "Mexican connection" is provided by XEW-TV, the Mexico City superstation. XEW-TV may be found on the American Westar 4 satellite, and utilizes this facility under a special Western Union tariff.

The Atlantic and Pacific Intelsat satellites carry international news feeds between the continents on a daily basis, and may be viewed along either US coast. The ABC Nightline news show, UPI International Television, and the British Visnews Organization transmit hours of programming via Intelsat each week. Intelsat satellites are often used to relay outbound programs from the US to Japan, Australia, Asia, South America, and Europe in tandem with a domestic US satellite which links the programming from US coast to coast.

A number of other nations also lease Intelsat transponders for exclusive use internally, several of these domestic television services of other countries can be seen by US Intelsat viewers. The Brazilian national television network beams its own shows and news from Rio via the Atlantic Intelsat satellites, and during the Faulkland Islands crisis, Argentine national television was a popular source of satellite tv viewing for many American observers.

The USSR operates several different and incompatible communications satellite networks which can be easily seen throughout the United States. Inter-Sputnik is a geosynchronous Soviet satellite system with a series of birds positioned over the Atlantic and Pacific Oceans. It is used to relay programming between the USSR and the Soviet-block countries, including Cuba. Cuba also uses the Russian satellites to beam television programming back to the Soviet Union, and to parts of Central and South America. Cuba occasionally runs unauthorized transmission of US television shows and first-run movies which are uplinked from Havana in the US NTSC Television Standard! A second USSR satellite network provides domestic telephone and television communications. The Molynia satellites are not placed in geosynchronous orbit; instead this system requires a number of satellites in polar orbits. Molynia satellites are continuously moving, and the TVRO earth stations, which are pointed northward to the pole must be motorized to track the satellites as they move. Because of the elongated elliptical orbits used, when they enter certain parts of their orbits, the Molynia satellites will appear to be semistationary from the Russian earth stations extending across the seven time zones of the USSR. The dramatic time differences between the US and the USSR make for interesting viewing of these satellites in the States. The 6:00 pm nightly news from Moscow can be seen in New York City at 11:00 am. And, the Russian satellites are almost directly over the United States when they beam back their high-powered signals to the Soviet Union! They can be readily picked up by a TVRO satellite terminal anywhere in North America. (At least one of the Big-3 US television network news departments regularly looks at the Russian domestic satellites to gleen a bit of news from that tightly controlled society.)

In Chapter 12, "DX'ing the International Satellites," specific information and transponder-by-transponder listings of the programming on the international birds is provided. Appendix B presents detailed programming information on the US and Canadian systems.

CHAPTER 4

The Home TVRO System: Watching TV Beamed From Space

The home satellite television system consists of five major components. The first is the television studio and master control facility from which the programmer originates his satellite television feeds. This programming is then delivered to the transmit/receive (T-R) earth station via video tape or by direct video circuit. The signal is uplinked to the geosynchronous communication satellite in the 6 gigahertz band. The satellite, often called the "space segment," electronically amplifies this incoming signal, shifts it downward in frequency to the 3.7 gigahertz band, and retransmits it earthward. Back on the ground, TVRO earth station terminals scattered throughout the country receive simultaneously the satellite feed. The complete system is diagramed in Fig. 4–1. A master satellite command earth station, operated by the satellite common carrier owner, monitors the "housekeeping" functions of the satellite such as battery charge, orbital positioning, etc., and controls the overall operation of the space vehicle itself.

The studio and master control facility of the programmer is often part of an affiliated terrestrial television station. For instance, Trinity Broadcasting Network uplinks live program feeds from their Los Angeles uhf station to form the basis of the TBN network. Programs can originate from separate studio and control facilities as is the case with the Cable News Network. They can also be uplinked by common carriers at the earth stations, utilizing banks of video tape machines to play back prerecorded tapes. The HBO movie service has used the RCA Americom New Jersey uplink facility for this purpose. A typical master control facility is pictured in Fig. 4–2.

The T-R uplink earth station consists, typically, of a 30-foot parabolic antenna (Fig. 4–3) and associated transmitter electronics. The uplink is often owned by the common carrier which also owns the satellite system. The common carrier T-R station receives the incoming video signal

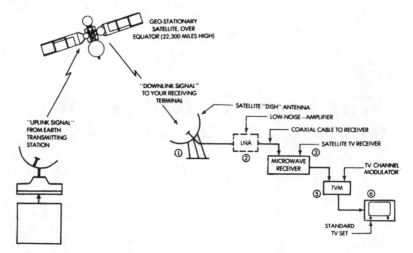

Fig. 4-1. A home satellite TVRO system.

Fig. 4-2. A typical master control facility.
(Courtesy Bell Telephone)

from the satellite programmers master control center via AT&T long lines microwave circuits. Some satellite programmers use their own portable 5-meter T-R earth stations mounted on mobile trailers; others use more permanent uplink antennas located at their studios. Each of the com-

Fig. 4-3. An "Antenna Farm" showing a 30-foot uplink antenna and two
TVRO antennas at the right. *(Courtesy Paradigm Manufacturing Inc.)*

munications satellites is fed by many independent programmers so it is
typical to have ten or more T-R antennas scattered throughout the
United States uplinking simultaneously to different transponders on the
same satellite.

The "space segment" satellite itself is located at a specific fixed "slot"
in the geosynchronous orbit. The satellite contains 24 transponders which
relay the uplinked signals back to earth via the 3.7 gigahertz C-Band.
The satellites are owned by RCA, Western Union, Comsat, AT&T, Telesat
Canada, and Intelsat. New players who will shortly be launching their
own birds include SP Communications, GTE, Hughes, and American
Satellite.

The TVRO television receive only earth station is typically located at
the headend of a cable television system, or in the parking lot of modern hotel. The TVRO satellite terminal has five major components: a
satellite dish antenna (Fig. 4−4), a low-noise preamplifier (LNA), a
satellite tv receiver, a vhf television modulator (which converts the video
signal of the satellite receiver to an ordinary vhf television channel), and
a television set (Fig. 4−5). Some TVRO configurations may also have a
motorized antenna mount, and a polarizing rotator which electronically
or mechanically shifts the satellite dish feedhorn to pick up either the 12
vertically or the 12 horizontally polarized transponder channels.

THE HOME TVRO SATELLITE TERMINAL

The typical private earth station feeds a single home television set.
The audio portion of the television program may also be fed into the

Fig. 4-4. A backyard TVRO dish.

home stereo system as well. Many TVRO satellite receivers are equipped to decode the special stereo audio which a number of satellite programmers are now transmitting, and these receivers include separate audio output jacks for connection to the auxiliary input of a stereo amplifier. Figure 4—6 shows the connections to a typical system.

Although the TVRO dish antenna is only one component, it is certainly the largest. TVRO satellite antennas come in two shapes: the parabolic dish, and the spherical antenna. Over 95% of the TVRO systems in North America use a parabolic dish antenna.

The typically TVRO parabolic antenna is between 8 and 15 feet in diameter, and is fabricated from metallized fiberglass or lightweight spun aluminum. It weighs between 150 and 500 pounds, including its support structure and base. There are three popular types of antennas—the single-piece spun aluminum (Fig. 4—7), the aluminum mesh (Fig. 4—8), and the molded fiberglass (Fig. 4—9). One of the most novel of the TVRO antennas is the Luly umbrella, named after Robert Luly the Southern California inventor and satellite tv manufacturer. The Luly antenna (Fig. 4—10) which comes in 8-, 10-, and 12-foot versions unfolds like a large beach umbrella. It consists of a special mylar surface impregnated with silver, and is supported by a precisely machined

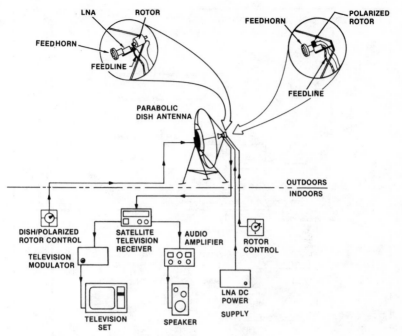

Fig. 4-5. A home TVRO earth station.

flexible rib structure. The entire antenna weighs under 25 pounds and can be set up in minutes on a modified heavy-duty 35-mm camera tripod. An optional permanent pier mount is also available, as well as a remote-controlled motorized satellite tracking mount. Popular with R-V and motorhome owners, the Luly umbrella is also useful as an instant site demonstration antenna for use by satellite TVRO dealers. The antenna should be enclosed to protect its structure from the elements when used in a permanent installation. This can be accomplished by placing it inside a large weather balloon or a small geodesic dome can be built around it. In mild climates, like those found in Southern California, the Luly need not be enclosed, but its fabric surface may have to be replaced after a few years of use. The author's own 8-foot Luly has been working quite well for a number of months on the rooftop of his Nob Hill home right in the heart of San Francisco (Fig. 4−11).

The spherical antenna experienced a brief flurry of sales activity. Now, however, except for the kit builder, it has almost completely disappeared from the market. The spherical is noted for its ability to receive television signals from several different satellites simultaneously. Unlike the parabolic, which focuses the energy of a single satellite on a single focal point located slightly in front of it, the spherical antenna

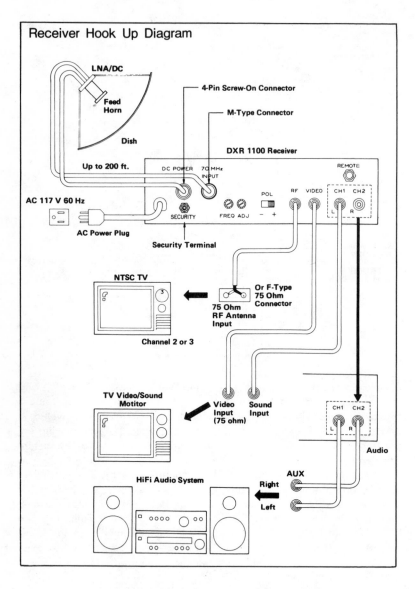

Fig. 4-6. Hook-up diagram for a TVRO system.
(Courtesy Dexcel Inc.)

Fig. 4-7. A 10-foot single-piece spun aluminum TVRO antenna with button-hook feedhorn.

Fig. 4-8. A popular 11-foot aluminum mesh antenna with remote-control motorized mount.
(Courtesy Paradigm Manufacturing, Inc.)

Fig. 4-9. A 6-foot molded fiber glass TVRO antenna. Receiver designer
Werner Vazken is at the right.

Fig. 4-10. The Luly umbrella antenna
(Courtesy Luly Telecommunications, Inc.)

Fig. 4-11. The author (left) and Lloyd Covens (publisher of Channel Guide) on the author's roof in downtown San Francisco.

focuses an arc of multiple satellites on five or six points of focus. Tv detectors or "feedhorns" are placed at each focal point, allowing the owner of a spherical antenna to pick up over 90 television channels simultaneously (Fig. 4-12).

Sphericals can be very inexpensive. Popular TVRO hobbyist units use woodstrips and window screening mesh as their major components, and a complete kit for the "do it yourself" experimenter (Fig. 4-13) costs under $700. By using construction plans available from several sources, the creative TVRO hobbyist can put together his own spherical antenna for under $200 in parts! In Chapter 9, basic construction details for building a "home brew" spherical are included.

The parabolic TVRO antenna must be properly positioned to point at the precise location of the satellite in space. The antenna is rotated right-to-left and up-and-down through adjustable mechanisms on its mount. The right to left rotation is known as the "azimuth" and extends from zero to 360 degrees. By convention, due north has an azimuth of zero/360 degrees, east is 90 degrees, south is 180 degrees, and west is 270 degrees. The up-and-down positioning of a TVRO antenna is called

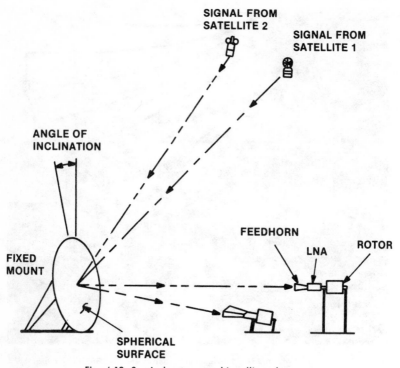

SIGNAL FROM
SATELLITE 2

SIGNAL FROM
SATELLITE 1

ANGLE OF
INCLINATION

FEEDHORN

LNA

ROTOR

FIXED
MOUNT

SPHERICAL
SURFACE

Fig. 4-12. Sperical antenna multisatellite advantage.

the "elevation." When the antenna points straight up, its elevation is said to be 90 degrees. When the antenna is aimed at the horizon, it has an elevation of zero degrees. When pointing a TVRO antenna at a specific satellite, the azimuth and elevation angles must be determined for the particular location on the face of the earth where the antenna is physically placed. (Appendix E gives several ways to calculate the proper azimuth and elevation (AZ/EL) angles including a computer program which will run on an Apple® II home computer.) The spherical has an advantage over the parabolic in initial alignment; once it has been positioned properly, its azimuth and elevation need not be changed.

There are two popular parabolic antenna mounts, the AZ/EL and the polar mount. (The less popular fixed mount is rarely found today.) In each structure, adjustments allow for the correct positioning of the antenna. The AZ/EL mount corresponds most readily to the azimuth and elevation angles normally supplied by the antenna manufacturers as an aid to locating the satellites in their geosynchronous orbital slots. The polar mount is recommended. With this mount, the azimuth and elevation adjustments are replaced with "hour-angle" and "declination" ad-

Fig. 4-13. A sperical antenna constructed from a kit. (The feedhorn for this particular antenna is located on a pole out of sight to the right of the picture.)

justments. By first setting the proper declination angle, the polar mount antenna is able to (approximately) sweep out the various satellite geosynchronous orbits by swinging its hour axis from left to right picking up satellite after satellite as the antenna is rotated through the arc. Minor adjustments are still necessary because the actual geosynchronous arc forms a true circle only for a viewer positioned directly below it on the Equator. From the higher North American latitudes, a slight angular error occurs, but this can be easily compensated by an offset adjustment. A simple polar mount is shown in Fig. 4—14. Notice the linear activator bar (motorized mount) at the left of the antenna pedestal.

Some manufacturers make motorized antenna mount systems which permit the user to control the antenna position by remotely dialing the correct azimuth and elevation angles via a small control console located in the living room. Other motorized antenna mounting systems use the polar configuration, and only rotate the antenna through its hour axis. These antennas cannot be positioned readily to look at satellites—such as the Russian domestic Molyna series—which are not placed into geosynchronous orbit. For 99% of the television satellites, however, the motorized polar mount antenna is probably the best choice. Some deluxe TVRO systems allow the user to type the desired satellite name into a home computer which automatically steers the antenna to the proper location. Some of the newer satellite receivers (Fig. 4—15) allow the user

Fig. 4-14. Rear of an 11-foot Paradigm antenna with a simple polar mount.
(Courtesy Paradigm Manufacturing, Inc.)

to simply dial the desired satellite by name or number; the proper azimuth and elevation angles are preset into the receiver mount controller module by the factory or installing dealer!

At the focal point of the antenna, the feedhorn is positioned to pick up the weak signals that the antenna has amplified through reflection and focusing. The feedhorn is a metal signal reflector and transmission waveguide which picks up and carries the microwave signals into the front end of the low noise amplifier. There are two popular types of feedhorns. The prime focus feedhorn (Fig. 4–16A) is placed at the focal point in front of the antenna, and the low noise amplifier (LNA) is bolted directly onto it. The cassegrain feedhorn (Fig. 4–16B) is located at the rear of the antenna, and the microwave signal energy, picked up by a small subreflector mounted at the focal point of the antenna, is routed through a metal funnel to the rear of the antenna, where the LNA is mounted. Both feedhorn assemblies are popular in TVRO antenna design, although the prime focus configuration is seen in greater numbers. A circular or "scalar" feedhorn (Fig. 4–17) provides a slightly better ability to collect stray microwave signals than rectangularly shaped

Fig. 4-15. A deluxe receiver with built-in controllers that activate the motorized antenna mount mechanism to effortlessly switch from satellite to satellite.
(Courtesy Microwave General)

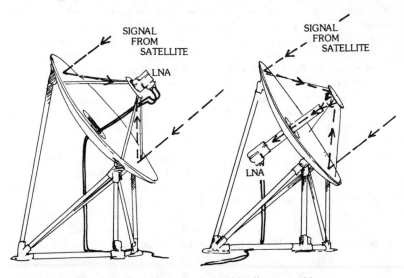

Fig. 4-16. The two popular types of feedhorn assemblies.
(Courtesy Luly Telecommunications, Inc.)

Fig. 4-17. The Chaparrel feedhorn bolted to an LNA for a KLM prime focus feed antenna.
(Courtesy KLM Electronics)

feedhorns, and can add as much as one-half decibel of increased signal power to the satellite reception. In marginal TVRO systems using 8- to 10-foot dishes, circular feedhorns will often make the difference between receiving poor pictures filled with "sparklies" and relatively clean noise-free ones.

The low noise amplifier (Fig. 4–18) is bolted to the end of the feedhorn. It is an extremely sensitive high frequency transistor circuit which can amplify the exceptionally weak signals received from the distant satellite, without swamping them with excessive noise generated in the amplifier itself. Because the microwave signal coming from the satellite transponder is so very weak (a typical satellite transponder has a power output of 5 to 11 watts), substantial preamplification is necessary. The received signal is less than one-millionth as strong as the local common carrier terrestrial microwave repeater output. At these low signal levels, even the natural noise produced by the radiated heat energy from the earth into space can interfere with the distance satellite signals. LNAs are, therefore, rated in terms of their noise figure represented by Kelvins (K). Noise becomes the dominant factor. The actual amount of noise the LNA generates is compared with that of a device operated at zero Kelvin (minus 270 degrees Celsius). LNAs with lower noise figures are better, and popular LNAs operate in the 80 to 120 K range. The output of the LNA is fed through a special low-loss large-diameter coaxial cable known as "Heliax" into the "down converter." The down converter shifts the 3.7 GHz signal to an intermediate frequency (i-f) signal usually in the 70-MHz range. This i-f signal can then be carried over ordinary inexpensive coax cable.

Until recently, down converters were located at or inside the satellite

Fig. 4-18. A typical low noise amplifier (LNA).
(Courtesy Avantek)

receiver, requiring the bulky and costly Heliax cable to bring the 3.7 GHz tv signals from the LNA to the receiver. Most cable television systems still use this technique. It is now possible, however, to mount the down converter at the antenna itself, and many TVRO satellite receiver manufacturers use this "dual conversion" technique by providing a separate waterproof down converter electronics package mounted near the LNA. One LNA manufacturer, Dexcel, introduced a combined LNA/ down converter in one tiny package. This device, called a Low Noise Converter (LNC), allows low-cost coax cable to be run for hundreds of feet from the antenna to the home satellite receiver. Other LNA manufacturers are now producing similar units.

The satellite receiver is the most important component of the home satellite television terminal. Six of the most popular receivers are pictured in Fig. 4–19. Unlike a tv set, the satellite receiver does not have a built-in audio amplifier or picture tube, and is more akin to a shortwave receiver or stereo component tuner. The satellite receiver takes the 3.7 GHz or i-f television signal (depending on whether the down converter is located at the antenna or built into the receiver itself), and converts the signal to baseband video and audio, which appear on output jacks at the rear of the unit. The baseband video signal cannot be fed directly into a home tv set; the signal is similar to one which comes from a television camera. To display this video on a tv set, it must first be run through an rf modulator to convert the signal to a vhf television channel. The output of the rf modulator is then fed through a lead-in cable to the vhf antenna terminals on the television set, and the tv is turned to the vhf channel used (typically 3 or 4).

The baseband video output of the satellite TVRO receiver can be fed

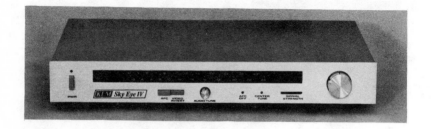

(A) KLM Sky Eye IV
(Courtesy KLM Electronics)

(B) Channel Master Model 6128.
(Courtesy Channel Master.)

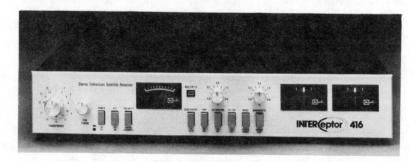

(C) Arunta Interceptor 416.
(Courtesy Arunta Engineering.)

Fig. 4-19. Six of the most

(D) Earth Terminals "Videophile."
(Courtesy Earth Terminals, Inc.)

(E) Birdview model 20/20.
(Courtesy Birdview Satellite Communications.)

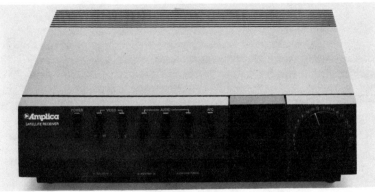

(F) Amplica receiver.
(Courtesy Amplica, Inc.)

popular TVRO receivers.

directly into a commercial tv monitor or projection television system, or into a video cassette recorder without the requirement that a modulator be used. This configuration improves the fidelity of the picture since the modulator, with its added noise and distortion, can be eliminated from loop. Most TVRO receivers provide built-in FCC-approved vhf modulators, and the audio output of the satellite receiver is heard through the tv speaker. The separate "audio-out" jack on the rear of the satellite receiver may also be connected via a shielded-wire audio cable to the auxiliary input of the home stereo hi-fi amplifier.

Dozens of companies manufacture satellite television receivers. The granddaddy of the home TVRO market is the International Crystal unit of Fig. 4–20. Their first model was introduced in 1979. Like most units, the International Crystal receiver tunes all channels on the 24 transponder birds, and allows the user to select between the various audio subcarriers which appear along with the video picture. Many satellite programmers transmit up to ten or more additional audio channels along with the tv program audio. By varying the audio subcarrier tuning control on the satellite TVRO receiver, these other audio channels may be heard as well.

Twenty-four transponder communications satellites utilize a frequency reuse technique known as "polarization." Twelve channels are transmitted using microwave signals which are polarized vertically with respect to the positioning of the satellite. A second group of 12 channels are transmitted using microwave signals which are polarized horizontally. The waves groups formed are orthogonal, or at right angles to each other, and they do not interfere with each other, allowing the 24 channels to be overlapped in frequency. Thus, each satellite will consume only half as much bandwidth as would otherwise be needed to transmit 24 simultaneous channels.

Fig. 4-20. The "Granddaddy" of the home TVRO receivers.
(Courtesy International Crystal Manufacturing Co.)

Back to earth, the TVRO feedhorn/LNA combination is rotated along its axis 90 degrees to pick up the separate vertical and horizontal 12-channel transponder groups. This is often accomplished by using a small television antenna rotor. The Chaparral "polar rotor" electronic polarizing device performs the same function much more elegantly, by electronically rotating just the front end of the feedhorn probe instead of the entire physical assembly.

The commercial TVRO satellite terminal solves the problem another way. Two separate LNAs, with separate cable feeds, are mounted on a special feedhorn that splits the signals into their corresponding vertical and horizontal components. The 12 horizontal transponders are shifted upward in frequency by 20 MHz (one-half transponder) from their corresponding vertical neighbors, but they all share the same 3.7 gigahertz band. Thus, any satellite TVRO receiver is capable of tuning the full 24 satellite transponder set by providing a fine tuning "frequency offset" control on the front panel of the receiver.

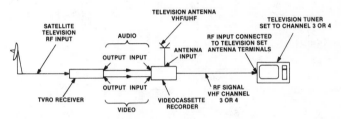

(A) Simple straight-through connections.

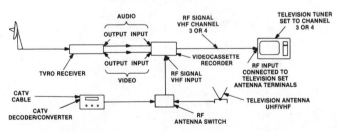

(B) Using a video cassette recorder with the satellite receiver.

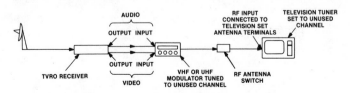

(C) System with TVRO, VCR, vhf-uhf local antenna, and CATV cable.

(Cont. on next page)

Fig. 4-21. Various home satellite tv configurations.

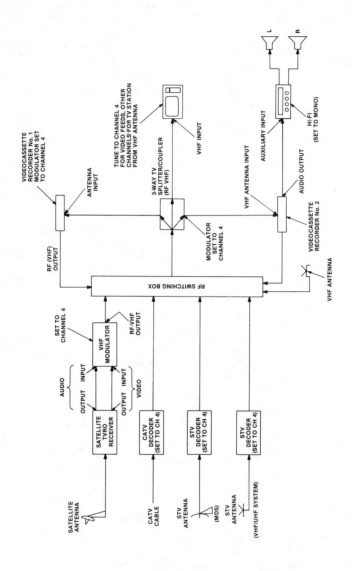

(D) System with two VCRs and other video services.

(Cont. on next page)

Fig. 4-21-cont. Various home satellite tv configurations.

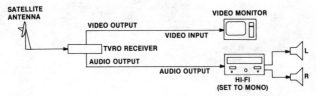

(E) System with video monitor and hi-fi sound.

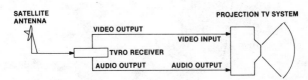

(F) System with projection-tv receivers.

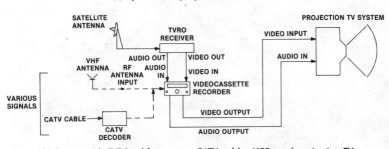

(G) System with TVRO, vhf antenna, CATV cable, VCR, and projection TV.

Fig. 4-21-cont. Various home satellite tv configurations.

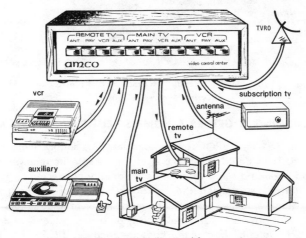

Fig. 4-22. An rf switcher to simplify connections.

Fig. 4-23. Combination satellite TVRO receiver and projection-tv system.
(Courtesy Microwave General.)

PLUGGING TOGETHER A HOME SATELLITE TVRO SYSTEM

The home satellite TVRO may be considered as simply another television source. The video and audio outputs of the satellite receiver may be connected through a small vhf channel modulator connected to the tv set antenna terminals, placing the signal on a regular tv channel (Fig. 4–21A). Many satellite receivers have built-in vhf modulators whose output signal appears on Channel 3 or 4, on a rear panel-mounted jack labeled "RF Output." If a video cassette recorder (VCR) is used, the receiver vhf channel modulator may be eliminated or bypassed; the VCR has its own built-in vhf modulator. In this case, a cable directly connects the audio and video outputs of the satellite TVRO receiver with their corresponding inputs on the video cassette recorder (Fig. 4–21B).

Over 25% of US homes are now wired for cable television. In Canada, over 50% of the households have cable. The TVRO satellite terminal will replace the premium channels, but the basic cable service will usually be retained to provide for clear reception of local television channels. The two systems can be used together in a complimentary fashion (Fig. 4–21C). If the TVRO satellite receiver rf output is set to a vhf channel different from the one the cable decoder uses, these two signals can be mixed together through a simple "vhf antenna combiner" and connected to the vhf input terminal of the video cassette recorder. In this configuration, the VCR becomes the master control for the entire television system.

The easiest way to simplify the jumble of connections that usually results is to switch the various video devices at their vhf channel rf signal feeds, using an inexpensive rf switcher. This configuration allows two or more video cassette recorders along with various other video signals to be routed both to the television set as well as to one or more VCRs (Fig. 4–21D).

The home satellite TVRO terminal works well with projection television systems and television monitors (Fig. 4–21E, F, and G). These devices directly accept baseband video and audio signals. The rf modulator is bypassed, improving picture quality. With this arrangement, most projection systems can provide big-screen studio quality pictures. The best (and simplest) configuration couples the home satellite TVRO terminal to a video cassette recorder, projection television system, and hi-fi stereo amplifier. With this arrangement, tv pictures received from the satellites can be razor sharp; better than the prerecorded movie tapes available in the stores!

Adding an rf switcher as shown in Fig. 4–22 simplifies the routing of the various tv signals and minimizes the jumble of wires and cables. A combination satellite TVRO receiver and projection-tv system is pictured in Fig. 4–23.

CHAPTER 5

SMATV and the Private Cable System: How to Cash in on the Booming Satellite TV Phenomenon

Cable television used to be a mom and pop business. Community antenna television systems were formed to pick up weak tv broadcasts from distant cities using a common antenna system mounted on a tall tower or a nearby mountain top. Thousands of these "minicable" systems still exist today, with hundreds of new franchishes being issued annually. Many are members of the Community Antenna Television Association (CATA), an Oklahoma based organization of small and independent cable television operators. There are still many opportunities open for entrepreneurs to form small cable companies, and the giant corporations continue to consolidate the industry by buying out the small systems when their owners put them up for sale. The new television satellites have made it easy to enter the mini-CATV business by providing a "big league" look with the fancy movie channels, sports and news networks, etc. available on the birds.

Many of the private CATV systems throughout the country service only a single housing development, apartment building, or condominium complex. Often they are connected to the existing master antenna television distribution system, the MATV cable which was originally installed by the building developer.

The differences between a small system CATV and an MATV system are often subtle, and they are easily confused. The nonfranchised private CATV systems, known as satellite-based MATV (or SMATV) systems are usually owned and operated by entrepreneurial organizations which are not the housing developers or management companies. Private CATV systems charge fees. MATV system costs are usually included in the monthly unit rental, and are provided by the building management.

Mini-CATV system cables often cross public property and right-of-ways, and thus require Public Utility Commission licenses. The MATV and SMATV or "private-cable" systems usually do not cross public property, and thus they are not so franchised.

New developments are rapidly occurring in the private CATV business. Office buildings are now being wired for business information and communications users to receive Reuters, Dow Jones, AP, UPI, and other satellite-based services (Fig. 5–1). Video conferencing may also be provided via private CATV systems. New two-way amplifiers allow computer communications and television programming to share the same coaxial cable. Viacom International (San Francisco) and Manhattan Cable are now jointly providing, along with Tymnet (the data communications common carrier), a New York-to-San Francisco high-speed computer data link. The cable companies utilize their local city distribution networks to distribute computer feeds to business data terminals within the two cities. Today, both private CATV and SMATV systems can easily plug into the communications satellites via their own TVRO earth stations (Fig. 5–2), providing both data communications and television services to their subscribers.

TECHNOLOGY OF THE PRIVATE CABLE SYSTEM

The commercial CATV system has an earth station which is similar to a home satellite TVRO terminal. At the headend, a receiving dish, LNA, and down converter are required. Where the usual home configuration

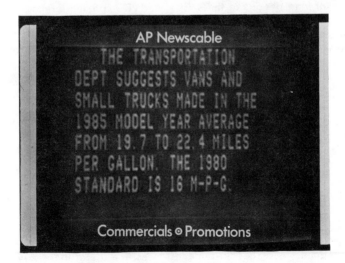

Fig. 5-1. Example of newswire service available on cable.
(Courtesy Associated Press)

Fig. 5-2. A fiber glass dish antenna feeding an SMATV system.

has a single TVRO satellite receiver, the commercial system will have a bank of independent receivers, each tuned to a different transponder. A downstream distribution system feeds these multiple channels to subscribers over a conventional coaxial cable. Existing MATV systems in multidwelling buildings can be upgraded to a "private cable" status by the installation of a new TVRO satellite terminal with the tv outputs tuned to unused vhf channels. A typical system is shown in Fig. 5–3.

The private cable system must function continuously on an unattended basis over long periods of time. Thus, its TVRO satellite terminal equipment should be installed to commercial specifications and standards. Often a larger antenna will be used to ensure that sufficient signal will be available to feed the bank of satellite receivers, providing broadcast-quality pictures. A parabolic antenna is most often used, but spherical antennas can enable the private cable operator to receive several satellites simultaneously. This gives the system the flexibility of choice, especially if special events (such as the world title boxing matches) are carried on a satellite other than Satcom 3. Sphericals are relatively maintenance free, and the typical open-mesh spherical provides less resistance to wind than does a solid-surface parabolic dish.

The TVRO antenna should be located to minimize interference from existing or future terrestrial microwave systems, especially those operated by the telephone companies which utilize the same 3.7 GHz band. Sometimes it is necessary to clear away trees which block the antenna (Fig. 5–4). Two engineering organizations, Compucon of Dallas, Texas, and ComSearch of Virginia, specialize in providing computerized data base frequency searches. In some instances, the use of these services is invaluable, especially when it is not possible to use a portable Luly umbrella antenna for a temporary site test.

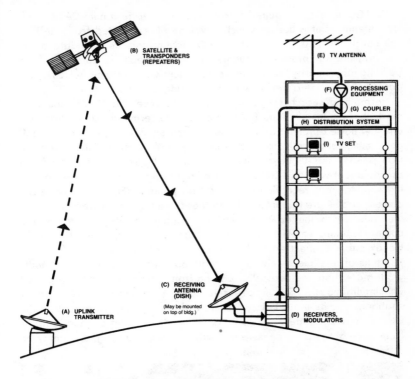

Fig. 5-3. Diagram of a mini-CATV or SMATV system.
(Courtesy Anixter-Mark, Inc.)

Fig. 5-4. Microwave engineering with a chain saw!
(Courtesy Channel One)

No FCC license is required to construct or operate a mini-CATV satellite antenna, but a license which is voluntarily requested from the FCC will cause the FCC to enter the TVRO location into its own data base. This registration process prevents the local telephone company from building any nearby future terrestrial microwave systems which could possibly cause interference with the existing TVRO antenna.

Normally a search for microwave interference (Fig. 5–5) should be made before installation. The home satellite terminal usually has only one low noise amplifier and feedhorn assembly which is either electronically or mechanically rotated, allowing both vertical and horizontal transponder sets to be received. Commercial systems use a feedhorn arrangement with two LNAs bolted at right angles to each other allowing simultaneous reception of all 24 transponders (Fig. 5–6). Separate coaxial cables are required to bring these two LNA feeds from the satellite antenna to the down converters (usually an integral part of the communications satellite receivers). Short runs of expensive and bulky Heliax cable connect each LNA to the receiver electronics.

Banks of multiple receivers are required, since private CATV systems want more than one vertical or one horizontal transponder to be picked up simultaneously. Each receiver is tuned to a separate transponder channel, and its input is isolated from the other receivers through the use of "power splitters" or dividers which deliver two or more outputs from each of the LNA feeds. This headend equipment (Fig. 5–7) may be

Fig. 5-5. A spectrum analyzer, feed, LNA, and scope camera assembled for microwave interference check at a suspected noisy site.
(Courtesy Channel One)

located in a small metal shed or enclosure adjacent to the TVRO antenna. The system pictured in Fig. 5–8 provides six simultaneous channels of satellite television. It can be remotely controlled through a dial-up telephone link.

The baseband video signal outputs from the multiple-receiver bank are then fed into separate vhf channel modulators, and the combined outputs of the modulators are mixed together through a signal combiner unit. This composite multichannel television signal is then fed into the cable distribution system of amplifiers and tv set couplers.

Cable television and master antenna tv systems also provide their viewers with ghost-free reception of the local television broadcast stations. Individually-tuned vhf and uhf antennas at the headend "antenna farm" (Fig. 5–9) feed banks of vhf and uhf tuners. Known as broadcast-tv signal-processing equipment, it is normally installed in a rack adjacent to the multiple-channel TVRO receivers. The signals are then likewise coupled into a cable distribution system via the signal combiner.

Many private CATV systems offer several tiers of programming. The basic monthly service will include the local television stations and one or more satellite feeds such as CBN, C-SPAN, etc. On a second tier, first-run, pay-television movies, such as those provided by SelecTv, are carried. These systems provide for a mechanism to lock out the subscribers who wish to obtain only the basic monthly service.

One way this is accomplished is by placing a small trap in series with the incoming cable at the customer's tv antenna leads to prevent the pay-television signal from reaching the tv set. This trap is removed when the subscriber takes the premium tier of programming. Several traps placed in series can block several premium channels in a multitier system. Fig. 5–10 shows a typical system.

Another technique places the premium services on midband cable channels. These cannot be tuned on a regular television set. A block converter connected to the incoming cable in the home translates these hidden channels onto unused vhf channels. Block converters (Fig. 5–11) are more expensive than traps, costing about $17 as compared to $1.00, but they provide for greater security.

The major drawback to block conversion techniques is that many of the newer television sets and video cassette machines can now tune these hidden channels (although this may not be stated in their instruction manuals). This is because the midband channels are located between vhf Channels 6 and 7 on the television dial while the superband channels are located between the upper vhf and lower uhf bands. If a private cable company wired up an apartment building of video enthusiasts using the block conversion technique, revenues would rapidly drop as the grapevine circulated the information on how to find the not-so-secret channels.

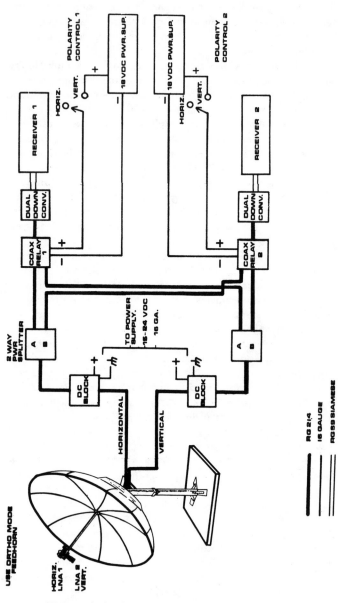

(A) Basic dual-polarized, two-receiver system.
(Courtesy Interset Corp.)

Fig. 5-6. Typical receive

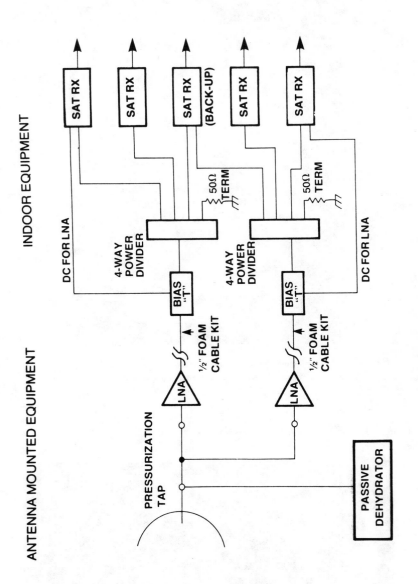

(B) Dual-polarized, multiple-receive system.
(Courtesy AM/COM Video Satellite Inc.)

terminal configurations.

Fig. 5-7. A self-contained "headend" system consisting of a bank of TVRO receivers,
a bank of signal processors, and modulators.
(Courtesy Channel One)

Fig. 5-8. Avantek AR1000 multichannel SMATV and CATV receiver system.
(Courtesy Avantek)

Fig. 5-9. A typical antenna farm for a CATV system showing TVRO satellite dish and an array of vhf and uhf antennas to pick up distant "off-the-air" tv station signals.
(Courtesy John Kinick, Kintech)

A full-channel CATV decoder is the most effective way of handling this problem. Many decoders have built-in descrambling systems which work in conjunction with channel scramblers located at the private CATV headend. Newer systems enable each box to have its own electronic serial number, and each decoder be turned on and off by remote control at the system headend—or through the use of a telephone modem from even another city. These addressable systems allow for the immediate upgrading of customer services, and can provide secure presentation of pay-per-view events. The unit in Fig. 5–12 also features a full-feature wireless remote control.

However, the most effective inexpensive way that a private cable system can block the pirate instinct of the viewer is to offer all subscribers a single monthly service charge which covers the cost of all services, including the pay movie channel(s).

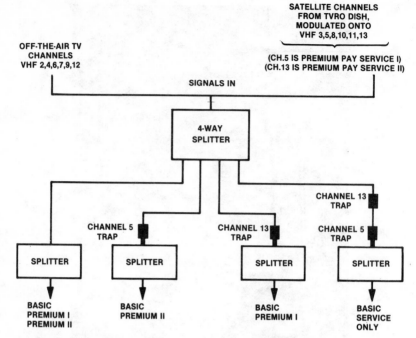

Fig. 5-10. A typical "home operated" SMATV system for Denver. It provides for 6 "off-the-air" channels and 6 satellite channels. Traps are used to block pay channels from unauthorized subscribers.

PRIVATE CATV PROGRAMMING AVAILABILITY

There are many satellite programmers whose services can be picked up by the private CATV company either for free or at a negligible cost. The most popular of these is the religious channel, CBN, which is not considered to be a broadcast television service under FCC rules, and thus carries no government restrictions. This is not true of the TBN religious network, which rebroadcasts the signal of a Los Angeles uhf station. Carrying TBN automatically causes the system to be defined as a CATV operation under FCC rules. In general, it is better to avoid regulation, at any governmental level, whenever possible. Carrying CBN requires that a one-page authorization agreement between the programmer and the mini-CATV system be signed. This agreement is readily attainable from CBN.

Cable News Network, ESPN, and Nickelodeon are popular services that are provided to cable systems for between $.04 to $.20 per subscriber per month. They are usually included by the cable company in its basic programming package.

Fig. 5-11. A block converter used with an SMATV system to decode a group of channels
for the local subscriber.
(Courtesy Oak Communications)

Fig. 5-12. The MAAST multiple-audio secure cable tv/STV decoder.
(Courtesy Telease)

The Modern Satellite Network, Satellite Programming Network, C-SPAN, and ACSN, are either provided for free to the cable company, or charge a few cents per viewer per month. They are often included in the basic-tier programming package that the subscriber receives as part of the CATV service.

The chief advantage of these services is their low cost of acquisition. However, they tie up many of the available tv channels on the cable system, since each requires a separate TVRO satellite receiver and headend modulator. Many subscribers prefer a second and third movie channel instead of a public service or news channel.

The "sweetener" for the private cable system is the pay-tv movie service. Movies entice the subscribers to take the overall service package. Although charges vary from supplier to supplier, all pay television satellite programmers have set sliding scale fees based on the price that the private cable system charges the home subscriber, and the total number of premium subscribers in the system. These rates typically range from $2.50 to $5.00 per-month, per-subscriber for each premium-tier viewer. Since the fees are determined in part on the retail rate the cable tv company levies to minimize these payments, most cable systems offer the movie services on a separately billed tier.

The three national superstations—WTBS-TV, WOR-TV, and WGN-TV—are available to FCC-recognized cable companies for from $.10 to $.15 per month per cable subscriber. To carry the service, another single-page authorization agreement is executed between the cable company and the appropriate "Resale Common Carrier" which feeds the superstation over the satellite (Fig. 5–13). Superstations are television broadcasting stations, regulated by both the FCC and the Commerce Department. Under the FCC Rules and Regulations, a private cable system which has 50 or more subscribers cannot regularly "import" the signal from distant superstations unless all local over-the-air television stations are first carried. (This rule is being gradually relaxed, with condominiums and housing associations now exempt.) The Copyright Tribunal imposes conditions on private CATV systems which carry local or imported tv broadcasting stations. Copyright fees begin at a few dollars per month and can range upward to several thousand dollars per year if the CATV system has thousands of subscribers. These fees are collected by the Federal Government and distributed to the actors, directors, and other artists involved in the production of the original television programs.

Local vhf and uhf television stations come under these same regulations which govern the superstations. Most private CATV systems carry one or more local tv stations. It is, therefore, important to note that the carriage of only "nonbroadcasting" satellite programs and networks by a private CATV system will effectively eliminate it being designated by any branch of the federal government as a CATV company. This automatically exempts it from any rules and regulations at the federal level. Conversely, to be classified by the FCC as a cable system, the private CATV or SMATV operation which has more than 50 subscribers need only pick up one superstation or local television station and distribute it over the cable. (An association of subscribers who jointly own the SMATV system is exempt.)

At present, it is not officially possible for a private CATV system to

SATELLITE SERVICE REQUEST AND AGREEMENT

1. This Agreement, made and effective this ___ day of _____, 19__, is by and between United Video, Inc., a communication common carrier, with its principal place of business at 5200 S. Harvard, Suite 215, Tulsa, Oklahoma (hereinafter referred to as "Carrier" and _____ (hereinafter referred to as "Customer").

2. To receive signals of WGN-TV, Chicago, Customer hereby orders satellite transmission service from Chicago, Illinois to Customers' receiving location(s). Customer agrees to payment of the monthly rates designated on Schedule 1 (reverse side) and further described in Carrier's Tariff #13, which is on file at the Federal Communications Commission (FCC). This Agreement and the obligations imposed hereby is and shall be governed by, subject to, and interpreted in accordance with all terms, rules and regulations, charges, practices and conditions of Carrier's tariff, including supplements or amendments thereto. Carrier and Customer shall comply with all provisions of that tariff, which is incorporated herein by reference and shall be part of this Agreement.

3. Customer shall be responsible for any necessary FCC and Copyright registration or authorization to permit Customer to utilize the service provided by the Carrier. Customer shall be responsible for construction and operation of earth station facilities necessary to receive Carriers' signal.

4. Since continued availability of this satellite transmission service is subject to the Carrier maintaining all necessary FCC, Copyright and other legal authority, as well as other factors, Carrier shall not be responsible for interruptions or discontinuance of satellite service. In the event of such interruptions or discontinuance of satellite service, the liability of the Carrier shall be limited to a pro rata refund of any service charges paid by Customer for the period during which service was not provided.

5. Customer agrees to receive this transmission service for a period of no less than three (3) years from the date service is implemented to the location(s) given on Schedule 1 (reverse side). Payment for service is due at the office of Carrier on or before the 1st day of the month for which service is provided.

WITNESS the due execution of this Agreement on the day stated above.

CUSTOMER

_____
Authorized Corporate Name

_____
Street Address

_____
City, State and Zip Code

(X)_____
Authorized Officer or Signer

_____
Typed Name of Signer

_____
Area Code and Phone Number

CARRIER

United Video, Inc.
_____
Authorized Corporate Name

5200 South Harvard, Suite 215
_____
Street Address

Tulsa, Oklahoma 74135
_____
City, State and Zip Code

_____
Authorized Officer or Signer

_____
Typed Name of Signer

1-800/331-4806
_____
Area Code and Phone Number

Fig. 5-13. A satellite service request and agreement.

carry the "Big 3" commercial networks and the Public Broadcasting Service directly from the satellite. Networks will not sign letters of authorization to allow cable companies to pick up their private feeds, because of prior agreements that they have entered into with their own local affiliates. Each of the three television giants have introduced new made-for-cable networks to infiltrate this market with new products. For example, ABC's ARTS service is available for carriage. Although the FCC restricts US cable companies from picking up over-the-air Canadian television and broadcasting stations, no official agreement between the two countries exists. Therefore, the nonregulated private CATV systems which are near the Canadian border might legally carry Canadian tv channels, complete with their movies and situation comedies, game shows, adventure stories, etc.

Canadian cable tv companies are restricted to carrying their own country's satellite feeds. However, hundreds of northern private CATV systems have for years aimed their TVRO antennas "south of the border," taking programming from the American communications satellites. Since recent Canadian government rulings have been in conflict on this issue, most of these smaller cable systems continue to pick up US satellite television. To counter this embarrassing situation in the marketplace, a number of Canadian pay-tv movie channels are being developed, and the Cancom four-city superstation network is now available on the Anik-D satellite.

For the private cable system, many programming alternatives are available. The typical operator will carry two to three local tv stations, a religious channel, the Cable News Network, a sports channel, C-SPAN, a pay-tv service, and one superstation. If money is no object, a private CATV system could easily carry over 30 satellite-fed television channels, and the number of new services is increasing every month.

A list of satellite-based program services which are officially interested in providing their feeds to the private cable operator is given in Table 5—1. A number of other programmers, including several of the cable-oriented pay-tv networks, are quietly providing their movie feeds to SMATV companies. As time passes, most of the satellite programs and channels available to the big CATV companies will also open up to the SMATV market. Most programmers have special requirements for rates, number of subscribers, and system size. Contact them for complete details.

RULES AND REGULATIONS GOVERNING PRIVATE CABLE

Private CATV systems fall under the rules and regulations of both the Federal Communications Commission and the Copyright Tribunal (Commerce Department). If a CATV system carries a local broadcast television station or a satellite-fed superstation, the operation is considered to be a cable system by the Copyright Tribunal. Such a cable company must complete a formal application and make semiannual payments according to a fee schedule based on the gross yearly revenues. If the minicable company has 50 or more subscribers, it also comes under the jurisdictional control of the Federal Communications Commission and a second set of forms must be filed with the FCC on an annual basis. CATV systems thus regulated also fall under the jurisdiction of all federal laws concerning Equal Employment Opportunities, as well as those regu-

Table 5–1. Satellite Services Open to Deal With SMATV (Private Cable)*

Service	Name	Address
ACSN	Appalachian Comm. Service Netwk (The Learning Channel)	1200 New Hampshire NW Suite 240 Washington, DC 20036 (202) 331-8100
	American Network (hotels, hospitals only)	735 Water St. Milwaukee, WI 53202 (414) 276-2277
CBN	Continental Broadcasting Network, Inc.	Pembroke Four Virginia Beach, VA 23463 (804) 424-7777
CNN	Cable News Network Cable News Network 2	1050 Techwood Dr. Atlanta, GA 30318 (404) 898-8500
CMTV	Country Music Television	Telstar Corp. 6455 S. Yosemite Englewood, CO 80111 (303) 850-9101 or (213) 659-4354
C-SPAN	Cable Satellite Public Affairs Netwk	400 N. Capital St Suite 155 Washington, DC 20001 (202) 737-3220
EROS	EROS	2 Lincoln Square Suite 18A New York, NY 10023 (212) 595-7900
ESPN	Entertainment/Sports Network	ESPN Plaza Bristol, CT 06010 (203) 584-8477
EWTN	Eternal Word TV Network	5817 Old Leeds Rd Birmingham, AL 35210 (205) 956-9535
FNN	Financial News Network	2525 Ocean Park Santa Monica, CA 90405 (213) 450-2412
GALA	Galevision	250 Park Avenue New York, NY 10020 (212) 484-1241
HTN	Home Theater Network	465 Congress St Portland, ME 04101 (207) 774-0300

*As of May 1, 1983

Courtesy Channel Guide

Table 5- 1-cont. Satellite Services Open to Deal with SMATV (Private Cable)

Service	Name	Address
TMC	The Movie Channel (hotels, hospitals only)	1133 Avenue of the Americas New York, NY 10036 (212) 944-5532
MSN	Modern Satellite Network	5000 Park St. North St. Petersburg, FL 33709 (813) 541-7571
NCN	National Christian Network	1150 West King St. Cocoa, FL 32922 (305) 632-1000
ON	ON-TV/Oak Media	16935 W. Bernardo Dr. Roncho Bernardo, CA 92127 (619) 485-9880
PTL	People That Love TV Network	Charlotte, NC 29279 (704) 542-6000
	Pleasure Channel	1888 Century Park East Suite 1106 Los Angeles, CA 90067 (213) 556-0680
SelecTv™	SelecTv™	4755 Alla Road Marina Del Ray, CA 90291 (213) 827-4400
SIN	Spanish International Network	250 Park Avenue New York, NY 10017 (212) 953-7500
SPN	Satellite Programming Network	P.O. Box 45684 Tulsa, OK 74145 (918) 481-0881
TBN	Trinity Broadcasting Network	P.O. Box A Santa Ana, CA 92711 (714) 832-2950
USA	USA Cable Network	1271 Avenue of the Americas New York, NY 10020 (212) 484-1866
	The Weather Channel (Requires local equipment)	Landmark Communications 2840 Mt. Wilkinson Pkwy Atlanta, GA 30339 (404) 434-6800
WGN-TV	Channel 9, Chicago	United Video 5200 S. Harvard Suite 215 Tulsa, OK 74135 (800) 331-4806

Table 5- 1-cont. Satellite Services Open to Deal with SMATV (Private Cable)

Service	Name	Address
WOR-TV	Channel 9, New York	Eastern Microwave P.O. Box 4872 Syracuse, NY 13221 (315) 455-5955
WTBS-TV	Channel 17, Atlanta	Satellite Syndicated Service Tulsa, OK 74145 (918) 481-0881

lations issued by other agencies that the FCC determines apply. These rules do not apply, however, to a master antenna television system which is included as part of tenant's monthly rent. Therefore, it may be to the advantage of a new private CATV system to install a separate cable feed into each apartment or dwelling unit and deliver the non-broadcast satellite television services physically over a different wire. The existing MATV system is retained for reception of local television broadcasts only.

Local building codes must, of course, be adhered to by the private cable operator. Some cities have established height limits which require that building permits be obtained before a TVRO antenna can be installed on a roof, or even in a parking lot. This process may involve retaining the services of a licensed professional engineer to verify that the roof structure can hold the several hundred pounds of additional weight. Insurance policies should also be reviewed if a 15-foot dish is attached to a building structure. An adjacent parking lot or field is usually a better location to place the TVRO and headend equipment. A small fence can be erected to protect the antenna and equipment shed.

Some cities and towns require business permits be obtained for any organization which deals with the public. These are usually issued automatically, and are concerned primarily with retail establishments rather than with the specific operation of a cable system. Specific laws concerning the operation of a CATV system fall within the jurisdiction of the city and state Public Utility Commissions (PUCs).

The PUCs define cable companies as being similar to the common carrier organizations and power companies who also string their cables across public property. Thus, in many cases, a private CATV system which restricts its operation to private property, and does not cross any public street or right-of-way, can usually avoid PUC jurisdiction. The owner of a multiple building complex can issue a private CATV system written authority to extend its cables from the TVRO headend to each apartment or dwelling unit. In a suburban housing tract, the developer may wish to retain rights to a narrow strip of access land to each house, for use by a homeowner association. The private CATV cable can be buried along this route, or mounted on poles on this private right-of-way, provided that a CATV clause has been written into the individual property titles, allowing future access to the buried cable as it passes from one private property to another.

Many private CATV owners incorporate each system separately. There are both tax and liability-protection advantages to incorporating. Selling the system in the future will also be easier since a transfer of ownership can be effected by a transfer of stock. The private cable operator will also find it easier to deal with cable television programmers and equipment suppliers who are geared to do business with corporations rather than individuals.

Condominium associations and housing communities which establish their own private cable systems may elect to operate in a nonprofit mode. If the private cable system is jointly owned by all subscribers, FCC regulations do not consider the organization to be a cable system, and thus this form of private cable operation is exempt from FCC regulations.

Whatever the actual structure chosen, it is always to the advantage of the private cable operator to look and act like a CATV company rather than an SMATV operator to the satellite program suppliers. The satellite networks, especially the pay-movie and commercially supported programmers, prefer doing business with cable companies. In many instances they will refuse to deal with SMATV operators, and in some cases even with "nonfranchised" cable systems. The name of an association can change its appearance. Thus, the "Eldorado Community Television Company sounds like a CATV system. The Eldorado Apartment Antenna Service definitely does not!

GOING INTO BUSINESS: PROJECTED INCOME AND EXPENSES

The cost to construct and operate a private CATV system will vary from location to location and system to system. Expenses depend on the number of channels carried and the number of subscribers whose homes must be wired. These figures define the amount of hardware which must be purchased and installed.

There are seven major costs to consider: physical hardware with installation; satellite programming fees; employee salaries; sales commissions; operating expenses (mail, telephone, etc.); ongoing repair and maintenance costs; and miscellaneous expenses (for professional engineers, independent consultants, etc.).

To illustrate, let us consider two typical private CATV examples. Chart 5−1 presents the Eldorado Community Cable Company, a minicable system located in a 20-story apartment complex. The building has 400 dwelling units and it is assumed that 50% of these subscribe to the cable system, which will offer a bundled service package of six satellite-fed channels. This example is conservative, as experience indicates 65% to 85% of multistory dwellers will subscribe to a private cable system when it is offered.

Chart 5—1. Example 1—Eldorado Community Cable Co.

CATV System is 20-story condominium or apartment building. (Based on 200 sub-scribers at $22/month each, with installation fee of $50/unit, representing 50% of total potential subscribers)

Income:

1) Monthly receipts:	$ 4,400	
2) Installation fees (one-time):	$10,000	

Expenses (Capitalized Costs):

1) Hardware (at dealer cost) TVRO and CATV equipment	$2,500.00	15' TVRO antenna with polar mount (Paraclypse)
	1,700.00	Dual 120 degree with feed-horns and cabling (DEXCEL)
	400.00	2 Four-way power splitters
	9,500.00	One 6-channel TVRO receiver with VHF Channel modulators (AVANTEK model ARC-1000)
	300.00	1 Signal Combiner
	3,000.00	6 Distribution amplifiers and 400 cable "taps"; miscellane-ous cable, parts, clamps, etc.
	350.00	1 Color TV monitor
Total Hardware costs	$17,750.00	
5-Year Lease-Purchase at 20%/year	$470.27	
Total Principal and Interest Payback	$28,216.20	
2) Installation:	600.00	TVRO Antenna/electronics (30 hours at $20/hour)
	12,000.00	CATV cable distribution net-work (400 wired units a 2 hrs/unit, $15/hour)
Total Installation Costs	$12,600.00	
5 year lease-Purchase at 20%/year	333.82	
Total Principal and Interest Payback	$20,029.20	

Chart 5-1-cont. Example 1—Eldorado Community Cable Co.

Expenses (Monthly Costs):

1) Programming: Free	$ 0.00	CBN (or other religious network)
200 × $0.04	8.00	ESPN Sports Channel
200 × 0.15	30.00	CNN (Note: CNN and WTBS when taken together qualify for lower $.20 rate—not used here)
200 × 0.10	20.00	WTBS (or other Superstation)
200 × 6.00	1,200.00	Premium Movie Service
200 × 0.01	2.00	C-SPAN (Public Service Channel) (Note that Movie Service Represents over 95% of programming expense)
	$1,260.00	
2) Salaries	600.00	Part-time technician
	600.00	Part-time bookkeeper/secretary
	$1,200.00	
3) Commissions	100.00	For selling expense-ongoing
4) Overhead	100.00	Rent
	60.00	Utilities
	25.00	Phone
	65.00	Miscellaneous
	$ 250.00	
5) Other	200.00	Professional engineers, consultants, insurance, etc.
TOTAL MONTHLY COSTS (First 5 Years)	$3,010.00	
Total Operating Expenses: (including lease payments)	$ 3,814.09	(if $50 installation fee is applied to reduce initial loan costs, this drops to $3,549.15)

The above assumes that the cost of equipment and installation was financed. If not, then total monthly operating expenses are $3,010, and the return on an initial cost for equipment and installation of $20,350 is 82% per year, assuming a monthly profit of $1,390!

Chart 5-1-cont. Example 1—Eldorado Community Cable Co.

Net Profit Per Month:	$ 850.85	(Total over 5 years =
(Before Taxes)		$51,050.75
Net Profit Per Month After		
5 Years:	$ 1,027.50	
(Assumes salary/ overhead increases of 25%)		
Therefore, from the 60th month onward, the yearly net profit becomes:	$12,330.00	

The six programming services are the CBN Christian programming channel, the ESPN sports network, Cable News Network, WTBS-TV, C-SPAN, and a premium movie service. An industrial-grade TVRO system with dual LNA feeds and six separate satellite receivers along with the necessary CATV distribution amplifiers, taps, and cabling will be required for the installation.

Assuming that the 200 subscribers each pay $22 per month, the total monthly receipts amount to $4,400. A one-time installation fee of $50 per unit will generate $10,000.

Total hardware costs come to $17,750. When financed on a five-year lease-purchase arrangement at 20% simple interest, the monthly payments for this equipment are $470.27. Related installation costs amount to $12,600, or $333.82 per month.

The programming expenses are the principal operating costs of such a system and amount to $1,260. Salaries, commissions, operating overheads, and other expenses total $1,750, for an overall monthly operating cost of $3,814.09.

This produces a net profit of $585.91 before taxes, or a total profit of $41,945.40 over the five-year period. However, if the installation fee is applied to reduce the initial installation loan costs, then the total operating expenses fall to $3,549.15 per month, increasing the actual net profit per month to $850.85, or a total income over the sixty months of $51,050.75, before taxes.

After the loan payments have been retired, the net profit per month increases significantly. Assuming that salary and overhead expenses go up by 25%, the annual net profit from the sixtieth month onward becomes $12,330.00. This does not consider any possible increase in monthly charges to the subscriber.

In the second example (Chart 5-2), the "American Private Cable Corporation" operates a similar service for a thousand-unit garden-type apartment or townhouse complex. Assuming 50% penetration, the 500

Chart 5—2. Example 2—American Private Cable Corp.

CATV System is a thousand unit garden-type apartment or townhouse complex. (Based on 500 subscribers at $22/unit for 50% penetration of 1,000-unit garden apartment/townhouse complex, with installation fee of $50/unit)

Income:

1) Monthly receipts:	$11,000	
2) Installation Fees:	$25,000	

Expenses (Capitalized Costs):

1) Hardware	$2,500.00	TVRO Antenna with mount
	1,700.00	Dual LNA's, etc.
	400.00	2 four-way power satellites
	9,500.00	Satellite multi-channel receiver
	300.00	Signal combiner
	8,800.00	20 distribution amplifiers with cable taps, cable, etc.
	350.00	1 color TV monitor
	1,000.00	Enclosure and Fence
Total Hardware Costs	$24,550.00	
5-year lease purchase at 20%/year	650.42	
2) Installation	650.00	TVRO Antenna/headend
	30,000.00	CATV distribution network (500 units wired at 3 hrs/unit; $20/ hr)
Total Installation Costs	$30,650.00	
5 year lease-purchase at 20%/year	812.04	

Expenses (Monthly Costs):

1) Programming Free	$ 0.00	CBN
	20.00	ESPN
	75.00	CNN
	50.00	WTBS
	3,000.00	Movie Service
	5.00	C-SPAN
TOTAL	$3,150.00	
2) Salaries	1,500	Full-time technician
	1,500	Full-time office staff
	$3,000.00	
3) Commissions	200.00	

Chart 5-2-cont. Example 2—American Private Cable Corp.

4) Overhead	500.00	
5) Other/Miscellaneous	500.00	
TOTAL MONTHLY COSTS	$7,350.00	
Total Operating Expenses (includes lease payments where installation fees have been applied to lower loan costs)	$8,150.11	($7,350 + $650.42 hardware + $149.69 installation)
Net Profit Per Month: (Before taxes)	$2,849.89	Total over 5 years = $170,993.40

The system is quite profitable both during its first 5 years of operation, and thereafter. Assuming indirect costs increase by 25% and monthly fees are raised gradually to $25, the system will continue to throw off a positive cash flow.

subscribers generate monthly receipts of $11,000 with a total installation revenue of $25,000.

In this example, it is assumed that only units which subscribe to the private cable system are wired at the time of service request, and total installation costs amount to $30,650. Associated hardware equipment costs come to $24,550. Monthly operating expenses amount to $7,350, with the operation supporting two full-time employees. When the cost of financing the hardware and associated installation expenses is included, less installation fees paid, the total operating expenses come to $8,150.11 per month. This provides a net profit per month of $2,849.89, or a total revenue of $170,993.40 over the first five years.

An entrepreneur considering entering the private CATV fiield will discover that a dozen similar systems will quickly produce a net income of well over $1 million per year! Doubling the headend equipment costs will only slightly lower this figure. There are over 25,000 apartment buildings, housing developments, condominiums and mobile home parks with 200 or more dwelling units in the United States. An additional 50,000 multifamily dwellings have over 100 units each. At present, the cable penetration for the United States has reached a little less than 30%, thus over two out of three American families do not currently receive cable television services. Half of these households are located in multidwelling buildings. The 1980s are a time of opportunity for the mini-cable entrepreneur to "cream skin" these profits from the giant cable companies.

THE NEXT STEP: GETTING INTO THE PRIVATE CABLE BUSINESS

The best way to minimize a business risk is to go with the pros. Local professional engineers and cable television installation companies may be found throughout the United States. Any consultant retained should be intimately familiar with the local government agencies and the building codes of his area. The National Society of Professional Engineers, headquartered in Washington, DC, may be able to provide or recommend a qualified professional in a specific area. The state agency licensing boards also maintain directories of professional engineers and consultants, and "business card" advertisements in the back pages of the cable and broadcasting magazines are helpful.

The private cable operator should consider joining two national trade associations. The Community Antenna Television Association (CATA), located in Oklahoma, is a national membership organization of small CATV and SMATV operators. The National Cable Television Association (NCTA) in Washington DC is the cable industry's voice, tending to be dominated by the 50 giant MSO cable companies. Both associations hold annual meetings and trade shows which are worth attending, as are regional meetings held by independent state cable associations. One of the major groups on the West Coast is the California Community Television Association (CCTA) whose offices are located in Castro Valley, California. The CCTA holds its annual convention and trade show each November in Anaheim, and thousands of people attend.

SPACE, the Society of Private and Commercial Earth Stations, in Washington DC, has an SMATV division which actively represents the interests of the private cable operator at the FCC and on Capitol Hill. The National Satellite Cable Association (NSCA), located in Houston, performs a similar service for the SMATV operator.

Cablevision Magazine, TVC Magazine, and the C-ED Journal can provide useful information to the cable television operator, and Satellite TV, a new Denver-based publication, specializes in the SMATV marketplace. Most CATV equipment manufacturers and distributors can be of service, as well as the home satellite TVRO manufacturers.

Finally, the Satellite Center located in San Francisco, California publishes a detailed private CATV and SMATV study package for the entrepreneur, and can provide engineering, financing, and discount equipment purchasing services to the start-up private CATV operator on a "turn key" basis! The report, Starting A Private CATV System is available for $95 from the Satellite Center.

Further information on the satellite programmers and equipment manufacturers can be found by referring to Appendices B, C, and G. Appendix F discusses in more detail the rules and regulations applicable for legally operating a private CATV business.

Private cable will explode into a multibillion dollar per year business over the next five years. This is a giant new pie that even the smallest entrepreneur can share a piece of. Good luck!

CHAPTER 6

Putting Satellite Television to Work in the Business and Nonprofit Organization

Satellite television can be a boon to the commercial business enterprise. The innkeeper, tavern operator, and restaurant owner have one common element: They all depend on patrons for their business. Satellite television can create a sizzle and excitement which attracts new customers. Likewise, the hospital, church, and school can all use satellite television for informational, educational, and instructional purposes. In addition, the hospital, like a hotel, can provide first-run movie service, via a satellite TVRO earth station, to its patients, helping to entertain them during their post-operative and recovery periods.

SATELLITE TELEVISION AND THE HOTEL

Ninety-nine percent of hotels have internal MATV systems which feed color televisions in each guest room. Most of these "minicable" systems use relatively new equipment, and are properly maintained. Many hotels generate a profit from this equipment by providing an in-room pay-movie service which is charged on a pay-per-view basis.

This movie channel is often provided by Spectradyne, a Dallas-based firm which installs a combination CATV-like decoder and electronic register adjacent to each television set. Used in conjunction with the central billing computer, when any room guest selects a desired movie, the billing will be automatically posted to the room account via the Spectradyne equipment. The movies themselves are provided on videotape cassettes which are continuously running in a multidrive video playback system located in the hotel.

Other hotels take a single movie channel feed delivered via an independent MDS common carrier that provides the programming to the hotel complex via terrestrial microwave service. These hotels have small microwave antennas which are mounted on their roofs and aimed at the local MDS transmitter located on a nearby mountain peak or building top. The pay-movie channel is converted by a small down converter electronics package into a vhf television channel, which is then fed into the hotel master antenna television system (Fig. 6—1). Hotels which provide this type of movie service usually do so on an inclusive basis, offering the movies free as part of the nightly room charge.

The hotel management which installs its own TVRO earth station can obtain movies directly from the satellite, bypassing these local suppliers. Other satellite-fed channels can be distributed as well, increasing the overall quality and value of the hotel room television service. For example, the Cable News Network, the Financial News Network, and C-SPAN can provide the professional and business guest with information services. The children's networks and shows can be carried to entertain the kiddies, while the sports channels can capture the avid sporting fan.

The TVRO earth station can also be used to plug the hotel into the rapidly expanding videoconferencing networks. Robert Wold, Vidsat, Videonet, Netcom, and others produce and distribute via satellite national video teleconferences, political fund-raisers, and Fortune-500 annual meetings (Fig. 6-2). Special events and sporting activities like championship boxing matches can be carried on a pay-per-view basis to the guest rooms and public-admittance areas of the hotel. The Reagan Administration often uses videoconferencing to bring together

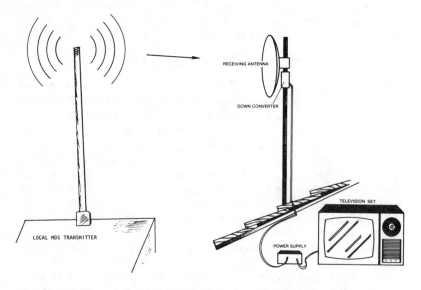

Fig. 6-1. The MDS service feeding pay-tv programs to a hotel or hospital from a central distribution point in the city.

(Courtesy Jim Lentz)

Fig. 6-2. The author and Robert Wold talk live from a recent New York conference to an audience in Los Angeles from the conference hotel using satellite facilities provided by the Wold Corporation.

thousands of people at gala fund-raising parties held in hotels throughout the country. The President can reach out to each of these locations electronically via satellite; two-way audio talkback channels are usually provided to allow people at the various hotels to be heard back in Washington.

Charity events can be staged in hotels with satellite videoconferencing facilities. Closed-circuit meetings can increase the room registrations, while saving the participants time and travel expenses. For the major corporation, annual video shareholders meetings, and video teleconferences tieing together the regional sales offices can save hundreds of thousands of dollars.

Many of the national hotel chains are now installing their own TVRO earth stations in their larger properties. The first and largest of these, the Holiday Inn group, now has several hundred five-meter dishes strategically located at hotels throughout the country (Fig. 6–3). The Hilton, Hyatt, and Marriott chains are following this lead, and by the end of the 1980s, most national hotels will have their own TVRO earth stations installed and operating. Some have even suggested developing their own private channels to present top entertainment acts originating in their Las Vegas and Los Angeles properties for nationwide satellite distribution to their affiliate hotels (Fig. 6–4).

Fig. 6-3. The author standing in front of a TVRO antenna in the parking lot of a Holiday Inn.

THE HOTEL ELECTRONIC ENTERTAINMENT SYSTEM

A hotel MATV system is similar to a private CATV network, using much of the same TVRO and cable equipment. Several national installation companies provide television services to hotels under contract, and can assist the hotel in upgrading its MATV system to a fully-automated, satellite-based operation. In some cases, the MATV system will need to be expanded to carry the increased number of channels that a satellite TVRO earth station can deliver. This may call for the installation of additional vhf channel modulators and an overhaul of the cable distribution plant. Some form of automated pay-per-view billing control also needs to be added so this function will not disrupt the existing staff operations. A number of manufacturers produce automated headend billing and control equipment for this purpose.

Fig. 6-4. A portable uplink used in teleconferences and special events. It can also be used to feed live entertainment to hotels and CATV systems nationwide. (The author recently used a similar T/R dish to transmit a 5-hour seminar on satellite television to TVRO enthusiasts and dealers throughout the country.)
(Courtesy Videostar)

The ultimate system consists of a microprocessor-controlled channel switcher and room-based tv decoder providing a two-way communications capability back to the headend. To view a desired movie, the room guest pushes a button on the decoder or dials a specific access code on the room telephone. This action sends a signal to the control microprocessor which captures the order information, printing out the order on a small terminal in the billing center of the hotel. In a simpler semi-automatic system, the guest dials the hotel switchboard "movie operator" who captures the order request manually, unscrambling the movie service for that particular room using a small operator console.

Although hotels can negotiate directly with a number of different satellite program suppliers, several organizations now specialize in providing satellite-fed movie packages to hotels (and hospitals). Appendix B describes the various satellite services that are available.

While buying programming is a straightforward process, determining how the room guest should be charged may be more difficult. Should the movie service be given away as part of the overall promotion of the hotel and its facilities? In this case, the additional charge must be included as part of the "bundled" daily room rate. Some customers who do not watch television may resent subsidizing the other guests; this could drive away visitors. Billing on a per-movie basis is more complicated, but usually more profitable. The best approach may be to provide four or five "free" channels as part of the basic television service, and to bill the movies on a pay-per-view basis. One motel on the Eastern seaboard reports that their occupancy rate increased by over fifty percent when satellite television was introduced. Many people, especially the frequent traveler, will consciously select hotels which actively advertise and provide satellite television services.

SATELLITE TV IN THE RESTAURANT AND BAR

Like the hotel industry, many restaurants and taverns have discovered the excitement of satellite television. Although one New England tavern made the news by running "General Hospital" straight from the ABC Television satellite feeds a few months back, it's best to stick with those programmers for which copyrights have been contractually obtained. Sports services like ESPN (Fig. 6–5) can attract large nightly audiences who gather around the big screen projection television set at the end of the bar. Adult entertainment and adventure programming can also draw large crowds. A local bar and hotel can share the same TVRO earth station if the two establishments are located in the same building or are near each other.

Tavern and restaurant installations can vary from commercial CATV-quality 15-foot parabolic dishes to consumer-quality 10-foot antennas using low-cost satellite receivers. A system costing $3,000 to $4,000 will provide excellent reception on a projection television system with electronics to accept the direct video output of the satellite TVRO receiver. This arrangement eliminates the need to convert the satellite signal to a vhf television channel along with the picture degradation caused by the channel remodulation process.

Some organizations rent both TVRO downlinks and uplink earth stations available for use throughout the country. Westsat Communications, Pleasanton, California, provides trailer-mounted ADM-3 11-foot TVRO terminals for use throughout the San Francisco Bay area. Many TVRO dealers operate a separate rental division for this purpose. The promotion of closed-circuit special events for which tickets are sold can bring in tens of thousands of additional dollars to the restaurant or bar equipped to handle such crowds. (Chapter 7 discusses the video teleconferencing rental business opportunity more thoroughly.)

SATELLITE TELEVISION IN THE HOSPITAL AND RETIREMENT HOME

Hospitals and rest homes provide a living facility similar to a hotel or apartment building. Most hospital rooms are furnished with television sets, or they are available for rent by the patients. Convalescent homes have community tv and family rooms. The TVRO satellite terminal can increase the choices that the patient has for entertainment. Programs

WEDNESDAY FEBRUARY 16

AM

12:30 **This Week in the NBA**
1:00 **SEC College Basketball:**
  Georgia at Mississippi State
3:00 **SportsCenter**
5:00 **SportsCenter**
6:00 **ESPN's SportsForum**
6:30 **This Week in the NBA**
7:00 **SportsCenter**
9:00 **ESPN's SportsWoman**
9:30 **Vitalis/U.S. Olympic Invitational
  Track Meet**

PM

12:00 **FIS World Cup Skiing:** Women's
  Downhill from Schruns, Austria
1:00 **Gymnastics:** USGF Single
  Elimination Championships
2:00 **Vic's Vacant Lot** (Children)
2:30 **ESPN's SportsWoman**
3:00 **Ski School**
3:30 **Fishin' Hole**
4:00 **Winterworld Series:**
  "Record of Time"
4:30 **SportsCenter**
5:00 **ACC College Basketball:**
  North Carolina at Maryland (L)
7:00 **Big 10 College Basketball:**
  Iowa at Indiana
9:00 **SportsCenter**
9:30 **Notre Dame Basketball:**
  Notre Dame at Pittsburgh
11:30 **SportsCenter**

THURSDAY FEBRUARY 17

AM

12:30 **ACC College Basketball:**
  North Carolina at Maryland
2:30 **ESPN's SportsForum**
3:00 **SportsCenter**
5:00 **SportsCenter**
6:00 **Gymnastics:** USGF Single
  Elimination Championships
7:00 **SportsCenter**
9:00 **ACC College Basketball:**
  North Carolina at Maryland
11:00 **Big 10 College Basketball:**
  Iowa at Indiana

PM

1:00 **Notre Dame Basketball:**
  Notre Dame at Pittsburgh
3:00 **F.A. Soccer:** "Road to Wembley"
4:00 **ESPN's SportsForum — Thurs. Ed.**
4:30 **SportsCenter**
5:00 **NFL Theatre:** Best-Ever Runners
6:00 **Big 8 College Basketball:**
  Kansas at Missouri (L)
8:00 **SportsCenter**
9:00 **Exhibition Baseball:**
  Los Angeles Dodgers vs. USC
11:00 **ESPN's SportsForum**
11:30 **SportsCenter**

FRIDAY FEBRUARY 18

AM

12:30 **Big 8 College Basketball:**
  Kansas at Missouri
2:30 **ESPN's SportsForum**
3:00 **SportsCenter**
5:00 **SportsCenter**
6:00 **ESPN's SportsWoman**
6:30 **ESPN's SportsForum**
7:00 **SportsCenter**
9:00 **Big 8 College Basketball:**
  Kansas at Missouri
11:00 **Exhibition Baseball:**
  Los Angeles Dodgers vs. USC

PM

1:00 **Vitalis/U.S. Olympic Invitational
  Track Meet**
3:30 **To Be Announced**
4:00 **College Basketball Report**
4:30 **SportsCenter**
5:00 **Winterworld Series:** "Born on Skis"
5:30 **Budweiser Presents Top Rank
  Boxing** (L)
8:00 **SportsCenter**
9:00 **Gymnastics:** USGF Single
  Elimination Championships
10:00 **F.A. Soccer:** "Road to Wembley"
11:00 **College Basketball Report**
11:30 **SportsCenter**

ESPN SPORTS GUIDE

Fig. 6-5. Portions of a typical ESPN programming day.
(Courtesy ESPN)

can be selected to match the patient's interests, from old to young, from movies to education.

Hospital administrations are revising their bureaucratic procedures and methods, working toward making their patients stay less traumatic and more pleasant. Gift shops, restaurants, and other community facilities have been added. Entertainment provided by television satellites is another patient service. The American Network, located on Satcom 4 beams their programming to hospitals throughout the country.

The TVRO earth station can also be used by the hospital staff to receive the educational, professional management, and medical information channels providing satellite feeds to hospital administrations. The number of out-of-town trips which the average medical professional must take every year can be reduced, and the staff can maintain its professional competence keeping up with the new medical and technological breakthroughs, while still being available and on call by the hospital. TVRO satellite antennas like the one in Fig. 6-6 can simultaneously receive two or more adjacent satellites and are now available at

Fig. 6-6. An antenna which can pick up several satellites simultaneously.
(Courtesy Antenna Technology Corp.)

relatively low cost. This means that the same earth station used to pick up a movie service for viewing by the patients can also receive a medical seminar or videoconferencing channel for use by the staff.

The convalescent or rest home can be thought of as similar to an apartment building, with all of the corresponding satellite-fed services and networks available for viewing. The facility can be treated as a private CATV system rather than just an MATV operation. This distinction will allow the rest home management to deal directly with many of the satellite program suppliers.

SATELLITE TELEVISION FOR THE SCHOOL

A wide variety of educational programming is available by satellite. C-SPAN, the Cable Satellite Public Affairs Network, covers the House of Representatives when they are in session, via the Satcom 3 satellite (Fig. 6–7). C-SPAN was formed in December 1977 on a nonprofit basis, with funding provided by twenty-five of the largest cable companies operating in the United States. C-SPAN is carried to over 11 million homes via cable. In addition to the gavel-to-gavel coverage of Congress, C-SPAN presents speeches by leading public figures at the National Press Club, personal interviews of people in the political arena, and informative call-in and interactive public affairs shows.

The Close-Up Foundation produced a series of civic-oriented programs for high school students, and in conjunction with C-SPAN runs week-long seminars and interviews on various topical subjects of public interest. During each school year, these satellite-fed seminars are

Fig. 6-7. C-SPAN delivers gavel-to-gavel coverage of the House of Representatives.
(Courtesy C-SPAN)

viewed by hundreds of high school classes throughout the United States, and a supplementary guide called "Issues and Answers" is published to support the public service. The goals of the project are:

1. Provide Social Studies teachers with timely and supplementary resources, creating, literally, a living textbook on Government.
2. Challenge and motivate students to examine current affairs and issues as well as the structure of the US Government.
3. Demonstrate through peer identification how young people can participate in the democratic process.

In 1982, C-SPAN offered over sixty one-half hour programs in four separate series over its public affairs channel. On most days, the Press Club series follows the House of Representatives coverage; the network operates from roughly 9:30 am to 6:00 pm around the House of Representatives schedule. C-SPAN is presently investigating the possibility of bringing live debates from the floor of the US Senate, and may expand to a second satellite channel if Senate approval is given.

A wide variety of educational programs are available for the school and institution from other sources as well. The Learning Channel (formerly the Appalachian Community Service Network) feeds dozens of seminars and continuing-education courses each week via Satcom 3. Originally established as an outreach program to bring high-quality education to rural areas in Appalachia, The Learning Channel has become a nationwide satellite network providing job training, quality educational and instructional programming, and cultural events via cable.

The Learning Channel courses include arts, music, humanities, and many others. A typical sampling presents "It's Everybody's Business," a show oriented to the business community; "Earth, Sea, and Sky," an introduction to earth sciences; "Designing Home Interiors"; "Applied Sketching Techniques"; and "Introductory Biology." Many courses grant under-graduate credits through participation in a local university; the television student attends an initial class orientation, and proceeds to do the remaining work at home.

Other special programs include major teleconferences at the USDA Extension School in Washington, and the annual conference of the American Council on Education. Community colleges and other educational institutions can negotiate directly with the local cable company to carry The Learning Channel programming, or can install their own TVRO earth stations for direct reception.

The American Educational Television Network (AETN) distributes non-degree continuing education programs to cable companies (Fig. 6–8). These seminars are offered in conjunction with national professional associations to fulfill the course requirements levied by the various state and regional regulatory bodies. Over 30 million professionals who must take some form of ongoing training live in the United States. They in-

Summer Program Schedule

August

Sunday				
			1 5:00pm LIVING ENVIRONMENT-10 *Engery Alternatives* 5:30pm CASE STUDIES SM. BUSINESS-5 *Running the Show*	
3 6:00pm LIVING ENVIRONMENT-11 *Conservation of Vital Resources* 6:30pm CASE STUDIES SM. BUSINESS-6 *The Balancing Act*	**5** LIVING ENVIRONMENT-12 *Economic Geology* CASE STUDIES SM. BUSINESS-7 *The Breaking Point*	**7** LIVING ENVIRONMENT-13 *Solid Waste* CASE STUDIES SM. BUSINESS-8 *Their Own Brand*	**8** 5:00pm LIVING ENVIRONMENT-14 *Wildlife Management* 5:30pm CASE STUDIES SM. BUSINESS-9 *Dealing and Wheeling*	
10 6:00pm LIVING ENVIRONMENT-15 *Forest and Man* 6:30pm CASE STUDIES SM. BUSINESS-10 *Taking Off*	**12** LIVING ENVIRONMENT-16 *Land Use in the City* GREAT PLAINS EXPERIENCE-1 *The Land*	**14** LIVING ENVIRONMENT-17 *Water Resources* GREAT PLAINS EXPERIENCE-2 *Lakota: One Nation*	**15** 5:00pm LIVING ENVIRONMENT-19 *Air Pollution* 5:30pm GREAT PLAINS EXPERIENCE-3 *Clash of Cultures*	
17 6:00pm LIVING ENVIRONMENT-20 *Impact of Political Science* 6:30pm GREAT PLAINS EXPERIENCE-4 *Settling of the Plains*	**19** LIVING ENVIRONMENT-21 *Impact of Economic Systems* GREAT PLAINS EXPERIENCE-5 *The Heirs to No Mans Land*	**21** LIVING ENVIRONMENT-22 *Myths of Technology* GREAT PLAINS EXPERIENCE-6 *Four Portraits*	**22** 5:00pm LIVING ENVIRONMENT-23 *Individual Involvement* 5:30pm GOING METRIC-201 *Measurement of Length*	
24 OFF	**26** OFF	**28** OFF	**29** 5:00pm LIVING ENVIRONMENT-23 *Individual Involvement* 5:30pm LOOSENING THE GRIP-4 *Signs & Symptoms*	
31 6:00pm LIVING ENVIRONMENT-24 *Solutions and Projections* 6:30pm LOOSENING THE GRIP-11 *An Ounce of Prevention*				

Time Zone Reference Tables

	Pacific DST	Mountain ST	Central DST	Eastern DST
	2:00 PM	2:00 PM	4:00 PM	5:00 PM
	2:30 PM	2:30 PM	4:30 PM	5:30 PM
	3:00 PM	3:00 PM	5:00 PM	6:00 PM
	3:30 PM	3:30 PM	5:30 PM	6:30 PM

Fig. 6-8. A typical AETN "University of the Air" schedule.
(Courtesy American Educational Television Network)

clude dentists, CPAs, attorneys, and nurses, among others. The seminars are often carried by a local cable company, and the participating professional is enrolled via his or her national association. The cost per seminar ranges from $50 to $150, and tests are issued. The final results are then relayed to the applicable licensing agency of the state in which the person resides.

Since AETN primarily distributes its programming through cable systems, there are many locations throughout the United States in which the service is not carried. This creates an opportunity for the professional organizations and associations to install their own TVRO earth stations at their regional offices in each city. Continuing education classes can then be held in conjunction with ongoing monthly meetings, etc. In many cases, the programming can be videotaped with permission, and made available for showing at the local public library.

Dozens of local and regional library systems are now installing their own TVRO earth stations at library buildings throughout the country. Many libraries have large audio-visual centers, television equipment, and public meeting rooms; the co-location of a TVRO satellite terminal can inexpensively expand the library services to the community, in an era when more labor-intensive projects are being curtailed. Services are available to all satellite programmers and news organizations. Fig. 6–9 shows the uplink and intracity facilities available in Washington DC from Metrosat.

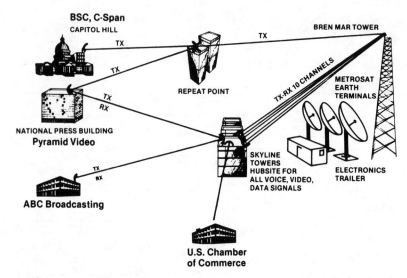

Fig. 6-9. Uplink and intracity Washington DC services available from Metrosat.
(Courtesy Communications Management Technology, Inc.)

THE ELECTRONIC CHURCH

Many religious groups have recently enlarged their congregations through reorganization and revitalization. The most successful of these have often tried new and unorthodox ways to reach out and share their message. Today, the television ministry is one of the most powerful tools at the disposal of the church. It began in the early fifties with Bishop Fulton Sheen, but the satellite television industry was trail-blazed by Pat Robertson, a former Wall Street stockbroker and founder of Christian Broadcasting Network (CBN). CBN reaches into the homes of over 17 million families via local cable and television stations, delivering its 24-hour a day Christian and family programming via the Satcom 3 satellite.

At least six major religious channels are now downlinked, including CBN, People That Love Television (PTL), the Trinity Broadcasting Network (TBN), Eternal Word Television Network (EWTN), the National Christian Network (NCN) (Fig. 6–10), and the National Jewish Television Network (NJT). Most other established churches, including the national Roman Catholic and Lutheran organizations are investigating establishing their own satellite channels. The Church of the Latter Day Saints (Mormons) will be delivering programming from its Utah headquarters to its congregations via Bonneville International, its Salt Lake-based tv broadcasting subsidiary. By 1985, perhaps a dozen or more churches and television ministries will be beaming their programming down to earth from the heavens.

Fig. 6-10. NCN, one of the many interdenominational religious programmers
used by churches and Sunday Schools.
(Courtesy National Christian Network)

Local churches and groups can obtain unused terrestrial microwave channels via the Instructional Fixed Television Service (IFTS) available to nonprofit organizations, to beam their sermons and educational programs throughout their region. Some Southern California churches now use these low-cost local video channels to provide Sunday School classes and large-scale worship events which are simultaneously projected on giant display screens in a dozen or more locations through Los Angeles and Orange Counties. By the 1990s, thousands of churches will be wired into these private television networks for local and nationwide prayer services.

A local church can install its own 10- to 15-foot TVRO earth station (Fig. 6–11), using less expensive consumer-quality electronics and a single LNA/down converter or LNC package. A combination Janeil dish/ Dexcel LNC package can be installed for under $3,500. For an additional $1,000 or less, the fixed-position antenna base can be upgraded to a remote-controlled antenna, which can be repositioned from satellite to satellite using a small control panel located in the church. With this arrangement, programs can be easily changed and quickly selected. To complete the installation, a consumer-grade video projection television

Fig. 6-11. A small TVRO antenna that can easily be mounted on the roof of a church or school building.

system such as those manufactured by the Sony or Kloss Videobeam Corporations can be purchased. Multiple display systems can be connected to the output of the TVRO receiver, depending on the size of the viewing audience and number of meeting rooms to be served. Finally, a video cassette recorder can be added to tape the best programming, creating a permanent church library. Other satellite channels such as ACSN and the children's networks can be provided for use at Sunday School services and evening "clambakes," "potluck dinners," and other social events.

FUNDRAISING AND CHARITABLE EVENTS

Satellite television is an exciting new tool for a fundraiser. Dozens of cities are now tied together for telethons and videoconferences, with many such events being telecast annually. It is easy to plug into a national telethon via a TVRO earth station; the addition of a telephone-based talkback audio teleconference unit provides two-way conversations with the guest stars or national officers in the distant tv studios.

A nonprofit or charitable organization can rent a TVRO earth station "on demand" for between $500 and $1,000. The purchase price of a system could easily be justified if only one or two such events are held per year. Appendix C provides a detailed listing of equipment manufacturers, dealers, and national distributors who can sell or rent such TVRO satellite terminals.

Of course a local nonprofit organization need not depend on a national telethon or videoconference as the basis of its fundraising activities. Other special events can be carried on a ticket sales basis. The World Championship Boxing matches have been particularly good attractions for use as fundraising events. Ticket prices run from $10 to $30 and the nonprofit organization which sponsors the local showing of the closed-circuit event will receive from 20 to 50 percent of the "gate"!

CHAPTER **7**

Video Teleconferencing via Satellite TV: How to Enter This Lucrative Business

According to a recent article in the *Wall Street Journal,* video teleconferencing is one of the most significant business tools to have been spun off from the NASA space technology. The development of the communications satellite has slashed the cost of providing a nationwide multipoint video channel. Before the advent of the domestic communications satellites, only the television networks could afford to lease the terrestrial television circuits provided by AT&T on an occasional-use basis. The costs and infrastructure required to fully utilize this technology were just too great. Today, hundreds of hotels, hospitals, office buildings, and theater complexes are equipped with TVRO satellite terminals. A dozen major organizations have appeared to provide "turnkey" video conferencing services. Users have included Hughes Aircraft, TRW, Ford Motor Company, The American Bar Association, Ciba-Geigy, The American Hospital Association, The AFL-CIO, and The Republican National Committee.

Videoconferencing is currently a 40-million dollar industry, and is predicted to reach 250 million by 1985. Fig. 7–1 shows a split-screen video conference where the second screen can also display graphics. A split-screen arrangement for a small conference room using two cameras is diagrammed in Fig. 7–2. Looking at the increased revenues that video conferences can generate, most of the major hotel chains have committed themselves to installing their own TVRO and associated video teleconferencing facilities throughout the country. The Holiday Inn Hi-Net System has over 200 earth stations in 120 cities, with almost every state represented. The Ramada Inn chain has equipped a number of its hotels with TVROs and the Hilton, Sheraton, and Hyatt Hotel conglomerates are in the developmental stages of installing their own networks. Where facilities are not yet available, temporary setups can be provided. For example, Fig. 7–3 shows a temporary antenna installed outside the Fairmont Hotel in San Francisco for a computer manufac-

Fig. 7-1. A split-screen videoconference room.
(Courtesy Satellite Communication Services)

turers national sales and marketing conference. Similar facilities were used at dozens of hotels throughout the country.

WHO USES VIDEO CONFERENCING?

Video conferencing services are now used for every imaginable purpose by corporations, nonprofit organizations, and government agencies in every state. There are at least eleven major functions for a video conference, which include:

1. Convention and Trade Show Meetings
2. Education Conferences and Symposiums
3. News Conferences
4. General Business Meetings
5. Stockholder Meetings
6. Sales and Marketing Meetings (Both National and International)
7. Fundraising Events for Political Parties
8. Charitable Events and Telethons

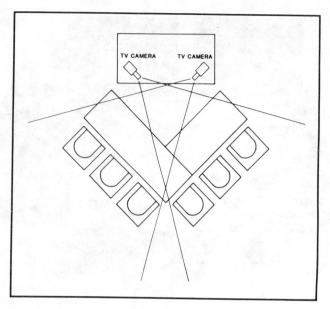

Fig. 7-2. A two-camera split-screen small video teleconference room.

9. New Product Promotions and PR Campaigns
10. In-House Corporate Training and Educational Seminars
11. Special Entertainment Events (Such as World Championship Boxing Matches, Private Shows from Hollywood, etc.)

The advantages of a custom video conference are numerous. First, the video conference can save substantial time on the part of all participants. Ciba-Geigy recently held a four-continent video teleconference business round table with participants in London, Rio de Janeiro, Tokyo, and 1500 attendees at a corporate convention in Atlanta. By tying these four cities together via videoconference, hours of travel time were eliminated, even at Concorde speeds.

Video conferences save money. By eliminating air fares, overnight hotel accommodations in distant cities, etc., periodic sales meetings can be held more frequently and at lower costs. Corporate productivity usually increases at a video conference, since by nature the event has to be better planned, and getting down to the "meat" of the matter is expected. Moreover, improved communications occur since everyone in the organization who attends the video conference will come away with the same information. More employees can be reached, eliminating the need to arbitrarily exclude entry-level and junior personnel from attending a corporate convention or meeting.

Fig. 7-3. A TVRO dish temporarily installed outside the Fairmont Hotel in San Francisco.

Scheduling major speakers for participation in video conferences is made easier by flexibility in location and timing. A famous or important speaker can literally be in two places at once through video teleconferencing. President Reagan has used video conferencing technology to speak to dozens of organizations and associations at their annual conventions, as did President Carter before him.

An organization can improve its public image when holding a video conference by being associated with this "leading edge" technology. Employee morale is also improved, especially for offices far removed from the national corporate headquarters. An increased sense of corporate identity and unity will often result.

Video conferences can be easily scrambled using the Oak Orion system or other similar equipment, and the entire event can be videotaped for future showings. Finally, the video conference can be held anywhere, in any city throughout the United States, as well as in most countries.

PLANNING FOR A VIDEO CONFERENCING EVENT

There are eight important aspects to the coordination and management of a video conference. First, the program must be carefully planned. How long will the meeting be held? Who will participate as speakers? Who will the attendees be in each city? Will the event be an electronic "round table of the air" held among various cities, a one-way broadcast, or some combination?

The meeting facilities must be considered. Where will the principal meeting place be located (if there is one)? From where will the satellite transmissions be uplinked? How are the video signals delivered from the meeting hall to the T-R uplink site (terrestrial microwave links, telephone company-supplied local loops, etc.)? Will the event be a one-way video broadcast, a one-way video/two-way audio, or a full duplex two-way video event?

Contracts must be negotiated for the communications facilities and equipment needed to make the video conference happen. This includes satellite time, earth station uplink usage, TVRO arrangements at the distant meeting sites, program and coordination/talkback audio circuits, and actual meeting room space. Hundreds of facilities equipped for one-way or two-way video conferences now exist nationwide. They include tv studios (both PBS and commercial stations), hotels, office buildings, theaters, and hospitals. Finally, the tv production equipment— including the cameras, lights, television sets, microphones, teleprompters, and video projection equipment—must be obtained.

The production crews and event director must be hired for each location. Fig. 7-4 shows the production crew and director reviewing an upcoming event with the sponsoring organization at the uplink studio. The program itself should be carefully scripted to fit the time restrictions of the satellite facilities which have been made available for use by the event. Graphics preparation including the production of slides, charts,

Fig. 7-4. The production crew and director review an upcoming event at the uplink studio.
(Courtesy Netcom International)

tape, and film should be prepared well in advance. Finally, the whole process should be rehearsed with the key players (or their stand-ins) before the day the video conference is held, and this includes the site coordination for potential microwave interference if portable TVRO trailer-mounted earth stations are used to bring the event into an unequipped meeting room. Finally, the event should be recorded for future showings.

VIDEO CONFERENCING ORGANIZATIONS

Luckily, there are many national organizations which specialize in the "turnkey" organization and production of a video conference or video event. The largest of these is the Robert Wold Corporation, the Los Angeles-based affiliate of Cox Broadcasting. Wold owns or leases eleven satellite transponders on a number of domestic communications satellites, providing over 80,000 hours per year of satellite transponder time to commercial organizations, television networks, cable programmers, and video conference organizers. Wold operates its own network of 45 TVRO earth stations, and uplink facilities at its Los Angeles, New York, and Washington, DC television operating centers (TOCs) which are interconnected via terrestrial microwave to AT&T and other common carrier facilities. Wold also owns two transportable uplink earth stations, the "flying dishes," which can be airlifted throughout the country, and operates a network facility in Honolulu, Hawaii. (Audio conferencing is also possible with Wold; the organization provides five SCPC 19 dBW high-power audio circuits on Westar 3, and shares the access rights to over 700 receive-only earth stations used by Associated Press for feeding their audio news network to newspapers and radio stations nationwide.)

Other major players include Netcom, based in San Francisco; Videonet, a Woodland Hills, California subsidiary of Oak Industries; Videostar, an Atlanta-based national video conference coordinator (Fig. 7–5); and PSSC, the Public Service Satellite Consortium which owns its own uplink facilities in Denver, Colorado.

To these major players are added Teleconcepts and Modern Telecommunications, (MTI), both of New York City, and Satnet. The Hi-Net Video Communications organization is the Holiday Inn subsidiary which can arrange for the use of any of its 200-plus TVRO-equipped hotels (all of which are configured for at least one-way video and two-way audio service). A list of the major video conference organizations is given in Table 7–1.

The first-time or smaller video conference user will often use the AT&T Picturephone Meeting Service (PMS), which operates dozens of two-way full-color video teleconference meeting facilities at Bell locations throughout the continental United States (Fig. 7–6). A number of major

SATELLITE TELECONFERENCING

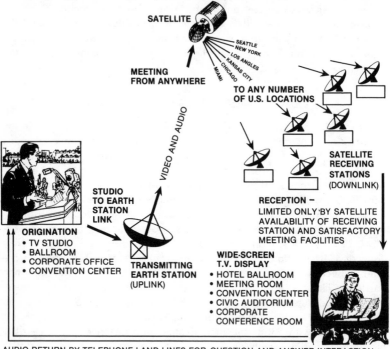

AUDIO RETURN BY TELEPHONE LAND LINES FOR QUESTION AND ANSWER INTERACTION

Fig. 7-5. Videostar Tele-Meeting™ network.
(Courtesy Videostar Atlanta)

organizations have contracted with AT&T to provide private PMS rooms for internal use only, and the PMS FCC tariff establishes hourly rates running from several hundred dollars and up. The advantage of using the Picturephone Meeting Service is that it can be quite cost-effective for a small and limited group, while allowing a sophisticated multilocation two-way video conference with minimal technical or management coordination required on the part of the user. Fig. 7–7 shows a closeup of the video displays at the PMS room. The cameras above the monitors automatically switch to pick up the person speaking.

TAPPING INTO THE VIDEO CONFERENCE PROFITS: BECOMING A LOCAL TVRO RENTAL SOURCE

As video conferencing expands, so does the need for the rental of portable TVRO earth stations to bring the one-time events to the desired

Table 7–1. List of Major Video Conference Organizations

Organization	Type	Location	Telephone
All Mobile Video	Video Production	New York, NY	(212) 757-8919
American Educational Television Network	Turnkey Producer	Irvine, CA	(714) 955-3800
American Hospital Television Network	Specialty Producer	Gates Mills, OH	(216) 951-6682
American Video Channels	Video Production	New York, NY	(212) 765-6324
Appalachian Community Service Network	Turnkey Producer	Washington, DC	(202) 331-8100
AT&T Picturephone Meeting Service	Turnkey Centers	Bedminster, NJ	(291) 234-7879
Austin Satellite Television	Turnkey Producer	Austin, TX	(512) 346-2557
Bell & Howell Satellite Network	Turnkey Producer	Washington, DC	(202) 484-9270
Bonneville Internat'l	Turnkey Producer	Salt Lake Cty. UT	(801) 237-2450
Center for Non-Broadcast Television	Turnkey Center	New York, NY	(212) 794-3260
Centro Corporation	Hardware Rental	Los Angeles, CA	(213) 203-8033
Chicago Video	Turnkey Producer	Arlington Hghts. IL	(312) 577-2430
Color Leasing Studios	Video Production	Fairfield, NJ	(201) 575-1118
Comart Aniforms	Turnkey Producer	New York, NY	(212) 867-7500
Compact Satellite Services	Video Production	Burbank, CA	(213) 840-7000
Confernet	Turnkey Producer	Sewickley, PA	(412) 771-4700
Digital Video Corporation	Video Production	Orlando, FL	(305) 425-1999
GBH Production Services	Turnkey Producer	Boston, MA	(617) 492-2777
General Television Network	Video Production	Oak Park, MI	(313) 548-2500
Hilton Communications Network	Turnkey Producer	Beverly Hills, CA	(213) 278-4321
Hi-Net Communication	Turnkey Producer	Memphis, TN	(901) 369-7539

Table 7-1-cont. List of Major Video Conference Organizations

Organization	Type	Location	Telephone
Hughes Television Network	Turnkey Producer	New York, NY	(212) 563-8900
Live-Video	Video Production	Plainville, CT	(203) 793-0587
Maryland Center for Public Broadcasting	Turnkey Producer	Owings Mills, MD	(301) 337-4086
Midtown Video	Video Production	Denver, CO	(303) 778-1681
Mobile Video Services	Turnkey Producer	Washington, DC	(202) 298-9100
MTI	Turnkey Producer	New York, NY	(212) 355-0510
Netcom International	Turnkey Producer	San Francisco, CA	(415) 921-1441
NET Telecon/WNET	Turnkey Producer	New York, NY	(212) 560-2067
Ohio University Telecommunications Ctr.	Video Production	Athens, OH	(614) 594-5244
Omnia Corporation	Video Production	Edina, MN	(612) 929-8228
One Pass Video	Video Production	San Francisco, CA	(415) 777-5777
Organizational Media Systems	Turnkey Producer	Fort Worth, TX	(817) 281-4126
Polycom Teleproductions	Video Production	Chicago, IL	(312) 337-6000
Producers Color Service	Turnkey Producer	Southfield, MI	(313) 352-5253
Public Service Satellite Consortium	Turnkey Producer	Washington, DC	(202) 331-1154
Pyramid Video	Video Production	Washington, DC	(202) 783-5030
Jim Sant'Andrea	Turnkey Producer	New York, NY	(212) 974-5451
Satellite Communications Network	Hardware Rental	New York, NY	(212) 466-0507
Satellite Syndicated Systems	Hardware Rental	Tulsa, OK	(918) 481-0881
SCETV Teleconference Design Group	Turnkey Producer	Columbia, SC	(803) 758-7261
Skaggs Telecommuni-cations Service	Turnkey Producer	Salt Lake Cty. UT	(801) 539-1427

Table 7-1-cont. List of Major Video Conference Organizations

Organization	Type	Location	Telephone
TCS Productions	Video Production	New Kensington, PA	(412) 361-5758
Teleconcepts in Communications	Turnkey Producer	New York, NY	(212) 355-7113
Telemation Productions	Turnkey Producer	Glenview, IL	(312) 729-5215
Telemation Productions	Video Production	Seattle, WA	(206) 623-5934
Television Production Services	Video Production	E. Brunswick, NJ	(201) 287-3626
TNT Communications	Turnkey Producer	New York, NY	(212) 644-0200
Vega Associates	Turnkey Producer	New York, NY	(212) 980-6668
Video Systems Network	Turnkey Producer	Los Angeles, CA	(415) 820-5595
Videonet	Turnkey Producer	Woodland Hills, CA	(213) 999-3113
Videostar Connections	Turnkey Producer	Atlanta, GA	(404) 257-0121
Webster Productions	Video Production	Chicago, IL	(312) 951-7500
West Coast Satellite	Video Production	San Diego, CA	(714) 272-0680
Western Union/Video Services	Turnkey Producer	Upper Saddle River, NJ	(201) 825-5000
WETACOM	Turnkey Producer	Washington, DC	(202) 998-2700
Robert Wold Company	Turnkey Producer	Los Angeles, CA	(213) 474-3500
WPHL-TV	Video Production	Philadelphia, PA	(215) 878-1700

The turnkey producers can organize and hold a complete videoconference from start to finish, negotiating for satellite rental time, arranging to hire a video production company and crew, and renting the necessary TVRO site hardware throughout the United States and elsewhere.

meeting rooms. The number of video conferences that can be held at any one time is limited by the availability of prewired conference rooms. As the cost for air transportation continues to increase and the number of video conferences rises, the demand for using trailer-mounted commercial-grade TVRO terminals is expanding rapidly. Some organ-

Fig. 7-6. A Bell Telephone Picturephone® Meeting Service (PMS) videoconference room.
(Courtesy Bell Telephone)

Fig. 7-7. Closeup of the video displays at the PMS room.
(Courtesy Bell Telephone)

izations, such as Westsat, specialize in renting portable TVRO facilities, along with operating personnel, for use by the video conference organizers and management companies. Typical TVRO terminal charges run from $500 to $800 per day, and up, depending on the geographic location and complexity of the temporary installation. This rate is applied whether the video conference lasts for one or eight hours. Many video conferencing events, especially sales meetings, new product announcements, and training seminars run for only half a day, since the three-hour, cross-country time zone changes will limit the number of productive hours to an afternoon-East Coast/morning-West Coast meeting.

The video conference entrepreneur who wishes to enter the rental business should be familiar with the TVRO marketplace distribution mechanism. There are six major components in the TVRO earth station distribution chain. The electronic suppliers provide integrated circuits, transistors, and other circuit components that are used in the TVRO receiver, LNA, and dish components. The products of these organizations are purchased by the TVRO component manufacturer.

The TVRO component manufacturer assembles these components into the complete system modules such as LNAs, down converters, receivers, and antennas. They are known loosely as the "TVRO manufacturers," and usually sell their products to national distributors or dealers. Since this is a new industry, the manufacturers will often give single-unit prices to consumers who contact them directly, and are knowledgeable about their product. The novice consumer, however, will be referred to an appropriate local dealer for further assistance.

National distributors buy TVRO system components in large quantities from the manufacturers, and often establish subdistributorships, providing an intermediate tier between them and the stocking dealer. This two-tier distributor arrangement will probably disappear as the dealers themselves become more sophisticated and the market expands. Moreover, many of the original distributors have begun to manufacture their own systems, especially the parabolic dishes, tending to blur the distinction between them and TVRO component manufacturers. Since the manufacturers also sell directly to dealers, this makes the distribution chain both extraordinarily flexible and very muddled at the same time.

The pure dealer buys his product either from the distributor or TVRO component manufacturer. Some installing dealerships, like Channel One, Inc., of Lincoln, Massachusetts, often put their own "house brand names" on products which they have purchased essentially off the shelf from the TVRO component manufacturers.

The "System Integrating Dealer" may be affiliated with a component manufacturing operation, usually a metal working shop which produced the private-brand parabolic dishes that the dealer sells on an exclusive basis. Microwave General (Mountain View, California) is such an organization, operating both a dealership and manufacturing facility. Muntz TV, the Los Angeles operation of Earl ("Madman") Muntz, produces its own private-label systems, including the dish and receiver components which it handles exclusively in its own stores. The TVRO marketplace distribution is confusing indeed!

Assuming that the entrepreneur wishes to purchase one or more TVRO systems for use as rental units, the equipment can be bought directly from the manufacturers themselves at considerable price savings. Chart 7-1 presents an example of the dealer costs for a commercial-grade system which will provide acceptable video conferencing service for most parts of the country. (A 13-foot antenna may be required in some parts of Florida and the extreme western edges of the United States.)

Chart 7.1 Example of Dealer Costs for a Commercial Grade TVRO System for Video Conferencing

1. ADM 11-foot parabolic antenna or equivalent (e.g. Paraclypse 12 foot)	$1,000.00
2. Dexcel 100 K LNA or equivalent	450.00
3. Avcom COM-12 receiver or equivalent	1,100.00
4. Chaparral "Superfeed" feedhorn assembly	150.00
5. 4-wheel trailer with polar mount	1,300.00
6. Motorized mount with controller	500.00
7. Miscellaneous filters for QRM (interference)	300.00
8. Portable television monitor (5")	400.00
9. Miscellaneous tools, compass, inclinometer, cables	300.00
TOTAL	$5,500.00

Assumes equipment purchased at wholesale through national dealer/distributor.

The basic system consists of an ADM 11-foot parabolic antenna (or equivalent) (Fig. 7–8), a 100-K Dexcel LNA, Avcom Model COM-12 rack-mounted receiver (Fig. 7–9), Sony color television monitor, trailer with trailer hitch, and associated rf channel modulators, power splitters, and cabling, etc. A deluxe system might also include a motorized polar mount with remote control facility to aid in rapid dish positioning and to recover from any accidental movement of the trailer support structure.

The Paraclypse, KLM, and Janeil antennas would make good substitutes for the ADM-3 11-foot dish; the Earth Terminals, Arunta, KLM Sky Eye IV, and Drake receivers could substitute for the Avcom unit. Both Amplica and Avantek manufacture 100-K LNAs which could be used in lieu of the Dexcel low-noise amplifier.

Another tack to take in the selection of an antenna might be to use the 12-foot fold-up portable Luly Umbrella parabolic. This extraordinary antenna will provide outstanding service for a temporary video conference, and can be set up on its specially modified heavy-duty camera tripod in a matter of moments. A semipermanent sturdy mounting base is also available. In mild weather, the Luly can be used directly without

Fig. 7-8. A trailer-mounted TVRO dish which makes setting up for a videoconference an easy matter.

Fig. 7-9. The Avcom COM-12 TVRO rack-mounted receiver.
(Courtesy Avcom of Virginia, Inc.)

a protective shroud. If it is reasonably sheltered from high winds, it will not be affected by the occasional gust or two. It can be easily transported in most station wagons, along with the receiver and LNA electronics. In a location which is subject to more severe climate conditions, especially where heavy snow or ice is present, the Luly would have to be protected by assembling a small geodesic dome structure around it. This can be constructed in a few hours, far less time than it would take to assemble an ADM-3 from its pieces (Fig. 7–10), if a mobile trailer mounting arrangement were not used with that particular antenna. The

147

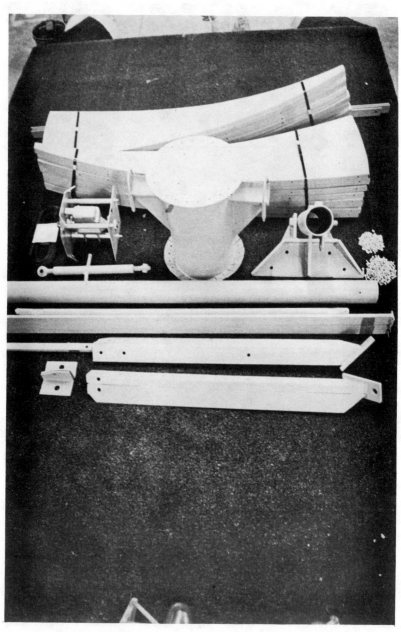

Fig. 7-10. The ADM-3 11-foot antenna broken down for shipping.
(Courtesy Antenna Development & Manufacturing, Inc.)

Fig. 7-11. The fully assembled ADM-3 antenna.
(Courtesy Antenna Development & Manufacturing, Inc.)

fully assembled ADM-3 antenna is pictured in Fig. 7-11. Note the polar-mount configuration with support pole.

On a rental system whose total cost will run about $7,000 to $8,000 (maximum), the video teleconference equipment rental operator should be able to charge around $500 per day. Assuming that the technician will receive $100 for his efforts, and that other overheads for running the small business will amount to an additional $100 per event, then an income of roughly $300 per day's rental should be realized. In the top fifty US cities, it should be possible to obtain two or three engagements per month on an "on-call" basis and one to two emergency rentals per quarter. These can be charged for at a premium of from 50 to 100 percent over the going rate. Thus, the video conference equipment rental business in the major markets should produce a reasonable income of from $9,000 to $13,200 per year, paying for the original investment within the first nine months or so of operation. Even if these figures are reduced by fifty percent, the business will still be profitable, as the number of video conferences increases.

In addition to renting the TVRO equipment, the video conference entrepreneur can also provide for industrial-grade video projection equipment and large-screen television systems. Table 7–2 summarizes a number of the popularly used commercial video projection equipment and features. Table 7–3 presents a similar list of selected consumer-oriented large screen television sets, which are quite adequate for the smaller and more intimate video teleconference.

Table 7–2. Commercial Video Projection Systems

Manufacturer	Screen Size	Light Output*	System Type	Resolution (lines)	View Angle (Degrees)
Aquastar III-C	4'–26'	500 lumens	one piece	300	50–120
GE PW-3000	40 in.	140	rear projection	420	80
Hitachi CT-5011	50 in.	120	one piece	400	50–70
Kloss Novabeam II	64 in.	200 lumens	one piece	400	50–120
Mitsubishi VS-322R	50 in.	180	one piece	400	90
North American Phillips	50 in.	120	rear projection	410	90
NEC PJ-6000	60 in.	80	one piece	400	120
Panasonic CT-4600	45 in.	115	rear projection	350	64

*In foot lamberts unless otherwise indicated

Finding customers for this fledgling business should not prove too difficult; simply contact the major video teleconferencing management organizations, such as Videostar, etc. Inform them of your new availability as an organization that will work closely with them, and one who understands the peculiarities of the video conferencing business and the

Table 7—3. Consumer-Oriented Large Screen Televisions

Manufacturer	Screen Size	Light Output*	Resolution (lines)	Price
Curtis-Mathes				
G-531R	60 in.	120	350	N/A
Fisher PT900	40 in.	70	330	$3,799
Magnavox 8505	50 in.	50	330	$3,495
Mitsubishi				
US-520 UD	50 in.	120	240	$4,000
Muntz 6346	63 in.	120	240	$1,695

*In foot lamberts

technology of the satellite television TVRO earth station. If you are contemplating establishing a TVRO dealership to carry these products for the consumer or SMATV/minicable markets, then the rental of TVRO satellite terminal equipment would provide a logical adjunct to the primary business. The Satellite Center, San Francisco, provides a series of two-day seminars for prospective TVRO dealers, which are held periodically throughout the country. A series of publications on establishing a video conferencing business, becoming an SMATV operator, and starting a TVRO dealership are available from this organization. By the middle of the decade, hundreds of video conferences will be held throughout the country each business day. It is literally a ground-floor business which will see explosive growth. And, video conferences are fun, too.

CHAPTER 8

How to Build Your Own Personal Earth Station

THE "EL CHEAPO SYSTEM": BUILDING YOUR OWN FROM SCRATCH

In the mid-1970s, Taylor Howard, Professor of Electrical Engineering at Stanford University, decided to construct his own backyard satellite tv earth station as an experiment in microwave engineering. Professor Howard modified an old military surplus parabolic antenna which had been rusting by the side of the house. Not wanting to spend the $10,000 that the professional TVRO satellite receivers were then selling for, Taylor designed his own "home brew" receiver circuit boards. By carefully scrounging the junk box for odds and ends, and using the hundred thousand dollars worth of radio alignment equipment to which he had access, Taylor was able to complete his first home TVRO earth station (Fig. 8–1) for only a couple of thousand dollars.

About the same time, Robert Coleman, an engineer in South Carolina, purchased an old surplus telephone terrestrial microwave parabolic antenna with electronics, which he modified using surplus military parts and electronic hobby components. By pounding out the dents in his antenna by himself, and working at night over several months, Bob was able to build his system for under $1,000. Clyde Washburn, Jr., out in the Midwest, and Nelson Ethier in Canada (two TVRO experts) put together a number of home TVRO systems using low-cost parabolic antennas and inexpensive components.

Together this group of pioneers proved that amateur radio operators and electronic hobbyists could tackle the heretofore elusive and forbidding territory of microwave communications, and construct a backyard dish inexpensively.

By 1978 Robert Cooper, Jr. had contacted these experimenters and others, creating an informal communications "grapevine" of practical tips and "how-to" information. Bob was the original editor of The Community Antenna Television Journal, based in Arcadia, Oklahoma, and had recently established Satellite Television Technology (STT) as a new

Fig. 8-1. The original backyard TVRO dish. Taylor Howard's military-surplus prime-focus feed 15-foot mesh-covered parabolic antenna has been in use since 1976.

consulting firm in the industry. STT published these developments, including the receiver schematics and antenna construction information, in a series of manuals, along with a small booklet on the potentials of home satellite television reception.

The response was overwhelming. Thousands of copies of these manuals were sold for $30 to $40 each, and a new industry was born.

The printed circuit boards for Professor Howard's TVRO receiver originally consisted of five functional components; modified boards are now available from Robert M. Coleman, Route 3, Box 58-A, Travelers Rest, South Carolina 29690. The latest-design boards sell for $99, less components (which can be obtained at most electronics parts stores). The insides of a "do-it-yourself" TVRO receiver is pictured in Fig. 8–2. Most such systems follow a form of the original design by Taylor Howard.

Similar receiver modules and kits are now available from a number of other suppliers including Sigma International, Inc., Box 1118, Scottsdale, AZ 85252 (Fig. 8–3), and NHz Electronics, 2111 West

153

Fig. 8-2. The insides of a "do-it-yourself TVRO receiver.

Camelback Road, Phoenix, Arizona 85015 which provides a set of five satellite receiver boards including dual-audio (stereo) boards and most components for $100. The receiver manufacturers, KLM and Sat-Tec Systems both sell low-cost kit versions of their highly successful consumer-grade TVRO receivers. In Chapter 10, detailed instruction plans and schematics covering the KLM Sky Eye-V Receiver are presented.

While Taylor Howard was at work on his backyard parabolic dish, Hayden McCullough was building his own *spherical* TVRO antenna in Salem, Arkansas. Based on an imaginative design engineered by Oliver Swan, Hayden set about to manufacture three different sizes of his "8-Ball" Antennas in 8-foot, 10-foot, and 12-foot versions. His company, McCullough Satellite Systems, Inc. sells these wood and metal-screening antennas for $650, $700, and $750 respectively (Fig. 8–4).

McCullough's spherical antennas can be assembled over a weekend by one person, and are some of the cheapest antennas on the marketplace. Using a wood and screen latticework-like assembly, the antennas look like they could be used to grow ivy as easily as receive satellite television pictures! The structures are built from 1½- by 1½-inch horizontal redwood strips spaced 8 inches apart. These strips, in turn, are attached by 1 by 3-inch cross strips. The redwood horizontal and vertical

154

Fig. 8-3. Advertisement for TVRO receiver circuit board.
(Courtesy Sigma International, Inc.)

Fig. 8-4. The inexpensive McCullough 8-ball spherical antenna.
(Courtesy McCullough Satellite Systems, Inc.)

strips are fastened to an angle-iron frame using adjustable bolts to establish the exact curvature of the reflective surface. This surface consists of inexpensive aluminum mesh which has been stapled to the horizontal strips. A separate feedhorn, positioned about 15 feet in front of the antenna, picks up the signals that have been reflected and concentrated by the aluminum screening.

Unlike the parabolic antenna, a spherical antenna can receive *several* different satellites simultaneously. This is accomplished by placing two or more feedhorns at the proper focal points for the desired satellites. The antenna is also available in kit form. Details on its construction have been published in several electronics hobbyist magazines. In Chapter 9, the McCullough spherical design is reviewed in somewhat more detail.

By using a do-it-yourself receiver, down converter, the LNA kits, along with a wood-and-screen spherical antenna, the home TVRO satellite terminal can be put together for well under $500! The LNA—the most sophisticated and expensive component—can be purchased assembled for about $300 from the national distributors. LNA kits can be obtained from Simcomm Labs, Box 60, Kersey, Colorado 80644. An LNA printed circuit board package with partial components is available from Gillaspie and Associates, Inc. for under $50.

Since the ultra-low-noise GaAsFET LNA Transistors now sell for under $50 apiece, an LNA kit in component form (not including the machined casing) should cost only $100 to $150. The LNA operates directly at the microwave frequencies. Sophisticated (and expensive) test equipment is required to properly align the LNA "front end" system (Fig. 8-5). It is not advisable for the microwave electronics novice to tackle this project. Instead, it would be better to buy an off-the-shelf 120-degree LNA, pur-

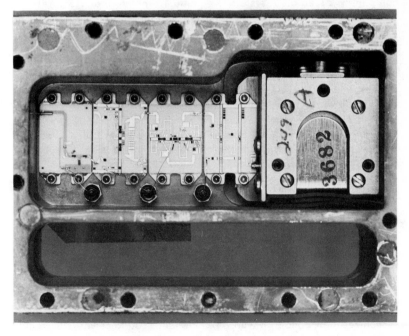

Fig. 8-5. Inside a commercial LNA.
(Courtesy Dexcel, Inc. © Mort Thomas)

chase a popular receiver kit with down converter, and build a preassembled, multisegment parabolic dish antenna which is shipped to the site in knocked-down pieces. Such a system can be obtained for $1,000 to $1,400, and can be assembled with a weekend of work.

In Chapter 9, the KLM parabolic mesh antenna kit is described. This TVRO dish can be assembled for under $200 using locally fabricated metal components, or for under $600 if purchased from the factory. It is a professional-looking system and the polar mount can be easily motorized in the future.

Optionally, the do-it-yourselfer might want to build the McCullough spherical antenna, using wood stripping and aluminum screening readily available at local building supply yards. For the enthusiast who is completely determined to build the entire TVRO system from scratch, the spherical antenna is much easier to fabricate.

THE COMPONENT SYSTEM APPROACH: PUTTING TOGETHER A MEDIUM-PRICED SYSTEM

Perhaps the best way of owning your own TVRO system is to buy completely assembled system elements from the manufacturers or dealers, taking the boxes home, and plugging them together. There are three major components: The antenna with base and feedhorn assembly, the receiver/down converter, and the LNA. By going this route, the backyard TVRO owner can set up his own system in a day or less.

The major distributors such as National Microtech can supply complete satellite TVRO terminal packages, shipped to the site in UPS-deliverable or truck-sized boxes. The ADM-3 11-foot parabolic antenna coupled with an Avantek 120-degree LNA and Sky Eye IV receiver with down converter provide a complete system. National Microtech has even produced a videocassette with step-by-step installation instructions. Longs Electronics, Birmingham, Alabama (800-633-6461), will sell a complete satellite system consisting of a four-section, 10-foot, metal-bonded fiberglass dish, polar mount, feedhorn assembly with polarizing rotator, HR101 "Entertainer" Receiver, Avantek 120-degree LNA, and Chaparral Feed Horn (cables, RF modulator, and shipping are separate). This system can be assembled in under three hours, is weather-resistant, and virtually maintenance-free. The Echosphere Corporation and Intersat—both national distributors—have similarly priced packages available for the do-it-yourself hobbyist.

Two outstanding low-cost TVRO packages are available for the middle-range "do-it-yourselfer." The Janeil baked enamel and mesh sectionized parabolic dish with Dexcel Model 1100 Receiver/LNC combination (Fig. 8—6) can be purchased directly from Janeil. The Paradigm aluminum mesh antenna (Fig. 8—7) with KLM IV Receiver and Avantek

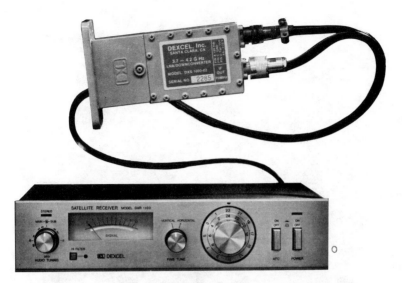

Fig. 8-6. The Dexcel 1100 LNC/Receiver combination with stereo sound.
(Courtesy, Dexcel, Inc.)

or Amplica 120-degree LNA package can be ordered directly from Paradigm for shipping overnight.

It takes from four to ten hours to assemble a metal-mesh parabolic dish antenna. This assumes a first-time user is working slowly and systematically. Following assembly, the antenna must be sited properly, setting the correct azimuth and elevation adjustments for the satellite desired. One advantage of a parabolic antenna is that its surface is formed at the factory and few adjustments are necessary in the field. Dish surfaces are precisely shaped; their tolerances can vary only by 1/16 of an inch or so over the entire face of the antenna. Surface variations greater than this limit the effectiveness and may completely prevent signal reception by a small-bore antenna. Spherical antennas, which are assembled piece-by-piece, require more care in aligning the surface shape. A spherical template or proofing instructions are supplied for this purpose. A parabolic dish is far better for the "plug-'em-in right away" buyer and they outsell the sphericals by 20-to-1. Appendix E describes the procedure for locating the birds in space, and how to aim your TVRO dish to the right point.

Videotapes and instructions for the TVRO dish assembly are available from Satellite Television Technology in Oklahoma as well as from many of the national distributors. With experience, aiming the antenna to the proper point in space is a very simple process. Some people have been known to assemble and install complete TVRO systems in under one

Fig. 8-7. The Paradigm 11-foot aluminum mesh dish comes knocked-down
for easy assembly.
(Courtesy Paradigm Manufacturing, Inc.)

hour, using antennas that come preassembled in one piece. A seasoned
installer can successfully aim a dish at the Satcom 3 satellite within a
matter of minutes. Your author has unpacked, set up, and successfully
pointed his lightweight 8-foot tripod-mounted Luly antenna (and Dexcel
receiver/LNC combination) in under ten minutes, most of which was con-
sumed by unfolding and leveling the camera tripod!

The Luly fold-up umbrella antenna (Fig. 8–8) can fit in the back of a
station wagon and can be taken to the vacation home or even used in
an RV. Set up, the Luly antenna will work anywhere. In Fig. 8–9, it is
used in conjunction with the *battery operated* Gillaspie TVRO receiver/tv
system for true portability.

Fig. 8-8. The Luly umbrella antenna folded up for transport.
(Courtesy Gillaspie & Associates)

With proper selection and preparation of the desired site, either on the roof or in the backyard, a TVRO video hobbyist can install his own satellite television earth station quickly and safely.

BUYING A DEALER-INSTALLED "TURNKEY" SYSTEM

The easiest way to install a backyard satellite TVRO dish is to have the dealer do it! The TVRO dealer will make a preliminary visit to the site, and recommend the proper antenna size to use, along with a receiver/LNA system.

Turnkey systems vary in price from $3,000 to $5,000 to $10,000 or more depending on the size of the antenna and the features selected. The typical price for a manually cranked polar-mount single-satellite antenna with a 24-channel tunable receiver and horizontal/vertical polarity rotator control, properly mounted on a concrete pad or prepared site

Fig. 8-9. The Luly antenna set up for operation with a battery-powered TVRO
receiver/tv receiver combination.
(Courtesy Gillaspie & Associates)

will run between $3,500 and $5,000. A fully remote controlled motorized antenna mount (Fig. 8–10) might raise this price by $1,000 to $2,000 or more. If two people in the household wish to view different programs simultaneously, then a dual LNA feedhorn assembly, signal power splitter, and two or more satellite receivers will need to be connected to the same antenna. Such a multireceiver system can increase the cost by $2,000 to $4,000 or more. This deluxe configuration begins to approach the sophistication of the commercial private cable system. Adding a deluxe receiver like the one in Fig. 8–11 which features a full-function infrared remote control for both channel selection and satellite selection increases the cost.

Not all TVRO satellite dealers are equal. Experience in the industry varies, as does the quality of workmanship and installation. Since the industry is still very new, even the most experienced organization will only have been in business a few years. The customer should check the TVRO dealer carefully. A visit to one or more existing TVRO installations should provide sufficient information to make a buying decision. Ask the current customers if they have had any trouble with the system components or the dealer servicing record. Suppliers who sell equipment to the dealer can be contacted directly, and their off-the-record opinions as to the reliability of the dealership can be solicited. Although by far and away the majority of satellite TVRO dealers are reliable and upstanding

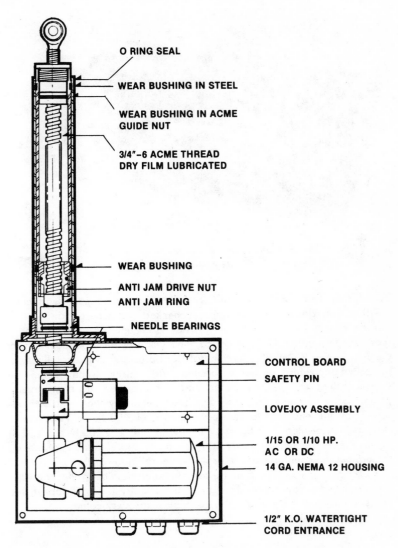

O RING SEAL

WEAR BUSHING IN STEEL

WEAR BUSHING IN ACME
GUIDE NUT

3/4"-6 ACME THREAD
DRY FILM LUBRICATED

WEAR BUSHING

ANTI JAM DRIVE NUT

ANTI JAM RING

NEEDLE BEARINGS

CONTROL BOARD

SAFETY PIN

LOVEJOY ASSEMBLY

1/15 OR 1/10 HP.
AC OR DC

14 GA. NEMA 12 HOUSING

1/2" K.O. WATERTIGHT
CORD ENTRANCE

Fig. 8-10. A linear-drive system used to drive a motorized TVRO dish.
(Courtesy Tel-Vi Communications)

businessmen in their communities, there will always be the occasional undercapitalized or shady fellow who "snookers" a customer or two. Luckily, this is quite rare!

The information provided in this book should give the reader a good handle on the industry and assist in the overall TVRO selection process.

Fig. 8-11. The Intersat deluxe remote-control receiver.
(Courtesy Intersat Corp.)

By referring to Appendix D, the size of the antenna recommended by the installing dealer can be quickly checked. In general, an 11-foot parabolic dish TVRO system will provide excellent consumer-quality tv reception throughout North America, and will provide broadcast-quality pictures in the "CONUS" (Continental US) boresight of the satellites.

Tens of thousands of home satellite TVRO terminals have been installed through the United States and Canada over the past several years. Many of these systems have been put in by the owners themselves. By the end of 1982 there were over 60,000 home TVRO earth stations in operation. This figure will grow to over 250,000 by the end of 1984! The burgeoning industry will see many more dealers spring up throughout the country over the next few years. Many of the "old-timers" like Channel One in Boston or Helfer's Antenna Service in Pleasantville, New York, who now travel throughout the country to install top-of-the-line systems, will be joined by local dealers in every community.

CHAPTER **9**

Construct a Backyard TVRO Antenna for Under $200

THE DO-IT-YOURSELF APPROACH

It is now possible to build your own backyard satellite antenna for under $200 using inexpensive designs and common materials. Or, by spending between $500 and $800, a commercial parabolic antenna can be purchased in knocked-down kit form for assembly in a few hours, saving from $500 to $1,000 in dealer installation costs.

There are three choices available to the energetic "do-it-yourselfer." The wood slat and mesh screen spherical, the open-face-form petalized fiberglass parabolic, or the expanded metal screen/rib parabolic can be constructed. All three antenna types require varying degrees of construction skill and expertise with different types of building materials. If you are good at working with wood, the homemade spherical may make sense. Anyone who has tackled building a fiberglass boat should find the fabrication of a parabolic antenna straightforward. The metalworker who wishes to construct a panelized metal-mesh antenna will find that detailed construction plans are available. Most people who frequently use small hand tools will probably be more comfortable with assembling a precut and pretested spherical kit from a detailed instruction manual. This chapter will review the spherical and parabolic designs and walk the reader through the assembly instructions for a typical aluminum rib and mesh 11-foot antenna.

THE HOME MADE SPHERICAL

The 12-foot wood-framed spherical antenna is one of the least expensive to build and is easy to align. In addition, the spherical design allows for the simultaneous reception of two or more satellites if separate strategically placed feedhorns are used. Once it has been set in place, the spherical antenna never needs to be moved to track a different Clarke-orbit satellite. A popular spherical antenna consists of two major pieces: a steel frame that provides a strong nonflexing support, and a wood-lattice assembly to which the screen wire reflective surface is stapled. One of the advantages to this type of structure is that the steel frame need not be formed to close tolerances, because the "proofing" adjustments are accomplished with adjustable bolts that distort the wood lattice strips.

The angle-iron framework is formed of five vertical and three horizontal members made from 1/2- × 1/2-inch, 1/8-inch thick galvanized angle iron shown in Fig. 9–1.

Five vertical lattice strips cut from 3- × 3/4-inch redwood are joined together by 19 redwood strips each 2- × 3/4-inch which become the horizontal ribs. After this wood framing has been bolted together, the lattice assembly is ready for attaching to the angle-iron support structure by 36 adjustable bolts. (See Fig. 9–2.) By setting the length of the

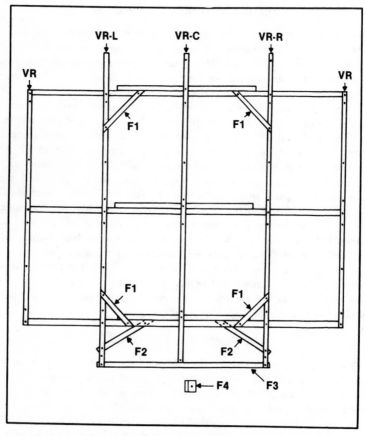

Fig. 9-1. Detail of the angle-iron framework for the spherical antenna.
(Courtesy Downlink)

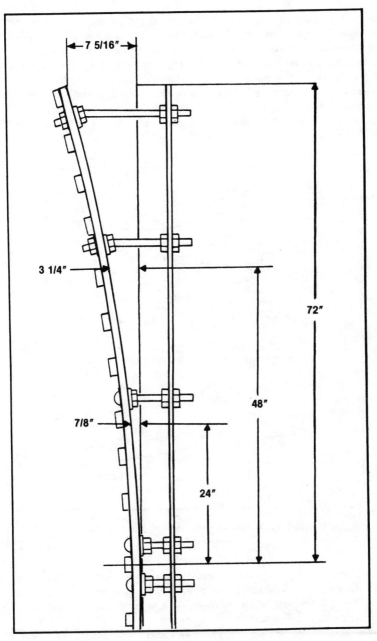

Fig. 9-2. Tuning the wood lattice frame to the proper spherical shape by adjusting bolts. (Courtesy Downlink)

bolts with ¼-inch "stop nuts," (Fig. 9-3) the proper curvature in the vertical direction can be achieved. This tuning is done to set the shape of the wood lattice to fit a spherical arc of curvature of 30 feet. By attaching a string to a radius point set precisely 30 feet away, the arc of the redwood lattice assembly can be slowly "proofed" as the 36 adjustment bolts are tightened. After this static tuning process has been completed, the spacing of the bolts will vary from 3 $^7/_{32}$ inches at 4 feet from each side of the center of the assembly to 7 ¼ inches on the left and right ends.

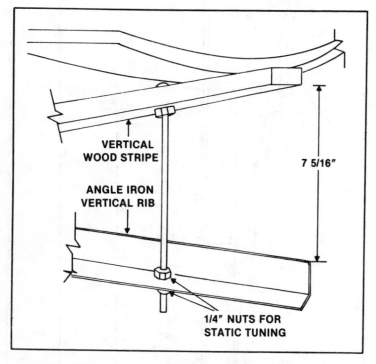

Fig. 9-3. Closeup of assembly after "static tuning" has been completed.
(Courtesy Downlink)

A heavy mesh screening of .025-inch diameter wire, ⅛-inch mesh is recommended to withstand the abuse that snow and ice can create. However, lightweight and inexpensive screening can be used in mild and temperate climates.

The screen must be carefully unrolled and stapled onto the wood lattice assembly tightly and accurately, with no gaps or spaces (Fig. 9-4). The process is similar to stretching a canvas painting; care should be

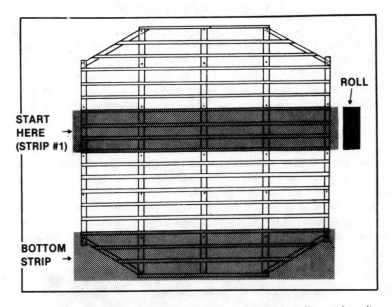

Fig. 9-4. Screening is applied to the wood frame by carefully unrolling and stapling.
(Courtesy Downlink)

taken to ensure that the screen is not stretched too tightly or puckers can result which dramatically distort the shape of the antenna surface.

Upon completion of the antenna dish, a support system of rear legs and braces should be built. This can be constructed of wood or angle iron, and should be placed on poured-concrete base pads to firmly support the heavy dish structure.

The antenna is aligned in a "polar" configuration to cause the geosynchronous arc of satellites to cut a corresponding arc across a horizontal line on the center of the dish surface. The feedhorn assembly consists of an LNA and Chaparral-type feedhorn mounted with a metal clamp to an aluminum pole which has been thrust into the ground. The feedhorn should be located 15 feet from the optical focal point of the spherical antenna corresponding to a specific satellite to be viewed.

By careful scrounging of parts, especially through the use of recycled angle iron and wood strippings, less than $200 can be spent in constructing this antenna—not including the cost of the spherical feedhorn assembly.

Complete assembly instructions are available from McCullough Satellite Systems, Inc. in Salem, Arkansas (see Appendix C for address). Dealers who wish to purchase this antenna can buy them in single quantities starting at under $500 for the 8-foot version.

ASSEMBLING THE KLM X-11 DISH

The KLM X-11 is a high performance, 11-foot parabolic antenna, suitable for most areas of the continential United States and Canada. Slide-in screen panels and reinforced aluminum support ribs ensure high strength, low weight, and wind load. Two people can assemble the X-11 and its associated Polar-Trak mount in about 2½ hours, and both prime and hub mounting feeds are available. The antenna, which weighs 125 pounds, is shipped via UPS from Morgan Hill, California.

The antenna (Fig. 9–5) is constructed from stainless steel hardware and corrosion-resistant aluminum alloys, and has a gain of 40.5 dB, representing a 55% efficiency. It can survive winds of 100 miles per hour, and has a wind load of 72 square feet. The focal length of the dish is 61 feet and its f/D ratio is .47.

In assembly, 24 ribs are bolted to the upper and lower hub plates using three bolts per rib. A stainless steel cable harness is attached near the perimeter and the turn buckle is tightened to set and hold the critical dish dimensions. Then the expanded-metal screen panels are slid in,

Fig. 9-5. The KLM X-11 aluminum mesh antenna assembled.
(Courtesy KLM Electronics)

and the perimeter strips installed. Assembly is complete! No exotic hardware, pop rivets, or baling wire are used.

Detailed Assembly Instructions

1. Use a support stand to permit dish assembly at a convenient height (3−4 ft). Clamp or bolt the square lower hub plate to the stand.
2. Attach the polarity rotator to the aluminum angle bracket using the ¼−20 hardware. Orient rotator so the weatherproofing collar is "up."
3. Bolt the angle bracket to the square lower hub plate using the ¼−20 × ¾-inch bolts. (Fig. 9−6)
4. Connect the rotator control cable to the terminals because they are difficult to reach once the dish is assembled. Terminals face the "down" side of the dish. Also loosely attach the rotator tube clamps.
5. Bolt the "down" rib to the right center of the side facing the rotator terminals. Bolt another rib to the opposite side of the plate (by angle bracket) to provide counterbalance. Secure with ¼−20 × ¾-inch bolts, nuts, and lockwashers (nuts placed inside rib channel). Finger tighten the hardware only. (See Fig. 9−7)
6. Place the 12-sided upper hub plate on top of the ribs and secure

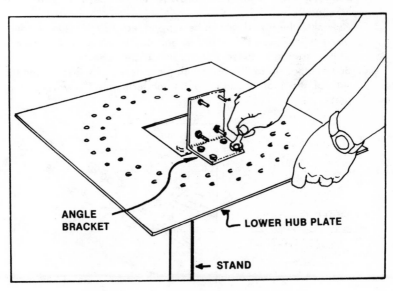

ANGLE BRACKET   **LOWER HUB PLATE**

← **STAND**

Fig. 9-6. Bolting angle bracket to lower hub plate.
(Courtesy KLM Electronics)

171

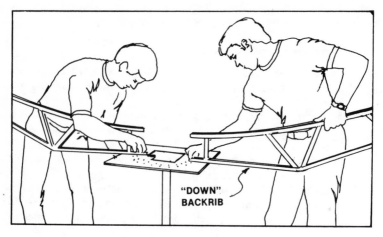

Fig. 9-7. Attaching the first two ribs to plate.
(Courtesy KLM Electronics)

it with ¼–20 × ¾-inch hardware. Position the nuts inside the rib channel, but use *no* lockwashers. Lightly tighten.

Sight along ribs and adjust so they form a straight line (see Fig. 9–8). Then securely tighten upper and lower hardware. *From this point on, finger tighten all hardware unless specifically directed otherwise.*

7. Assemble the tripod feed-tube support legs to the nylon bearing. Use the undrilled tabs on the legs positioned at 120° intervals (drilled tabs are bolted later to hub plate). See Fig. 9–9.

8. Bolt another rib to the *lower* hub plate next to the first one on the side of the plate facing the rotator terminals ("down side" of plate).

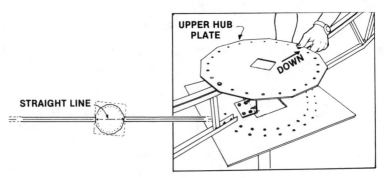

Fig. 9-8. Attaching upper hub plate.
(Courtesy KLM Electronics)

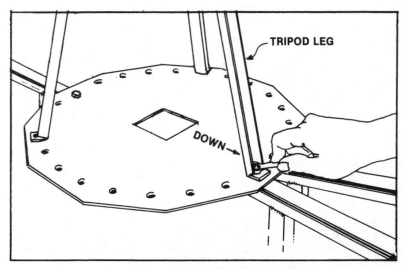

Fig. 9-9. Attaching tripod feed-tube support legs.
(Courtesy KLM Electronics)

9. Put a ¾-inch bolt through one of the feed support leg tabs and then use this to secure the third rib to upper hub plate. This is the "down" leg.
10. Add another rib to the opposite side of the hub as a counter-balance. Continue installing ribs to the opposite sides of the hub until *all* are in place. All hardware is finger tightened only! Every 8th upper hub plate hole is also used to secure one of the tripod legs.
11. If a two piece feed support is supplied, insert the crimped end of the lower tube into the upper tube (with LNA mount) and secure it with $8-32 \times 1¾$-inch bolt.
12. Bolt the LNA to its mount, then attach the feedline and LNA cover. Route the feedline out of the channel and down through the support tube. Allow enough length to reach the down converter. Secure the feedline with harness ties to keep the installation neat.
13. Install the feed assembly. Guide the feedline carefully through the rotator (Fig. 9–10). Loosely secure the clamps.
    The distance from the feedhorn to the upper hub should be 61 inches (as shown in Fig. 9–11) when using a KLM-type feedhorn.
14. Open the cable harness by removing one eyebolt. Adjust the other until no threads show inside. Position the turnbuckle between two ribs facing the rotator terminals (down), and string the cable around the dish on the outer rib support. Before placing the last four ribs, reattach the eyebolt. Then stretch the cable up over

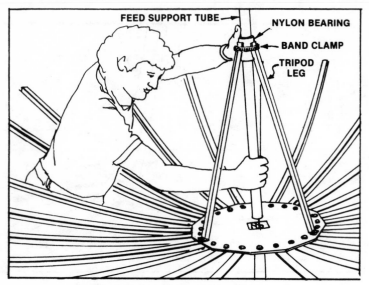

Fig. 9-10. Installing the feed assembly.
(Courtesy KLM Electronics)

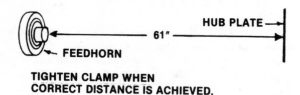

**TIGHTEN CLAMP WHEN
CORRECT DISTANCE IS ACHIEVED.**

Fig. 9-11. Measuring distance from feedhorn to upper hub.
(Courtesy KLM Electronics)

the last four supports, one at a time. Retighten the turnbuckle as needed to keep the cable from slipping off. (See Fig. 9–12.)
CAUTION: Leather gloves are recommended for working with the expanded-metal screen panels. Trimmed edges can be knife-sharp!

15. Install the first screen panel to the firmly bolted and aligned "down" rib and an adjacent loose rib. Work the narrow end firmly into the channels nearest the hub (Fig. 9–13) and then guide the long side into the bolted rib all the way to the perimeter. Adjust until the outer edge of the screen is flush (even) with the outer end of the rib.

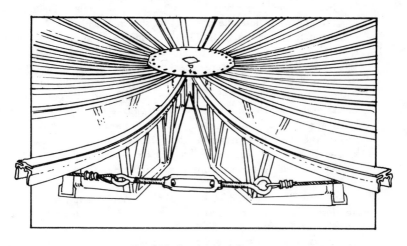

Fig. 9-12. Turn buckle adjustment.
(Courtesy KLM Electronics)

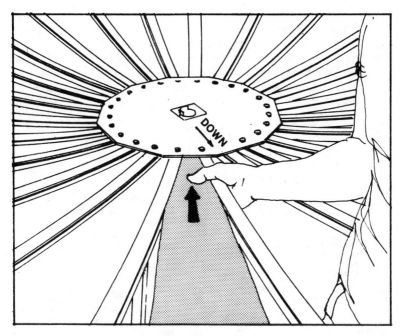

Fig. 9-13. Sliding screen panel into ribs.
(Courtesy KLM Electronics)

16. Finish by pulling the ribs together gently and guiding the other long screen edge into the channel as shown in Fig. 9–14.

Do not trim screens to fit. Screen panels are subject to rigid quality checks for size and fit and must not be trimmed. If the screen does not appear to fit, check to see that it is fully seated in its channels. Also check that the ribs are properly aligned and radiate from the center of hub plates. Misaligned ribs can cause the screen to have an uneven fit as the rib ends.

17. When the screen is well seated, install the perimeter channel between the rib ends using the No. 10 × ⅜-inch sheet-metal screws (Fig. 9–15). Do not tighten at this time.

18. Continue installing screen panels in a circular pattern as outlined in Steps 15, 16, and 17.

19. When installing the *last* screen panel, loosen the turnbuckle as needed to provide adequate clearance. Before installing the panel, step into the open wedge and tighten *all* the upper rib to hub bolts. Then, install the screen and tighten its turnbuckle until the perimeter channel can be screwed into place.

20. To ensure that the antenna is assembled to the correct parabolic curve, it is important to check the depth of the dish as illustrated in Fig. 9–16.

The easiest way to check the depth is to use a 13- or 14-foot length of strong string with a loop at one end. Measure 5½ feet from the loop and attach a short length of soft string that will hang straight. Tie a small paper clip to the loose end and adjust until the total length is 17¹/₁₆ inches.

To use, place the loop on a perimeter screw and stretch the main length across the dish to the other side (keep clear of feed

Fig. 9-14. Finishing the insertion of the first screen panel.
(Courtesy KLM Electronics)

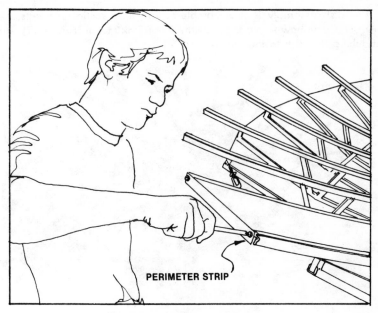

Fig. 9-15. Installing perimeter channel.
(Courtesy KLM Electronics)

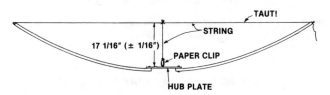

Fig. 9-16. Checking for correct depth of dish.
(Courtesy KLM Electronics)

hardware). Pull it taut and tie it around a nearby perimeter screw.

If the clip is more than $1/16$-inch away from the hub plate, loosen the cable turnbuckle until it is almost touching.

If the clip is laying on the hub plate, tighten the turnbuckle until it is almost touching (make sure that the string stays taut).

After any turnbuckle adjustment, "bump" all the ribs to help the cable take the new set.

When the correct depth is achieved, safety-wire the turnbuckle so it cannot unwind.

21. Firmly tighten *all* the No. 10 × ⅜-inch perimeter screws.
22. Tighten *all* the ¼-20 × ¾-inch lower hub/rib bolts.

The dish assembly is now complete. The next section presents the instructions on how to put together the Polar-Trak mount (Fig. 9–17) and attaching the dish to the mount.

Fig. 9-17. Close-up of Polar-Trak motorized drive for the X-11 antenna.
(Courtesy KLM Electronics)

ASSEMBLING THE POLAR-TRAK KLM MOUNT

1. The Polar-Trak base is 44 inches square and 24 inches deep. A round base, 48–50 inches in diameter, is also acceptable. About 1 cubic yard of concrete is used.

   The 3-inch × 72-inch AZ (azimuth) post is set perfectly vertical into the center of the base to a depth of 24 inches. No other orientation of the base or post is necessary. (See Fig. 9–18)

2. Install the ¼-13 × 1-inch bolts into the nuts welded to the 4-inch square × 60-inch pedestal column. Slide the pedestal over the AZ post. Orient the two tabs on the pedestal roughly to the north and tighten the locking bolts.

   With the pedestal bottomed on the base, dish clearance is provided to 15 degrees above the horizon. Should greater clearance be required due to terrain, obstacles, or location, use a short block of wood to support the pedestal while the bolts are tightened.

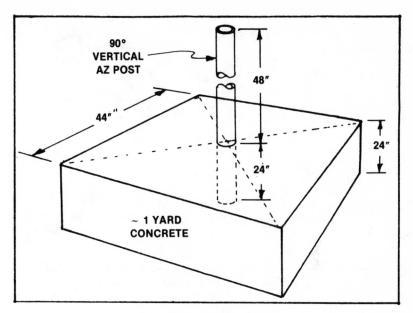

Fig. 9-18. Setting the aximuth post.
(Courtesy KLM Electronics)

3. Install the large right-angled polar/EL (elevation) assembly onto the upper end of the pedestal, with the long arm oriented to the north (the same side as the tabs on pedestal). Secure with ¾-10 × 6-inch bolt, flatwashers, lockwasher, and nut (Fig. 9–19). Do not tighten at this time.

4. Apply grease to one end of the 1-inch (od) × 20-inch polar axis shaft. Use a rubber mallet to tap on one of the large plastic bushings (flanged edge out) until it is flush.

   Install the shaft/bushing into the polar axis assembly (Fig. 9–20). Grease the other end of the shaft, and tap on the remaining bushing (Fig. 9–21).

   Grease the end faces of both bushings.

5. Attach the tangential-drive ring to the polar axis shaft with a ¾-16 × 1-inch bolt, flatwasher, and lockwasher.

   Run a ¾-10 nut about 1 inch onto each of the ¾-10 × 10-inch threaded rods. Install the rods on the tangential-drive channel and secure them with nuts and lockwashers (Fig. 9–22).

   Run another ¾-10 nut onto each of the rods and position it about 2 inches above the angle iron. (These two nuts will adjust the declination offset.) Then add a flatwasher to each rod. (See inset in Fig. 9–22.)

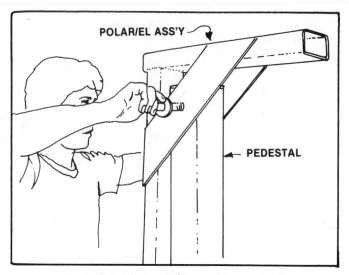

Fig. 9-19. Installing polar/EL assembly to pedestal.
(Courtesy KLM Electronics)

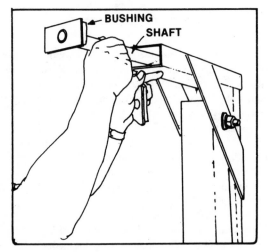

Fig. 9-20. Installing shaft and bushing into polar axis assembly.
(Courtesy KLM Electronics)

6. Install the square antenna-mount frame onto the threaded rods and secure it with two more ¾-10 nuts and lockwashers. (Fig. 9–23)

7. Secure the frame to the lower polar axis shaft with a ¾-16 × 1-inch bolt, flatwasher, and lockwasher. (Fig. 9–24)

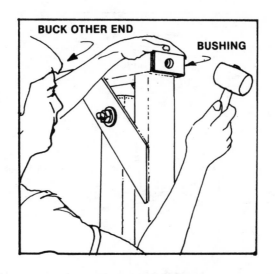

Fig. 9-21. Tapping bushing into place.
(Courtesy KLM Electronics)

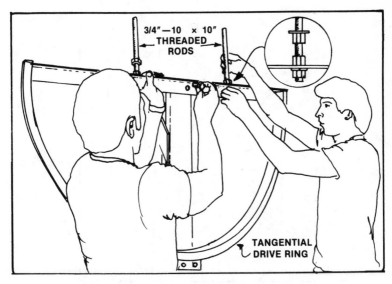

Fig. 9-22. Attaching tangential-drive ring to polar axis shaft.
(Courtesy KLM Electronics)

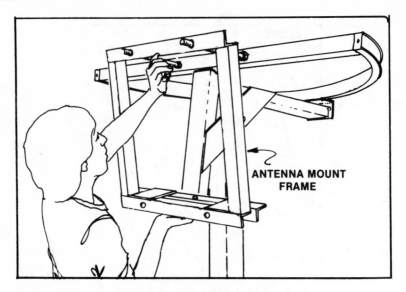

Fig. 9-23. Installing antenna mount frame.
(Courtesy KLM Electronics)

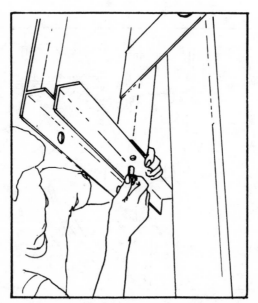

Fig. 9-24. Securing antenna mount frame to lower polar axis shaft.
(Courtesy KLM Electronics)

Manual Mount Only

8. Install the lock-down angle over the tangential ring. Secure it to the center hold on the bracket with a ½-13 × 1-inch bolt and lockwasher (see Fig. 9–25).

   CAUTION: The locking-angle prevents polar movement of the dish. *Always hold the dish firmly when loosening the lock-down angle to rotate dish or track other satellites.*

   NOTE: *The motorized module (option) interchanges with the manual lock-down without modification.*

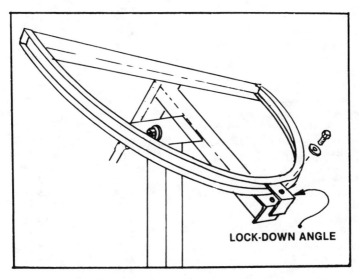

LOCK-DOWN ANGLE

Fig. 9-25. Installing lock-down angle on manual mount.
(*Courtesy KLM Electronics*)

Declination/Off-Set Adjustment

9. The farther from the equator a polar mount is located, the more it requires a small additional "tilt" adjustment to accurately track the tv satellite orbit. This is *not* an elevation setting for the polar axis, but a declination or offset angle between the dish and the polar axis. In Florida or Texas this declination is about 4.5 degrees. For a dish located at the Canadian/US border, the declination increases to a little over 7 degrees. Polar mounts that lack this adjustment can accurately track only a small portion of the satellite belt and the quality of reception suffers accordingly.

   The KLM Polar-Trak mount is fully adjustable for declinations; therefore, it can provide accurate tracking and excellent reception from horizon to horizon. Table 9—1 lists the angle of offset in degrees based on the latitude at your location.

Table 9—1. Declination Angle at Various Locations

Latitude (degrees)	Declination (degrees)	Dimension A (inches)
25	4.23	4.48
26 Brownsville, TX	4.38	4.53
27	4.53	5.58
28 Tampa, FL	4.68	4.64
29 Daytona Beach, FL	4.82	4.69
30 New Orleans, LA	4.96	4.74
31 Hattiesburg, MS	5.10	4.79
32 El Paso, TX	5.24	4.84
33 San Diego, CA	5.38	4.88
34 Atlanta, GA	5.51	4.93
35 Albuquerque, NM	5.65	4.98
36 Tulsa, OK	5.77	5.02
37 Joplin, MO	5.90	5.07
38 San Francisco, CA	6.02	5.11
39 Washington, DC	6.15	5.15
40 Denver, CO	6.26	5.20
41 Omaha, NE	6.38	5.24
42 Chicago, IL	6.49	5.28
43 Casper, WY	6.60	5.32
44 Eugene, OR	6.71	5.35
45 Minneapolis, MN	6.82	5.39
46 Butte, MT	6.92	5.43
47 Tacoma, WA	7.02	5.46
48	7.12	5.50
49 Vancouver, BC	7.21	5.53
50	7.30	5.56

The small changes in declination are difficult to measure accurately with the standard angle finders or inclinometers. To overcome this, "dimension A" is provided as a reference measurement between the polar assembly and the dish mount assembly. See Fig. 9—26. This declination adjustment can be done at any time without leveling or mount movement, but it is easiest at this point in the assembly process. Fig. 9—26 also shows the method for setting declination with an angle finder, and can be used for rough settings or doublechecking dimensional settings.

10. Set Dimension A by adjusting both sets of nuts that capture the dish mount angle iron. Measure at both ends of the angle iron to keep the setting even. Measure between the *inside* faces of the upper and lower angle iron as shown in Fig. 9—26. Firmly tighten these nuts when finished.

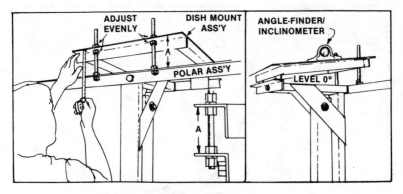

Fig. 9-26. Adjusting declination angle.
(Courtesy KLM Electronics)

Attaching The X-11 Dish

11. Raise the dish and carry it to the mount (remove the feed assembly, if possible, to reduce the dish weight and improve its balance). Place the dish on the mount, centering the square hub plate inside the dish mounting frame as shown in Fig. 9–27.

12. Install the upper rib-clamp angle onto the threaded rods (Fig. 9–28) and secure it with ¾-10 nuts and lockwashers.

13. Install the lower rib-clamp angle over the rib braces, and secure it to the mount frame with ¾-10 × 1½-inch bolts, lockwashers, and nuts. (See Fig. 9–29.)

14. Even up the eyebolts in the large turnbuckle. Install the turnbuckle between the pedestal tab and the polar assembly arm tab as shown in Fig. 9–30. Install the cotter pins. Use the upper pedestal tab if located in the upper US, etc.

Setting the Elevation

15. Fig. 9–31 shows how to use the latitude at your location to determine the correct elevation setting for the mount. The latitude figure may be used directly for degrees from vertical, or subtracted from 90 for degrees elevation from horizontal. Use an inclinometer or angle finder, and adjust the turnbuckle as shown in Fig. 9–32 to achieve the desired elevation.

16. Firmly tighten the ¾-10 × 6-inch (pedestal to polar assembly) pivot bolt and nut (installed in Step 3).

17. For sustained, long-term performance and mechanical integrity, it pays to make sure all hardware is thoroughly tightened. If possible, allow the dish and mount to temperature-cycle overnight. Over this period, hardware and assemblies will take a "set" and

185

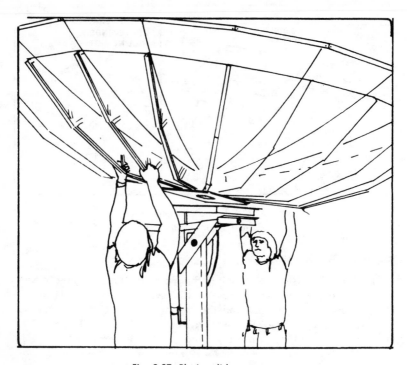

Fig. 9-27. Placing dish on mount.
(Courtesy KLM Electronics)

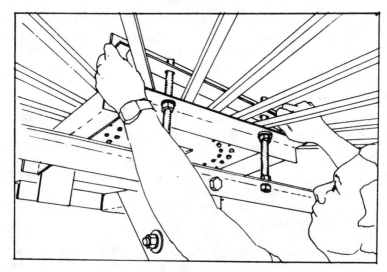

Fig. 9-28. Installing rib clamp angle.
(Courtesy KLM Electronics)

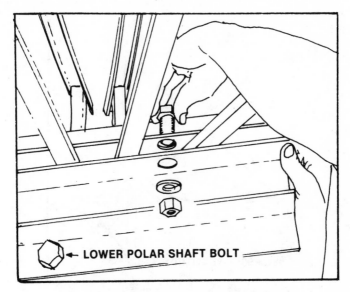

← LOWER POLAR SHAFT BOLT

Fig. 9-29. Securing rib clamp angle to mount frame.
(Courtesy KLM Electronics)

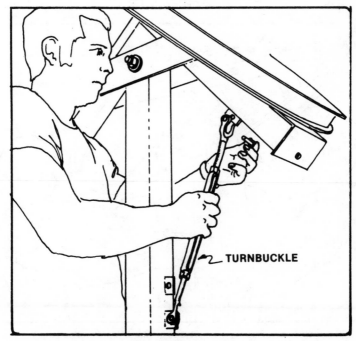

TURNBUCKLE

Fig. 9-30. Installing turnbuckle between pedestal and polar assembly arm.
(Courtesy KLM Electronics)

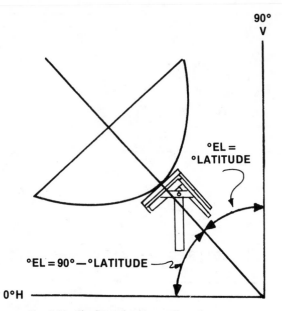

Fig. 9-31. Checking elevation setting of antenna.
(Courtesy KLM Electronics)

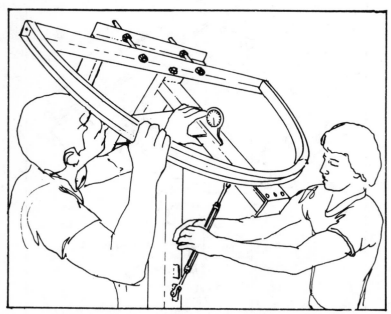

Fig. 9-32. Adjusting the turnbuckle to obtain correct elevation setting.
(Courtesy KLM Electronics)

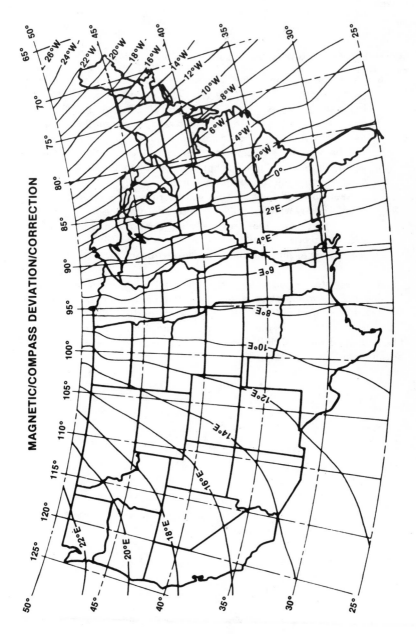

MAGNETIC/COMPASS DEVIATION/CORRECTION

Fig. 9-33. Magnetic compass deviation throughout the US.
(Courtesy KLM Electronics)

relax slightly. Then check and retighten all accessible hardware one more time.

Orienting the Mount to True South

18. "True" South follows the longitude (vertical) lines on a map or globe. A magnetic compass will deviate from "true" south by as much as 22 degrees in the continental US. The map in Fig. 9–33 can be used to determine the average magnetic correction needed in your area. For instance, in central Oregon "true" south will be found 20 degrees to the east of magnetic or compass south. In Maine, "true" south will be found 20 degrees to the west of magnetic or compass south.

Keep in mind the compass will be affected by nearby iron or steel object, including the mount, so adjusting the mount correctly may take several tries. For the initial setting it may be helpful to use the compass to find a landmark on the horizon (or at least in the distance) that the mount may also be pointed at.

With the system in operation, check the accuracy of *all* the settings. Declination, elevation, and true south, can be verified by clean, on-target, reception of satellites at the far ends and middle of the chain. Since true south is the most variable setting, start the accuracy and tracking adjustments with it. Then, if necessary, check the elevation and declination.

CHAPTER 10

The TVRO Satellite Receiver You Can Build for $300

In the old days of 1977, the process of constructing a receiver to successfully operate in the 3.7 gigahertz (GHz) range was quite a task. Before the advent of saw filters, "image reject mixers," and the new integrated circuits, discrete-component circuits requiring hours of workbench tuning were the only receiver systems available.

These days have long since passed for most satellite receiver manufacturers, and in the near future, we can expect to see receivers coming off the Japanese assembly lines consisting of one or two large-scale custom ICs!

For the electronics hobbyist who is interested in constructing his own satellite receiver "from scratch," an excellent multiprinted circuit board do-it-yourself receiver called the Lite Receiver IV™* has been designed by Dwight Rexroad. With careful parts scrounging, the receiver can be built for under $100. The secret of this receiver is to make the overall design noncritical by using inexpensive parts that can easily be obtained anywhere. Every component is "off the shelf." The front end of the receiver which operates in the high intermediate frequency (i-f) range (500 to 1,000 MHz) is simply a modified commercial uhf tv tuner module, used in the hundreds of thousands of television sets. The i-f amplifier is a MC1350 intergrated circuit, another product found by the millions in consumer televisions. By relying heavily on the Motorola handbook, Rex was able to use bargain-basement ICs such as the MC1357 quadrature detector (instead of the more commonly used phase lock loop circuits) as the video demodulator.

Known as the "cheap trick" receiver, the basic circuit consists of seven building blocks. The high i-f input is converted by the modified uhf tuner to a 70-MHz second i-f which is passed through an i-f filter with a bandwidth of 30 MHz. This output is fed into an i-f amplifier and de-

---

*Trademark by Martcomm, Inc.

modulator module, whose three outputs drive the video amp-filter-clamp module (for baseband video output), the audio subcarrier filter/demodulator (for audio output), and the afc tune module (for fine tuning across the transponder channel). The afc output is coupled back to the uhf tv tuner. A final module consists of a well-regulated power supply delivering voltages at +12, +18, and +30 volts.

The down converter is built in a separate waterproof box located at the TVRO dish antenna. It consists of a balanced mixer and local oscillator with an NEC MC5121 broadband amplifier providing the necessary gain. The down converter converts all twelve transponders in a group (3.7 to 4.2 GHz) to the 500 to 1,000 MHz region for delivery to the front end of the receiver. This double-conversion technique has two advantages. First, inexpensive RG-59U coaxial cable may be used to run the signal from the down converter out of the dish into the receiver, which might be up to several hundred feet away inside the house. Second, by using a "power divider" at the end of the cable run, two or more receivers can be connected to the same down converter for simultaneous reception of two or more transponders. This allows a receiver to be placed at every television set in the house for independent operation.

A complete set of printed circuit boards and construction details are available from Martcomm, Inc., P.O. Box 74, Mobile, Alabama 36601. (The receiver construction project has been described in a series of articles from May through November 1982, by *Seventy-Three* magazine. Copies may be found in most libraries)

BUILDING A "STATE OF THE ART" RECEIVER KIT

The KLM Sky Eye V TVRO satellite receiver (Fig. 10–1) is a dual-conversion single-board design used in conjunction with a dish-mounted 70-MHz out down converter which has been preassembled and aligned by the factory.

Fig. 10-1. The KLM Sky Eye V receiver kit which comes with a fully assembled downconverter. *(Courtesy KLM Electronics)*

The Sky Eye V circuit design uses new bandpass saw filters and non-critical IC modules to eliminate complex alignment of the i-f stages. Fig. 10−2 presents the basic simplified block diagram of the receiver. The complete circuit is given in Fig. 10−3.

When shipped from the factory the kit (Fig. 10−4) includes all necessary components, including the cabinet with front panel and knobs, down converter, power supply, etc. A separate rf modulator will need to be provided, and an FCC type-approved fully assembled vhf modulator which fits neatly into a space provided for it inside the receiver, is available from KLM as an optional component. Information on building an rf modulator is also provided. A number of commercial kits are available on the market for under $15.

Theory of Operation

The mixer module is assembled at the factory. The 3.7 to 4.2 GHz frequency band from the system LNA is routed to the mixer input connector. A factory-supplied coaxial transition section is used between the antenna transmission coaxial cable and the connector. Regulated voltage for the LNA is connected to the center conductor of the coax cable through a capacitor and rf choke combination. This LNA voltage connection is used for KLM-supplied LNAs. Other LNAs may require a different arrangement for power input. In this case, the 15-volt dc may be disconnected from the feedthrough capacitor inside the down converter module.

The input signal is routed to a pair of mixers through strip-line hybrids. The input signal is phase-shifted by 90 degrees to each mixer. The local oscillator frequency in the 3.630 to 4.130 GHz band is supplied by the voltage-controlled oscillator (vco). The pair of mixers and the hybrids comprise an image rejection mixer. The output frequency is the difference frequency of 70 megahertz, an i-f frequency which is rapidly becoming the industry de facto standard. This output is routed to a non-limiting amplifier consisting of several discrete transistors and associated components. The tuning voltage for the vco is obtained from the automatic frequency control (afc) circuits of the video demodulator, and routed to the mixer module via a separate input coax connector mounted on the down converter case.

At the receiver, the incoming 70-MHz i-f frequency is amplified by a four-stage discrete-component circuit, and fed into a diode limiter. Voltage to operate the tuning meter is amplified through one-half of an LM358 integrated circuit tapped into the collector output of the transistor Q3. The output of the limiter feeds the 70-MHz FL1 filter to precisely bandwidth-limit the signal. Another stage of 70-MHz i-f amplification restores levels to the discriminator circuit, provided by an LM1496 IC. After the complex signal has been recovered to baseband, the output of

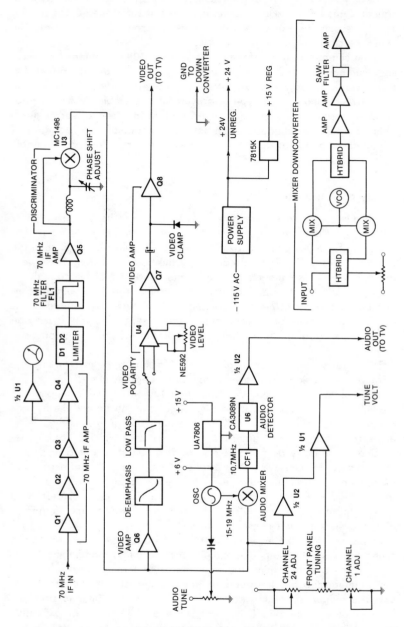

Fig. 10-2. Simplified block diagram of the Sky Eye V receiver kit.
(Courtesy KLM Electronics)

the discriminator is fed to three circuits: A final video amplifier with necessary de-emphasis and low-pass filtering, the audio demodulator (oscillator/mixer/detector/amplifier), and the afc/tuning voltage control circuit whose output voltage is fed back to the down converter through a final control circuit which varies the transponder tuning from the front panel tuning knob complete the circuit.

ASSEMBLY AND ALIGNMENT PROCEDURES

The KLM Sky Eye V can be assembled in a few evenings work, and can be aligned without the use of an expensive sweep generator or oscilloscope.

The down converter (with mixer) module is supplied by the factory preassembled and tested. Although the receiver unit needs no special equipment for tuneup or alignment, a vtvm or dmm is handy for troubleshooting or exact calibration of the tuning (vco) and output voltages.

Board Assembly

All parts should be mounted flush with the board, and soldered on both sides except as noted. The board is divided into four areas, in order to avoid overly long and complicated assembly and instructions. The board is coded with part numbers for easy reference (R1, R2, C1, C2, etc.). The layout of the circuit board is given in Fig. 10-5. Table 10-1 lists the parts used in the Sky Eye V.

First, install the resistors in areas marked A, B, C, and D. The chokes and diodes should then be installed in areas A, B, C, and D, noting the polarity or tab on the diodes. Trim pots are next to be inserted, followed by the voltage regulator (area A), the bridge rectifier (area C) and all capacitors for all areas, taking care to insert the capacitors with proper polarity as noted.

By this point, the board has been almost completely assembled. The bandpass filter FL1 is next to be inserted, along with the discriminator trimmer capacitor and ICs. Follow this with the push-button switches, meter light standoff, and light power leads. This completes the first stage of board assembly. It is important at this point to check for solder bridges, cold solder joints, touching components, etc. with a careful visual inspection of the entire board on both sides.

Chassis Assembly

The rear panel hardware (area "F") consisting of an RCA connector, terminal strip connectors, ac cord, and fuse holder are first to be installed. Next, the regulator socket and regulator are attached to the rear panel. The tantalum capacitor and diode are then wired to the regulator socket. Solder RG-59U coaxial cable leads to the RCA connectors from the points indicated on the "F" area of the circuit board. Attach the power transformer to the chassis with the green ac lead soldered to one of the mounting screws. Solder the transformer leads to the white ac power lead and the rear-panel fuse. Attach the ceramic resistor to the chassis, and connect its leads to the terminal strip and regulator socket. This completes the rear panel work.

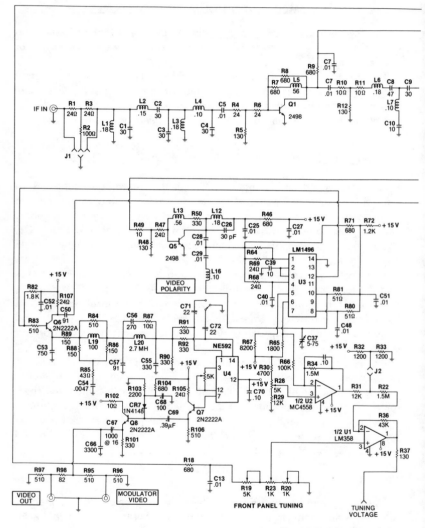

Fig. 10-3. Schematic of the
(Courtesy KLM

At the front of the receiver, bolt the tuning meter and solder its leads to the circuit board. Add the push buttons to the switches and install the audio and channel potentiometers. Run their leads to the circuit board but do not solder. Insert the board on the chassis with the four stand-off screws, feeding the pot shafts through the chassis face. Do not tighten the screws yet. Install the front panel to the chassis (the pot hex nuts and an additional screw firmly support the assembly), aligning the panel to

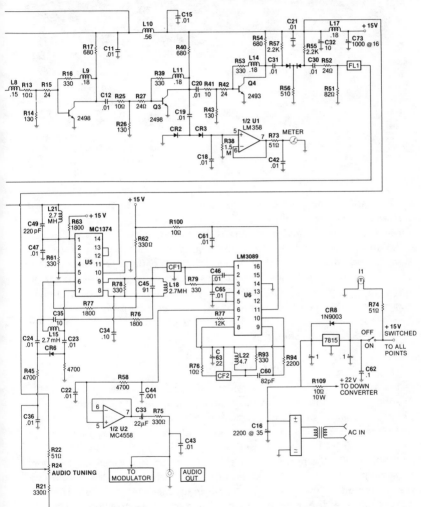

Sky Eye V receiver circuit.
Electronics)

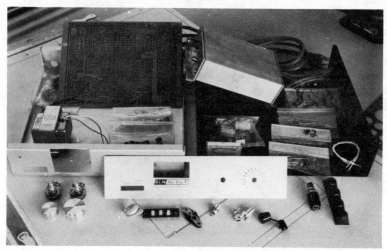

Fig. 10-4. Components of the Sky Eye V kit unpacked from the shipping carton.
(Courtesy KLM Electronics)

make it square and even. "Fiddle" with the board to provide the best push-button clearance through the front panel, and tighten all mounting screws. The front panel assembly is almost complete.

Solder the meter leads and the tuning pots to the printed circuit board, and the rear-panel coax leads to the chassis. Solder the regulator and terminal strip leads to the rear chassis, and solder the terminal strip lead to the ground lug. Finally, install the knobs to the tuning pot shafts. This completes the basic receiver assembly (Fig. 10—6). Check for correct placement of hardware and parts, and recheck for solder bridges, cold solder joints, etc.

TUNEUP AND FINAL ASSEMBLY

For tuneup, the remainder of the Sky Eye V receiver (dish, LNA, feedhorn, etc.) must be installed and functioning. The receiver down converter should be installed at the LNA, and all necessary cabling completed back to the receiver. Three conductor low-voltage unshielded lines and RG-59U coax are used.

The output of the receiver should be connected to a television monitor, or through a modulator to a standard tv set. Optionally, the output can be connected to the video input of a videocassette recorder. It is assumed that the dish is properly aimed and an active transponder signal is being processed by the down converter and this signal is being carried to the receiver. If this is not the case, then the dish and receiver must alternately be adjusted until a picture is available that will make it

198

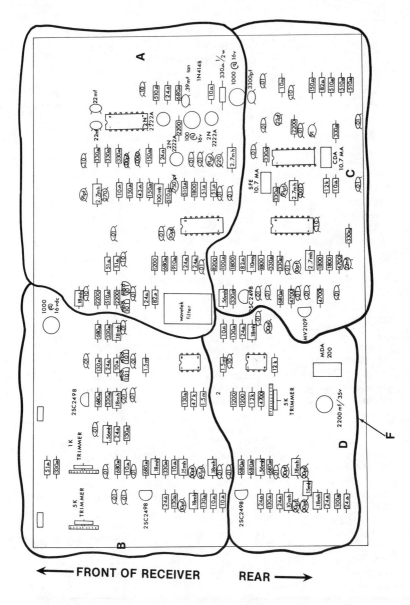

Fig. 10-5. Parts layout on the Sky Eye V circuit board.
(Courtesy KLM Electronics)

Table 10-1. Sky-Eye V Parts List

Item	Quantity	Item	Quantity	Item	Quantity
Resistors		Capacitors (Cont)		Sheet Metal	
10 Ohm	11	0.1 μF, Film	6	Chassis	1
24 Ohm	7	0.39 μF, 16 V, Tan	1	Cover	1
43 Ohm	1	1.0 μF, 35 V, Elect.	2	Front Panel	1
51 Ohm	4	22 μF, 16 V, Tan	4		
82 Ohm	3	47 μF, 25 V, Elect.	2	Switches	
130 Ohm	6	100 μF, 16 V, Elect.	1	On/Off	1
150 Ohm	3	470 μF, 15 V, Elect.	1		
220 Ohm, ¼ W	2	1000 μF, 16 V, Elect.	2	Plugs	
330 Ohm, ¼ W	8	1000 μF, 50 V, Elect.	1	F-59	1
330 Ohm, ½ W	1	5-80 pF, Variable	1	RCA	2
510 Ohm, ¼ W	7	100 pF, DM	2		
560 Ohm	1			Connectors	
680 Ohm	11	Inductors		3-Terminal strip	1
910 Ohm	2	.15 μH	4	F Chassis	1
1 kΩ, Variable (panel)	2	.56 μH	7	RCA Panel	2
1 kΩ, Variable (PCB)	1	.18 μH	3	Fuse Holder	1
1.2 kΩ	2	0.1 μH	1	TO-3 Socket	1
1.8 kΩ	2	2.2 μH	1		
2.2 kΩ	4	2.7 μH	3	Prntd Ckt Boards	
4.7 kΩ	3	4.7 μH	1	PCB Main	1
5 kΩ, Variable (PCB)	1	100 μH	1		
8.2 kΩ	1	115/25 Xfmr	1	Wire/Cable	
12 kΩ	5			18 AWG, Red	1
47 kΩ	2	Semiconductors		18 AWG, White	1
100 kΩ	4	2SC2498	5	18 AWG, Brown	1
1.5 Meg	3	MBD 101	4	18 AWG, Black	1
15 Meg, ¼ W	1	MC 4558	1	18 AWG, Blue	1
10 Ohm, 10 W	1	MC 1496	2	Power Cord	1
		MC 7806CT	1		
Capacitors		MC 1648	1	Miscellaneous	
10 pF	10	MV 2109	1	SFE 10.7 MABA	1
30 pF	4	MDA 200	1	CDA 10.7 MABA	1
47 pF Ceramic	1	2N2222A	4	Fuse 0.5 A, SB	1
82 pF, DM	1	1N4148	1	Dial Lamp	1
220 pF, DM	1	NE 592	1	Knob, Large	1
300 pF, DM	1	CA 3089	1	Knob, Small	1
330 pF, Ceramic	2	MC 7815KC	1	Solder Lug	2
750 pF, DM	1	1N4003	1	Rubber Feet	4
4700 pF, DM	1	LM 358	1	Meter	1
.001 μF, Ceramic	1			Serial No. Label	1
.0033 μF, Ceramic	1	Filter		Test Label	1
.01 μF	34	FL	1	Misc. Screws.	
				Washers, Nuts	

Fig. 10-6. Chassis layout showing location of the major parts of the Sky Eye V receiver
(Courtesy KLM Electronics)

possible to "zero in" the dish, optimize polarity, etc. The best way of accomplishing this is to borrow a known working satellite receiver with a 70-MHz i-f input to make sure the dish/LNA/down converter components are functioning and a picture is present.

To tune up the receiver, adjust the discriminator trimmer capacitor to its midpoint (see Fig. 10–7) and then rotate it, alternating with the channel knob tuning to obtain the best picture. Correct channel alignment is not important at this time.

Adjust the "HI" and "LO" channel alignment trim pots on the lower right-hand edge of the circuit board for correct channel alignment. Start with "LO" and transponder 4 and "HI" with transponder 19, working back and forth, while viewing the received pictures until both are accurate. Then check transponders 1 and 24, readjusting as necessary. Adjust the video level trim pot (front left-hand portion of the circuit board) to one volt peak-to-peak with a vtvm or dmm. If a meter is not available adjust the pot visually for the best picture. As a final check, the vco and LNA voltages should be measured on the terminal output strip. By varying the front panel channel tuning control through its range, the vco voltage should change from 2 to 6 volts. The LNA voltage should run between 22 and 30 volts. Install the chassis cover with the necessary screws, and sit back and enjoy your satellite television picture.

The KLM Sky Eye V is available in complete kit form from the factory

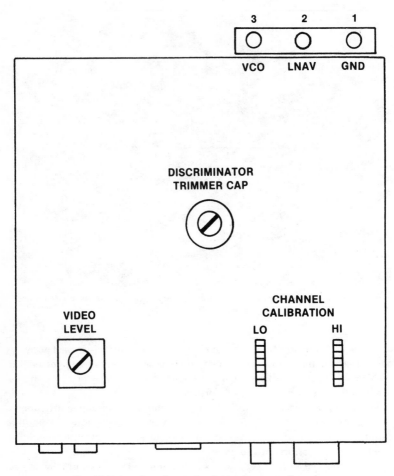

Fig. 10-7. Location of parts for final adjustment of receiver.
(Courtesy KLM Electronics)

at P.O. Box 816, Morgan Hill, California 95037. Optionally, the assembly manual may be purchased as a separate item allowing the "dyed-in-the-wool" electronics experimenter to use his own components. If the complete kit package including assembled and pretested down converter is purchased from the factory, a full 90-day limited warranty is provided. If any problems arise, the completed unit can be returned to the factory for alignment or repair for a nominal service charge.

By combining the KLM Sky Eye V receiver with the KLM X-11 Dish and a 120-degree LNA, a complete high quality TVRO earth station system will have been built for under $1500—a bargain indeed!

CHAPTER 11

Finding the Hidden Satellite Signals: Audio Feeds, Teletext, and News Wire Services on the Birds

CARRIERS AND SUBCARRIERS

A satellite transponder can carry a color television picture. It can also carry hi-fi audio channels, radio circuits, newswire feeds, special news teletypewriter channels, high-speed stock market and commodity exchange data feed, and teletext data services (the electronic newspapers of the air). Depending on the transponder, one or more of these auxiliary services may be riding along "piggy back" with the video and audio television programming feeds.

The basic NTSC (National Television Standards Committee) color television set requires a channel bandwidth of 6 megahertz to provide sufficient room to carry all of the frequently-changing color and gray-scale picture information and the accompanying sound. Ordinary telephone conversations require a bandwidth of only 3,000 hertz (3 kilohertz). It is possible to squeeze up to 2,000 simultaneous telephone conversations into a single video circuit through a process known as multiplexing.

Communications satellites share the same set of microwave frequencies that the telephone companies use for their terrestrial communications relay systems that form the backbone of the transmission network for the nation's telephone system. Because tens of thousands of these relay towers dot the country, an unacceptable amount of interference with the television satellites would occur if some special techniques were not used to effectively separate the two conflicting microwave services. The low-power transmitter of the communications satellite is 22,300 miles away; the nearest telephone company microwave transmitter may only be two blocks down the street. A TVRO antenna pointed upward can still pick up an undesired microwave signal from a nearby terrestrial tower.

The satellite common carriers disperse the 6-MHz video signal over a 36-megahertz-wide bandwidth, spreading out the 6-MHz tv signal so that it takes a greater bandwidth than would ordinarily be required. By using this technique, the terrestrial microwave interference which might be present in any 6-MHz portion of the total 36-MHz spectrum will be ignored in the TVRO receiver signal detection process. Although the satellite transponder channels appear to be 36-MHz wide, they are spaced at 40-MHz intervals to include a 4-MHz guardband between the upper and lower channel neighbors. The effective bandwidth of the transponder is limited to less than 8 to 10 MHz of usable spectrum. Since ordinary NTSC terrestrial color television pictures require only 6 MHz of space, additional "secret" signals can often be found between the 6- and 10-MHz portion of the spectrum.

Audio and data signals take up far less than a full video channel's spectrum. Because they consume a small portion of the available carrier bandwidth and can be considered as separate signals, they are known as subcarriers. Usually frequency modulated, these carriers may or may not be found in combination with a host video carrier signal. As Fig. 11-1 shows, the normal sound portion of a television program fed via satellite is placed on an fm subcarrier in the 5.8 to 7.4 MHz range (usually either 6.2 or 6.8 MHz).

In North America, the NTSC color television format uses a special 3.58 MHz color-burst subcarrier on which the color video signal information is placed. This differs from the European PAL or SECAM systems, which place their color information on a 4.33 megahertz subcarrier. Ordinary television sets have color decoder circuits that are tuned to the color burst signal to extract the necessary coloring information and recombine it with the black-and-white picture to produce the final color display.

Since the NTSC television system requires only a portion of the effective 8 to 10 MHz transponder spectrum available for use, any subcarrier frequency located above 5.8 MHz, or so, can be used to carry addi-

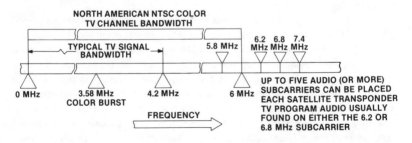

Fig. 11-1. A satellite tv carrier with multiple audio carriers.

tional unrelated audio and/or data channels. In the RCA Satcom satellite systems, many transponder owners feed separate audio subcarriers at 5.8, 6.2, 6.8, and 7.4 MHz. On Satcom 3, transponder 3 carries the program video and audio portions of the WGN-TV feed from Chicago. In addition, the classical stereo radio WFMT-FM is on the 6.3 and 6.48 MHz subcarriers. Also, the Seeburg "Lifestyle" music service may be found on 7.695 MHz. The WTBS-TV transponder is even more heavily loaded with independent audio feeds.

Most home TVRO satellite receivers have a front panel switch or tuning knob which can vary the audio subcarrier tuning from 5 to 8 MHz, allowing the various fm audio subcarriers to be picked up.

A transponder may be used to carry multiple audio and data subcarriers in place of a wide-band video carrier. By using frequency-modulation multiplexing techniques, hundreds or thousands of unrelated audio and/or data signals may be stacked one atop each other in ascending frequency. Fig. 11−2 illustrates this concept.

A basic telephone channel bandwidth is limited to 4,000 Hz. This single telephone conversation is modulated onto a carrier, and combined with eleven other similar conversations to form a carrier group. In telephone company parlance, five groups are multiplexed to form a 240-kHz wide supergroup. Five supergroups of 60 voice circuits each are combined to form a 1.2-MHz master group. Up to six or more master groups may be stacked together to form a wideband channel which fills the satellite transponder's 10.75-MHz frequency spectrum. Since each telephone channel carries one half of the telephone conversation, two separate telephone circuits must be utilized to provide a single full-duplex telephone conversation.

This ingenious technique of multiplexing channel upon channel of telephone conversations onto the same transponder was copied from the terrestrial microwave channelizing schemes, and it gives the telephone company tremendous flexibility in arranging and routing telephone circuits throughout the country.

By carefully engineering these circuit groups, and assigning specific master groups or supergroups to geographic regions or cities, a complex web of communications circuits consisting of point-to-point connections can be built up. For example, a sixty-voice-channel supergroup dedicated for use between Los Angeles and Dallas might be immediately adjacent in frequency to a supergroup carrying conversations between San Francisco and New York. Significant equipment savings can be realized, since the cost to multiplex, modulate, and demodulate sixty voice channels simultaneously is only several times higher than the cost of manipulating a single voice circuit.

In some cases, it is to the advantage of the common carrier to uplink a single audio telephone channel to a specific bird, and handle this feed independently, rather than as one of a group of signals. This process,

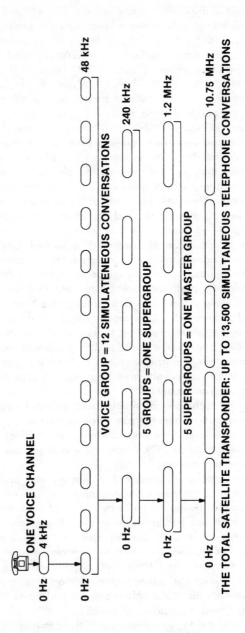

Fig. 11-2. The multiplexing of many audio channels onto one satellite transponder using the "telco-type" multiplexing scheme.

called single channel per carrier (SCPC), is used by operators of oil rigs in the Gulf of Mexico to provide telephone service to and from their offshore locations.

Wherever a single voice channel or audio service is desired, the SCPC transmission scheme can be used. SCPC transponders do not utilize the frequency spectrum as efficiently as the single-sideband fm multiplex schemes. A transponder can usually handle only several dozen or so SCPC channels when operating in this mode but the SCPC signals are transmitted at higher average power levels by the satellites. This allows smaller receive-only antennas to be used in picking up the SCPC signals. The newswire agencies use special-tariff SCPC channels to deliver high-quality audio news feeds to radio stations nationwide. The radio stations can then use inexpensive three- to five-foot rooftop-mounted, receive-only, parabolic antennas. Some SCPC channels carry high-speed teletypewriter and data signals for the national newspaper chains and news agencies relay services. Both the Commodity News Service (CNS) and Bonneville Satellite Corporation use SCPC techniques to move data between the various commodity floors throughout the country.

DECODING THE HIDDEN SIGNALS

Appendix B provides detailed information on those satellites and transponders which carry television programming. By the process of elimination, it can be reasonably concluded that the other transponders and satellites primarily carry multiplexed telephone channels, program audio channels, and data circuits.

The typical satellite TVRO receiver can tune in standard format NTSC video, and corresponding audio subcarriers. It cannot directly receive these other "hidden" signals. If the receiver is tuned to a nonvideo transponder, however, a 6-MHz band of frequencies will appear at the video-output jack on the rear panel of the receiver. (This assumes that the receiver does not have an internal audio subcarrier noise filter, or if one exists, that it has been bypassed.) This electronic spigot (the video output) can be connected to a shortwave communications receiver that tunes from 100 kHz to 10 MHz.

To detect these narrowband audio signals, the antenna input of the communications receiver is connected to the baseband video output of the TVRO satellite receiver as diagramed in Fig. 11–3. The TVRO receiver is then tuned to the desired transponder. When the communications receiver is switched to the single-sideband mode and set to the "lower-sideband" or "upper-sideband" position, the audio signals will be decoded. Beginning with the lower sideband, set the receiver to its lowest frequency (typically 100 to 200 kHz) and tune slowly upward over the dial. Every 4 kHz or so, a carrier should appear. These audio carriers contain telephone conversations, national radio network feeds, and

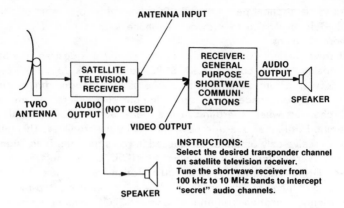

Fig. 11-3. Configuration to intercept the "secret" satellite audio signals.

private voice "leased-line" circuits. Multiplexed teletypewriter data channels will also be heard as high-speed varying audio tones.

To convert these data tones into readable teletype signals that can be printed out, an rtty demodulator must be connected to the audio output of the communications receiver as shown in Fig. 11—4.

A number of organizations sell rtty demodulators. They are often used by amateur radio operators to monitor the shortwave bands. These companies often advertise in ham radio magazines such as *QST, CQ, Seventy-Three,* and *Ham Radio*. An rtty demodulator kit, basically a variable-frequency fsk-modem, varies in price from $100 to $200. Commercially assembled units cost about $1,000.

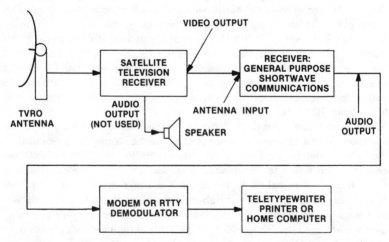

Fig. 11-4. Configuration to intercept the hidden teletypewriter and news wire channels.

If the TVRO television receiver includes an audio subcarrier filter, the multiplex carriers above 4.2 MHz will probably not be present at the video-out jack. Manufacturers build in these special circuits to minimize the possible noise interference that an inexpensive television set might experience if the program audio subcarrier is not filtered (i.e., removed) from the television video. These audio filters are considered to be a positive feature by TVRO television receiver manufacturers. Since nonvideo-mode transponders will often carry information stacked up to 10 MHz and beyond, a number of interesting data feeds and audio signals will be lost if this filter is not removed. Some manufacturers provide a second video output jack which will bypass the internal filter, allowing the shortwave communications receiver to pick up the full transponder spectrum of channels. Most dealers can modify a TVRO television receiver by installing a new "raw" transponder output jack which bypasses the filter.

The Communications Act of 1934, in particular Section 605—the Secrecy Clause—establishes rules and regulations for the interception of private communications which may be broadcast via radio transmissions. Appendix F discusses this in further detail. In general, good sense would suggest that anyone who listens in to these private conversations should *never* divulge the conversations or their content to anyone else. The same rule applies to the million-plus scanning receivers which monitor police and fire department channels and mobile and marine telephone conversations. The concept of listening in on private communications, whether coming from a terrestrial transmitter or a satellite transponder is the same.

RADIO STATION AND AUDIO FEEDS

Most audio programming handled via satellite is delivered through the SCPC or frequency division multiplex (FDM) multicarrier multiplexing schemes. A few exceptions, such as WFMT-FM, the classical music station from Chicago, operate on audio subcarriers of transponders transmitting in the conventional video mode. Several of these networks operate in stereo, using two different audio subcarriers to carry the separate left and right-channel information.

The Seeburg Music Service "Lifestyle" may be found at 7.695 MHz on Transponder 3 of the Satcom 3 satellite. The CNN radio network provides an all-news radio feed at 6.3 MHz on Transponder 14 of Satcom 3, and the Satellite Radio Network provides a contemporary and religious music format at 6.2 MHz on Transponder 2, Satcom 3. Transponder 3 of the Satcom 3 satellite is loaded with audio subcarrier services; over thirteen fm subcarriers are present! The Moody Broadcasting Network operates a stereo religious programming service at 5.47 and 7.92 MHz. Satellite Music Network (SMN) feeds its stereo country music service at 5.94 and 6.12 MHz. The SMN adult contemporary "rock" music network is located at 5.58 and 5.76 MHz. Stardust, a traditional middle-of-the-road popular music service, may be found at 8.055 and 8.145 MHz. Finally, Bonneville's "Beautiful Music Service" is located at 7.38 and 7.56 MHz.

Audio services are also found on Satcom 4, Transponder 7, (the Family Radio Network's stereo East and West Coast feeds), and on Satcom 1 (NBC Radio Programming). The Anik D satellite provides five outstanding services of fm stations located from Montreal to British Columbia, including French language, progressive rock, and all-news and information formats. Table 11−1 lists some of these radio services, along with other audio programming transmitted in a nonstandard digitized format.

Nonvideo transponders provide dozens of audio programming services operated in the SCPC and FDM modes. These require the use of a shortwave communications receiver to decode their signals. Perhaps the most interesting bird is the Satcom 2 satellite located 118.9 West Longitude. On Transponder 7 at 294 kilohertz (lower sideband mode) The Alaskan Forces Satellite Network (AFSN) is operated on behalf of the US Armed Forces stationed in the northernmost state. Anchorage country and western station KYAK may be found at 393 kHz on the same transponder, and Anchorage station KBYR is located at 333 kHz. At 958 kHz, the Alaskan Satellite Network broadcasts "Soul music" and black programming, while at 1190 kHz (upper sideband mode) a second audio channel is fed by AFSN.

The Mutual Radio Network operates a feed in upper sideband mode at 1702 kHz, and the ABC Radio Network feeds a 2358 kHz carrier in lower sideband mode.

Over on Transponder 11 of the Satcom 2 satellite, National Public Radio (NPR) can be found at 134 kHz (lsb mode) with a second channel at 602 kHz (lsb). At 2802 kHz, the signal of Anchorage radio station KBYR is duplicated during the day, and during the evening a Christian-oriented audio service takes over this frequency. Anchorage station KFQZ operating in the lower sideband mode can be found at 2854 kHz.

Moving from the odd to even transponders on Satcom 2, transponder 2, CBS Radio Sports feeds a usb-mode carrier at 421 kHz, and NBC News feeds are at 469 kHz (usb mode). On transponder 10, the Mutual Radio Network is carried at 2082 and 2454 kHz (both in lsb mode). Transponder 16 has some very interesting narrow band audio and data services present, and the reader is invited to discover these for himself. . . .

The Alaskan Forces Satellite Network (AFSN) may be found again on transponder 16 at 166 kHz (lsb mode) and CBS Radio News feeds at 238 kHz (lsb) are common. Sports feeds (lsb mode) are located at 310 kHz, and Philadelphia radio station KYW can even be heard at 552 kHz, operating in the lower sideband mode. NBC Radio News secondary

Table 11-1. The "Hidden" Audio Networks

Service	Satellite/ Transponder	Transmission Type	Program Type
ABC Radio Networks	Satcom 1/T23	Digital	All networks
Associated Press	Westar 3/T1	Analog, FDM	News
Bonneville Broadcast Systems	Satcom 3/T3	7.38/7.56 MHz Analog	"Beautiful music"
Bravo	Satcom 4/T6	6.8 MHz Analog	Stereo Multiplex Bravo program audit
CKAC-AM	ANIK D/T8	5.41 MHz Analog	Montreal French MOR/rock music
CITE-FM	ANIK D/T8	6.17 MHz Analog	Montreal French traditional MOR
CKO-FM	ANIK D/T14	6.17 MHz Analog	Toronto all-news
CIRK-FM	ANIK D/T18	6.17 MHz Analog	Progressive rock multiplex stereo
CFMI-FM	ANIK D/T22	6.17 MHz Analog	Adult rock music
CBS Radio Network	Satcom 1	Digital	News/Sports/Features
CNN Radio	Satcom 3/T14	6.3 MHz Analog	News & Information
Country Coast-to-Coast	Satcom 3/T3	5.94/6.12 MHz Analog	Country music in stereo
ESPN Affiliate Informational Network	Satcom 3/T7	6.2 MHz Analog	Affiliate information Program schedules
Global Satellite Network	Westar 4/T2	Analog	Talk/Music
HTN Plus	Satcom 3/T16	6.8 MHz	Multiplex stereo audio
Moody Broadcasting Network	Satcom 3/T3	5.47/7.92 MHz Analog	24-hour religious
Music in the Air— Broadway	Satcom 3/T6	5.40/5.94 MHz Analog	Broadway/Hollywood
Music in the Air— Country	Satcom 3/T6	5.58/5.76 MHz Analog	Country Music
Music in the Air— Comedy	Satcom 3/T6	7.785 MHz Analog	Comedy
Music in the Air— Big Band	Satcom 3/T6	7.695 MHz Analog	Big Band music
Music in the Air— 50s/60s Rock	Satcom 3/T6	6.435 MHz Analog	50s & 60s hits
Music Country Network	Westar 3/T1	Analog/Digital	Overnight country music, Interviews/ News/Sports/Weather Talk/Music

Table 11-1-cont. The "Hidden" Audio Networks

Service	Satellite/ Transponder	Transmission Type	Program Type
Mutual Broadcasting System	Westar 4/T2	Analog, FDM	News, Sports
NBC Radio Network (Interim System)	Satcom 1/T12	5.8 & 6.2 MHz Analog	News/Sports Features (2 feeds)
National Public Radio	Westar 4/T2	Analog, FDM	News/Talk/Music
RKO Radio Networks	Westar 3/T1 & T4	Analog, FDM	Talk/Music News/Sports
Satellite Music Network—Stardust	Satcom 3/T3	8.055/8.145 MHz Analog	MOR, popular music
Satellite Music Network—Star Station	Satcom 3/T3	5.58/5.76 MHz Analog	Adult contemporary "rock" music
Satellite Radio Network	Satcom 3/T2	6.2 MHz Analog	Contemporary/ Religious music
Seeburg/Lifestyle Music	Satcom 3/T3	7.695 MHz Analog	MOR light music
Sheridan Broadcasting Network (Family Radio Network)	Satcom 4/T7	5.58/5.76 & 5.94/6.12 MHz Analog	Religious News/Sports (East & West feeds)
Transtar	Westar 3/T2	Analog or Digital	Young Adult Music
United Press Int'l	Westar 3/T1	Analog/Digital/ Time Cues	News, Data
Wall Street Journal/ Dow Jones	Westar 3/T1	Analog, FDM	Business Financial News
WFMT-Chicago	Satcom 3/T3	6.3/6.48 MHz Analog	Classical music

feeds are at 694 kHz (lsb mode) and additional programming from the Voice of America and a number of independent radio feeds may be found scattered throughout this transponder.

Satcom 2 transponder 20 carries yet another NBC sports feed at 190 kHz (lsb mode), and way up at 7382 kHz (upper sideband), NBC Television operates a talkback coordination channel which plugs tv news crews out in the field back to the NBC New York Broadcasting Operations Center for those special sporting events which NBC TV delivers via one of its contract video channels.

NATIONAL NEWSWIRE CIRCUITS AND SPECIAL CHANNELS

There are dozens of unusual telephone and audio channels operating on the birds in FDM format. On the Satcom 2 satellite, the AP radio newsfeeds can be found in lower sideband mode at 950 kHz, transponder 10. Transponder 16, 682 kHz (lsb mode) feeds the UPI audio news service. The National Black Radio Network operates several audio channels on the same transponder.

Weather and airport control channels also use Satcom 2; the Anchorage Airport tower is located on transponder 7 at 1938 kHz (lsb mode) and transponder 11 at 1234 kHz (lsb). Valdez Weather Advisory Service for the Bering Sea, Alaska and the USSR is located at 2366 kHz, Transponder 11 (lsb). The Kenai Flight Service Weather provides daily summaries and 24-hour forecasts for the entire state on transponder 7, at 1946 kHz (lsb). Switching satellites, the National Public Radio Service (NPR) uses single channel per carrier techniques to deliver twelve separate audio channels via Westar-4. These SCPC transponders are easy to locate, since fewer audio channels are present with significantly higher power levels delivered to the small-bore receive-only earth stations.

THE TELETYPEWRITER NEWSWIRE SERVICES

UPI, AP, Reuters, and the other major news services provide dozens of teletypewriter channel feeds via satellite (Fig. 11–5). Every type of news service is represented, including stock market reports, regional weather, sports and national news, gold and silver quotes, and com-

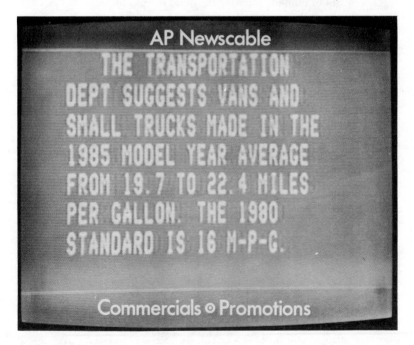

Fig. 11-5. One of the dozens of major news services on the birds.
(Courtesy AP Newscable)

modity exchange information. Fig. 11—6 shows the setup for a unique satellite service by Newstime. This service transmits a slow-scan tv picture (Fig. 11—7) along with the corresponding audio. The Satcom 2 satellite, Transponders 7 and 16 are the most heavily used, with transponders 4 and 10 following closely behind. To decode these teletypewriter channels, an rtty demodulator and printer must be used. In a deluxe configuration, the output of the rtty demodulator could be fed into the RS-232 serial data input channel of a home computer system, and with the appropriate programming the computer could become a highly sophisticated satellite-fed business terminal.

THE VERTICAL BLANKING INTERVAL AND THE NATIONAL TELETEXT SERVICES

Perhaps the most exciting data and text services which can be found on satellite are snuggled away in the vertical blanking interval (VBI), an unused portion of the tv picture. Teletext is an electronic "newspaper of the air" invented in England in the early 1970s. Fig. 11—8 shows a teletext slide, or page, being prepared at the teletext newsdesk. A computer terminal or camera digitizer (or both) is used to prepare the slide. Teletext has rapidly spread to dozens of countries worldwide, and at least three different incompatible global standards now exist. Versions of all three may now be found operating on the US domestic satellite systems. Teletext enables a home viewer equipped with a tv set-top decoder to display several hundred pages or more of alphanumeric text and 16-color graphics. The European television manufacturers have begun to build teletext decoders directly into their television sets, allowing small hand-held keypads to be to be used by the viewer to select the desired teletext news, weather, and other information pages.

The British version of teletext was soon followed by the French Antiope system and Canadian Telidon technology, both of which also allow pages of information to be delivered over the VBI or via a telephone line, when operated in a "videotext" mode.

In the teletext systems a high-speed digital data stream is sent during the time between the tv picture frames when the electron gun in the television set is momentarily switched off. Known as the vertical blanking interval or VBI, this period occurs 60 times each second. When the electron gun has successfully scanned to the bottom of the tv frame, it must be "retraced" back to the top of the picture tube to begin the next scan sequence. During this retrace or vertical blanking interval, when the electron gun is turned off, the teletext signal sent by the television station or programmer will be visible to the viewer.

In operation, a stream of teletext pages are sent over and over again in a round-robin teletext "wheel." When a desired page has been selected for display by the viewer, the teletext decoder strips off and

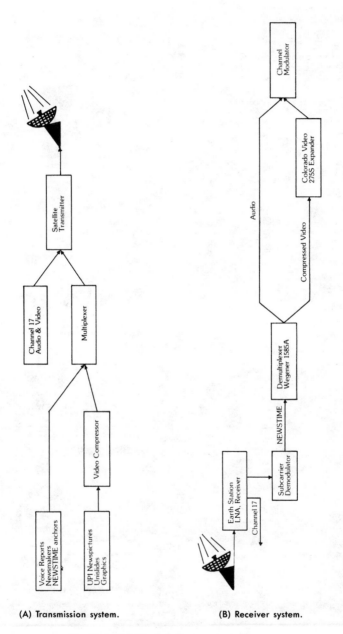

(A) Transmission system.   (B) Receiver system.

Fig. 11-6. Satellite-fed slow scan tv service.
(Courtesy Newstime)

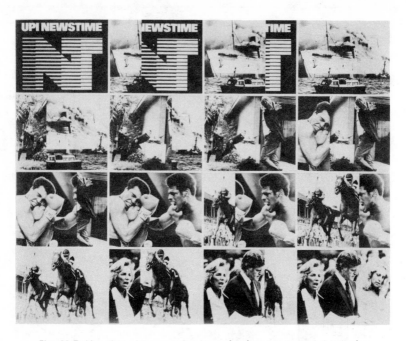

Fig. 11-7. Newstime pictures coming in on the slow-scan transmission mode.
(Courtesy Newstime)

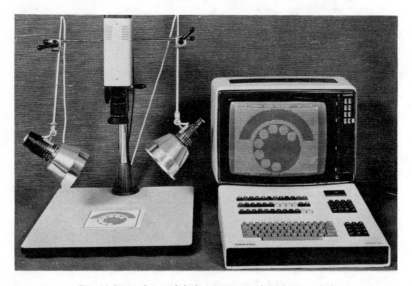

Fig. 11-8. A teletext slide being prepared at the newsdesk
(Courtesy Antiope)

captures this page as it flies by on its next round. Refresh memory electronics in the teletext decoder continues to display the retained page on the screen. Several hundred pages take ten to 20 seconds to transmit in this manner, and frequently selected pages, like the index page, are either resent several times during the same teletext wheel, or are captured and stored in a frequent-page memory in the teletext decoder, for instant display upon query.

Many leading broadcasting organizations in the United States are experimenting with teletext feeds. CBS has adopted the French Antiope System for a series of tests being conducted jointly in Los Angeles by network-owned KNXT-TV and KCET-TV the PBS station (Fig. 11−9). Since teletext allows for the closed captioning of shows for the deaf, the PBS Close Captioning Center at WGBN-TV in Boston is also participating in the CBS/Antiope experiment. This Southern California teletext system includes weather maps, sports summaries, news briefs, airline schedules, children's puzzles, emergency medical procedures, and a number of other intriguing pages of news and information.

Another system, called Keyfax (Fig. 11−10), uses the British teletext system and is being developed in Chicago for national satellite distribution. The service is provided by the Field Communications group, using their independent Chicago television station. KPIX-TV, Westinghouse Broadcasting's ABC affiliate in San Francisco, is also running a teletext trial using the French Antiope scheme. Still other television stations are experimenting with the Canadian Telidon System which is being actively developed for national satellite operation in Canada.

Several cable tv satellite programmers also use the vertical blanking interval to deliver national teletext newspapers. WTBS-TV, transponder 6, Satcom 3, has both UPI and Reuters cable news services piggybacked onto its retrace interval. These services provide a continually changing screen of news headlines which are decoded by headend special processors located at many cable companies throughout the the country. The decoded English text display is then inserted onto a spare cable channel for view by the CATV subscribers. WGN-TV, Satcom 3 transponder 3, shares its vertical blanking interval with the Dow Jones Cable News Service. Several combination slow-scan tv/text/audio feeds produced by the North American Newstime organization may be found on subcarriers of WTBS-TV at 6.2 and 7.4 MHz.

Bell Telephone and CBS have entered into an agreement to explore the mutual development of a national interactive teletext and videotex service based on a modified-Antiope scheme. In Canada, dozens of tv stations and telecommunications organizations are testing the Canadian Telidon teletext system with local and satellite feeds over the Anik birds (Fig. 11-11). As cable companies begin to provide more value added and multitiered services, dozens of satellite programmers will begin to feed new teletext newspapers over the vertical blanking intervals.

Fig. 11-9. Teletext pictures beng
(Courtesy

HOW'S YOUR HISTORY?

Our 16th president was assassinated on April 14, 1865. Who was he? What was he doing that evening? Who shot him?

Press REVEAL for the answer.

Abraham Lincoln was shot while attending a play at Ford's Theater in Washington, D.C. by John Wilkes Booth.

A SINGING TRIBUTE

MARK TAPER FORUM

The CENTER THEATRE GROUP provides a fond portrait of turn-of-the century America in TINTYPES. A musical celebration of how our melting pot heritage shaped the country's character.

DATE AND
SEAT INFO   PAGE 39

transmitted by KCET in Los Angeles.
CBS)

Fig. 11-10. The Keyfax television service from Chicago.
(Courtesy Keycom Publishing)

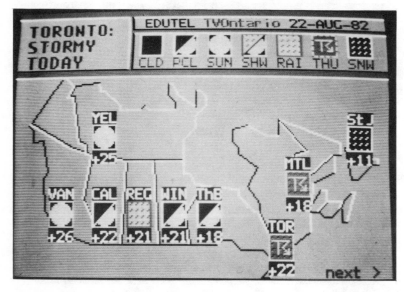

Fig. 11-11.. The Canadian Telidon teletext "electronic magazine" in Toronto.
(Courtesy TV Ontario)

A number of firms currently manufacture subcarrier decoders for use by cable tv companies, and teletext decoder kits, chip sets, and complete systems are now available for North American NTSC operation. Subcarrier text and data decoders for CATV systems are available from Wegener Communications in Atlanta. Southern Satellite Systems, the resale common carrier which delivers WTBS-TV via Satcom 3 provides a special "Cable Text" decoder for its teletext service. The Zenith Corporation has recently introduced a new cable television converter/teletext decoder package for testing by MSO cable operators. Satellite enthusiasts who are interested in building their own teletext decoder kits which function on the British Standards, should review the series of articles on the subject published in the late 1970s by the British magazine, *Wireless World*. With minor modification, this PAL-system decoder will function on the US NTSC standard. The most popular "hidden" data channels carried on the Satcom 3 satellite are summarized in Table 11—2.

MONITOR, REUTERS WORLDWIDE FINANCIAL NEWS SYSTEMS

On transponder 18 of the Satcom 3 satellite, Reuters operates the "Monitor Service," a high speed, digital commodity and stock market information service which utilizes the entire video bandwidth of the transponder. This service is fed to cable companies and business subscribers who have been equipped with a special microcomputer-based terminal. With this terminal, the Monitor user can immediately gain direct access to the data coming from dozens of global stock and commodity exchanges. Money market rates from the world banks are also updated continuously. This information is collected by the international network of Reuters correspondents who provide Euro-deposits, foreign exchange, Eurobond and CD rates, news reports, and in-depth analysis of the business world. All of these items are retrievable from the special Monitor terminal. Without the terminal the screen appears like that of Fig. 11—12A; decoded, the picture of Fig. 11—12B appears.

Manhattan Cable was the first to carry the Monitor service over a spare cable channel. Throughout mid and lower Manhattan office buildings, the incoming coaxial cables of the company are connected to the Monitor video terminals installed by Reuters personnel. Small desk-top printers can provide a hardcopy record of the information displayed on the screen. Thousands of pages of information are transmitted each minute. Fig. 11—13 (page 224) shows a listing of the individual services.

Individuals and businesses who own satellite TVRO earth stations can contact Reuters directly to obtain the Monitor microcomputer terminal for rent on a monthly basis. Using the Monitor Service in conjunction with a small TVRO terminal, a person living in even the remotest part of the United States can plug into the world's largest and most sophisticated financial news network. In addition, money market brokers and dealers

Table 11-2. Data Services on Satcom 3 Satellite

Service	Program Type	Transponder	Format*
Dow Jones Cable News	News Wire Feed	6	VBI—B&W Text
Electronic Program Guide	Graphics & Test "TV Guide" Satellite-Feed Channels	3	5.4/4.9 MHz Subcarrier-Text
KEYFAX National Teletext Magazine	General Teletext Service	6	VBI—UK Standard Graphics & Text
North American Newstime	News, Science & Travel Pictures	6	7.695 MHz Subcarrier—Slow-scan Color TV
QUOTRADER	Commodities News & Quotes Service for home computer	6	VBI—B&W Text
Reuters News-View	Financial & Sports Wires (2 feeds)	6	VBI—B&W Text
SSS Cable Text	General Press News Wire Feeds	6	VBI—Color Text
VPI Cablenews	English & Spanish Headline News	6	VBI—B&W Text

*VBI = Vertical blanking interval.

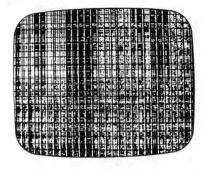

REUTER MONITOR
STOCK QUOTATIONS
NEWS - WEATHER - FOREX
COMMODITY PRICES
MONEY MARKETS - SPORTS
FINANCIAL/INDEX FUTURES
DOMESTIC - WORLDWIDE

(A) Undecoded.   (B) Decoded.

Fig. 11-12. The Reuters Monitor Service.
(Courtesy Reuters)

can now make personal inquiries anonymously, instantly interrogating any financial market without revealing this fact to anyone. Because of this advantage, and the ease of obtaining the service, there are now several thousand Reuters Monitor Service subscribers throughout North America who receive their daily financial data beamed directly from space.

INDIVIDUAL SERVICES AVAILABLE

Individual financial and commodity services are available within the system. Each such service includes news, market reports, preformatted quotations, and tickers where applicable. Each category of service has a Group number. Each display within that Group has a Page number. The groups are as follows:

GROUP	SERVICES
0	INFORMATION, OPERATIONAL AND ENHANCEMENT DATA
1	FINANCIAL AND OPTIONS NEWS
2	GRAINS/OILSEEDS
3	LIVESTOCK
4	COFFEE/COCOA/SUGAR
6	BUSINESS/MARKET NEWS
7	METALS
8	MONEY NEWS
9	INTERNATIONAL MONEY/FOREX RATES
12	CBT TICKER
13	CME TICKER
14	COFFEE/COCOA/SUGAR TICKER
16	ARBITRAGE — WHEAT/CORN/SOY
17	COMEX TICKER
18	ARBITRAGE — COCOA/COPPER/SILVER/SUGAR/COFFEE
20	COMBINED OPTIONS TICKER
22	CASH GRAINS
24	REUTERS NEWS-VIEW*
26	LONDON (WORLD) COMMODITIES TICKER
27	LONDON METALS QUOTATIONS
28	LONDON SOFT COMMODITIES QUOTATIONS
29	MONEY RATES INDEX/MONEY AND FINANCIAL FUTURES
31	IMM TICKER
32	KANSAS CITY TICKER
33	N.Y. MERCANTILE TICKER
34	N.Y. COTTON TICKER
35	NASDAQ TICKER
36	NYSE BONDS
37	WINNIPEG TICKER
38	MINNEAPOLIS TICKER
94	HEADLINES AND COMMODITIES STATISTICS

Fig. 11-13. The wealth of services available on the Reuters Monitor System.
(Courtesy Reuters)

CHAPTER 12

DXing the International Satellites

INTRODUCTION

Satellite television in the United States is delivered through low-power C-band satellites, originally built to carry telephone calls for RCA, Western Union, and the other common carriers. In the halcyon days of the early 70s, no one imagined that private organizations like Time-Life would be uplinking their own feeds. Few people dreamed that in a few short years cable companies would own their own headend TVROs. Fewer still imagined thousands of backyard dishes picking up these birds. US domsats were a spinoff of Intelsat, the giant global network of giant satellites. Intelsat dishes were large, running from 30 to 50 feet in diameter (or larger).

Because of the Intelsat C-band "big dish" mentality, a new higher-frequency 12 GHz "Ku-band" was opened up allowing a new generation of high-power satellites to provide direct broadcasting (DBS) of programs to the home. The American C-band system would be temporary. Ku satellites would take over the world.

The World Administrative Radio Conference (WARC), an arm of the UN International Telecommunications Union, met in 1977 to establish the WARC-BS broadcast satellite standards. WARC reserved the 11.7 to 12.5 GHz band for global satellite television broadcast use. Forty television channels were created. Each would be assigned a separate transponder. Countries were authorized to build their own satellite systems of up to 5 channels each. The channels would be divided into lower and upper portions of the band, and the satellites would be spaced at 6-degree intervals around the globe (Fig. 12−1). Power levels would be high, permitting 2½-foot home terminals to be used to pick up that country's DBS satellite feeds).

The multilateral satellite DBS agreement was to run until the end of 1993. The new DBS satellite standards will ensure dozens of television satellites to be built throughout the 1980s as country after country launch their own satellite television service.

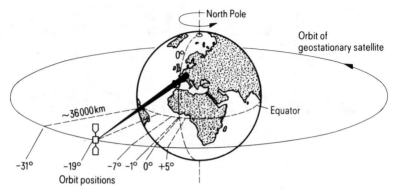

Fig. 12-1. Broadcasting satellites in geostationary orbit (European Region).
(Courtesy Telcom Report)

In the meanwhile, domestic C-band satellites have been built by a number of countries. Intelsat, the "granddaddy" of the communications satellite business, maintains an ever-growing global network of satellites. Stationed over the Atlantic, Pacific, and Indian Oceans, any of these Intelsat satellites can be seen from the United States.

Intersputnik, the international Soviet-bloc geostationary communications satellite system, is an up-and-coming junior version of the Intelsat network. Its satellite television feeds can also be picked up by a home TVRO terminal within view.

Other regional consortiums—the Arabsat and Indonesian Palapa systems are examples—operate C-band satellites similar to the American domestic birds.

Because so many of the smaller Soviet Union communities are located at extreme northern latitudes, Russian domestic television cannot reach them using the Intersputnik satellites. These geosynchronous satellites are literally "over the horizon." Therefore, the Soviet Union runs a second domestic satellite system called Molniya (Russian for "lightning"). Molniyas are all placed into the same nongeosynchronous U-shaped polar orbit, spaced roughly six hours apart. The Molniya orbit has apogees over the nothernmost territories of the Soviet Union and just above Hudson Bay, Canada, as shown in Fig. 12–2. The Molniya satellites, which must be actively tracked using motorized TVRO antennas, relay network television service feeds from Moscow to the Russian hinterlands. Luckily, Molniya satellites provide extremely powerful signals which can easily be seen throughout most of North America and Europe, bringing the TVRO enthusiast rather interesting political television programming, as well as international sports and cultural events.

The most readily observable non-US C-band satellites which can be seen throughout the United States are the Canadian C-band Aniks. The

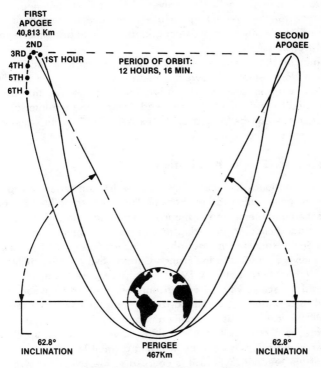

Fig. 12-2. Orbit of the Molniya satellites.

Anik satellites deliver the Canadian Broadcasting Corporation (CBC) feeds and other television channels to cable companies and television stations scattered throughout the Dominion. The Anik B and D satellites are identical to their US Westar cousins. They are stationed in the geosynchronous belt in the middle of the American satellites, and viewers in the United States can tune into Anik B or D using their standard TVRO earth stations. A second Anik series provides Canada with its own Ku-band DBS service to the Northwest Territories. (See Chapter 14.)

TECHNICAL DIFFERENCES

Depending upon the international satellite, the "North American type" C-band TVRO earth station will require various modifications to receive the international broadcasts.

Ku-Band Feeds

For satellites operating in the Ku band, a new LNA and down converter assembly must be used to convert these higher frequencies to the standard i-f input of the TVRO receiver. C-band TVRO antennas using a wire screen for their reflecting surface may have to be retrofit with a finer mesh, since Ku-band wavelengths are one quarter the length of C-band transmissions.

Commercial manufacturers are now selling small Ku-band TVRO terminals. In North America, however, there are few Ku-band satellites in operation; C-band systems dominate the hemisphere. International feeds to and from the United States and Canada use the Intelsat 4-GHz satellites; these can be picked up by existing TVRO earth stations with relatively minor modification.

Orthogonal vs. Circular Polarization

Most international satellites, including the Intelsat series, use a circular form of polarization rather than the linear horizontal/vertical polarization scheme implemented in the North American birds. Both left-hand and right-hand circular polarizing modes are in common use. The Intelsat and Soviet satellites use right-hand circular feeds; left-hand circular polarization is standard on the Franco-German Symphonie satellites. The Intelsat V birds use both. A TVRO system provided with a feedhorn configured to receive vertical or horizontally polarized transponders can be used to pick up circularly polarized signals, but two to three dB of gain will be lost. This is a significant amount for most systems. The scalar feedhorn can easily be modified, however, to pick up circularly polarized signals. This is done by inserting a small Teflon plate inside the mouth of the feedhorn where it is coupled to the LNA as shown in Fig. 12–3. The Teflon dielectric plate should be positioned at a 45-degree angle to the LNA probe. When used in conjunction with an electronic polarizer, this modification will allow both left-hand and right-hand circular polarized signals to be received. Note that simply rotating the LNA feedhorn assembly with a mechanical motor drive (found in lower-cost TVRO systems) will *not* work. The popular Chaparral "Polorotor," however, will receive all four types of polarization over a 135-degree probe angle range when a Teflon dielectric plate is installed as shown in Fig. 12–4. Note in Fig. 12–4 the plate is shown positioned for a right-hand circular polarized signal; for a left-hand polarized signal, the Teflon plate should be rotated 90°. Chaparral now manufactures an international version of the "Super Feed II" feedhorn, which includes such a built-in dielectric plate.

Program Audio Transmission

The North American C-band satellites use the standard FDM audio subcarrier scheme to deliver the program audio with the video over the same transponder. Many of the Intelsat television feeds use the same convention. However, in some cases the program audio is placed on a separate transponder using the single carrier per channel or FDM multi-

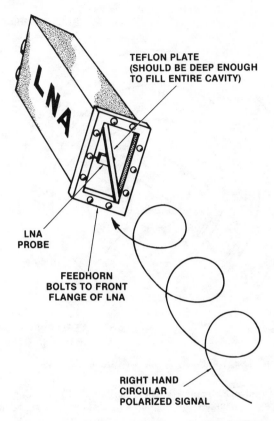

TEFLON PLATE
(SHOULD BE DEEP ENOUGH
TO FILL ENTIRE CAVITY)

LNA

LNA
PROBE

FEEDHORN
BOLTS TO FRONT
FLANGE OF LNA

RIGHT HAND
CIRCULAR
POLARIZED SIGNAL

Fig. 12-3. Teflon dielectric plate for rectangular feedhorn.

plexing scheme. This makes finding the "hidden audio" an adventure. The Russian Molniya satellites use a "sound-in-sync" audio multiplexing scheme in which the horizontal blanking period of the video signal is modified by the uplink station using a form of pulse width modulation (PWM). In the Molniya system two additional sound pulses are added and the sync pulse is both displaced. and narrowed. The pulse amplitudes are maintained at constant levels while the pulse widths are varied in relation to the program audio signal. The pulse frequency is 15,625 hertz, allowing a maximum audio bandwidth of 7.7 kHz. Two separate audio channels or a stereo sound track can be transmitted using this technique.

Low Power vs High Power Transponders

The typical North American C-band communication satellite transponders have power outputs from 5 to 12 watts, providing a footprint of 32

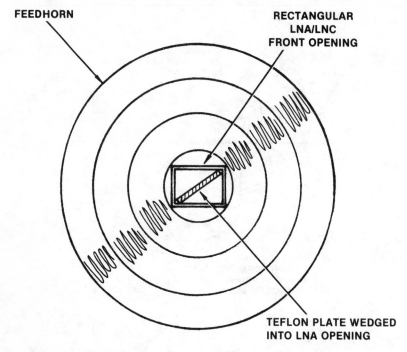

FEEDHORN

RECTANGULAR LNA/LNC FRONT OPENING

TEFLON PLATE WEDGED INTO LNA OPENING

Fig. 12-4. Teflon dielectric plate for scalar feedhorns.

to 40 dBW effective isotropic radiated power (EIRP). Although some of the international satellites deliver similar or even higher EIRPs, the Intelsat satellites are designed to operate with global and hemispheric beams, which cover a much wider geographic area. Thus, using the same power traveling wave tube (TWT) transponders, the EIRP values for the Intelsat satellites vary from 22 dBW for global beams through 26 to 29 dBW for hemispheric beams, and up to 41 through 44 dBW for steerable spot beams. To receive decent Intelsat pictures from hemispheric beams, a TVRO antenna size of 5 to 6 meters should be used. Global beam reception will require antennas in the 7 meter class or larger, although successful reception using much smaller size antennas has been reported throughout the world. Table 12–1 gives a comparison of antenna sizes vs EIRP thresholds.

The Intersputnik satellites are also designed to operate with global and hemispheric beams, and are limited to 6 channels each (as compared to the 24 to 50 or more channels delivered by the Intelsat birds). Since there are fewer channels, the Russians can use higher-power TWTs, driving their global and hemispheric beams at 15 watts and their Eurospot and other spot beams at 40 watts. Thus, the Intersputnik satellites can be seen with much smaller dishes. The Eurospot beam for the

Table 12-1. Antenna Size vs EIRP Threshold

Antenna Size (Meters)	LNA Noise Temperature (Kelvins)	Threshold EIRP (dBW)*
2.0	120	36.9
	100	36.4
3.0	120	33.3
	100	32.7
	85	32.3
3.7	100	31.0
	85	30.4
	75	30.0
4.5	100	29.2
	85	28.6
	75	28.2
6.0	100	26.6
	85	26.2
	75	25.7
7.5	85	24.1
	75	23.7
	65	23.4
10.0	75	21.1
	65	20.6

*For a TVRO antenna with elevation of 30 degrees or greater—add 1 to 2 dBW threshold for elevations from 15 to 10 degrees.

Gorizont (horizon) satellite at 14° W longitude can be received by 8-foot parabolic antenna in England, and a dish as small as two meters in the Scandinavian countries.

Polar Orbit Satellites

The majority of the world's communications satellites are parked directly over the equator in the Clarke orbit. These satellites cannot be seen from the surface of the earth above approximately 83° N. latitude (or below 83° S. latitude). At these extreme latitudes the satellites are below the local horizon. To solve this problem, the Russian Molniya domestic satellites use a polar orbit. To receive these satellites, continuous tracking of the satellites is required by the TVRO earth stations as the satellites move through the orbit. To find the Molniya satellites, a TVRO terminal must be aimed northward toward the Pole rather than southward toward the equator. As each Molniya passes below the terminating horizon, the next one in orbit is turned on, and the TVROs are quickly repositioned back to the rising bird. This process is repeated every six hours.

In Russia, computer-generated tables of "pointing angles" are produced on a daily basis to enable the antennas to be properly positioned. These angles change every few minutes, or so, and to complicate matters the Molniya satellites recede in their orbit several minutes each 24-hour earth day. Thus, a Molniya satellite won't be in the same position in its orbit at the same time each day. To locate the transmitting Molniya satellite when it reaches its 40,813 kilometer apogee over the Hudson Bay proceed as follows:

1. Align polar mount axis of dish with center of Hudson Bay (instead of true north). See Fig. 12–5 for positioning angle to set azimuth for your state.
2. Set the elevation for your site (Fig. 12–5).
3. The "on-air" Molniya satellite will be between the elevation given and elevation + 20°, depending on the position of the bird in its orbit.

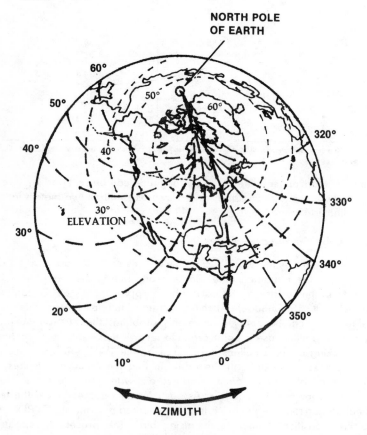

Fig. 12-5. Molniya pointing angle locator map.

4. Beginning with the listed elevation slowly rock the dish through a ± 15° azimuth change, looking for a signal on transponder 9.
5. If no signal is found, return to the given azimuth and increase the elevation by 1°.
6. Repeat Step 4 with new (higher) elevation until the bird is located.
7. When the "on-air" bird reaches its highest elevation at the end of the 6-hour transmission period, it is shut off and the next bird (lower elevation) is turned on. Return to Step 4 to find new bird.

A special Apple II® Molniya Tracking Program Disk and instructions on how to modify TVRO receivers is available from The Satellite Center for $30 ($40 overseas). (See Appendix G for address information.)

Transponder Schemes

Every designer of a satellite system uses a different transponder bandwidth and modulation scheme. In the North American C-band satellites, 36-MHz channels which are separated by 4-MHz guard bands (40 MHz per transponder) and using an orthogonally polarized frequency-reuse scheme are employed to squeeze 24 channels into the 3,720- to 4,220-MHz band. The Soviet Intersputnik birds have 6 channels spaced in 50-MHz increments from 3,675 to 3,975 MHz, using right-hand circular (RHC) polarization. The Molniya satellites operate three transponders (with bandwidths of approximately 34 MHz) and use 50-MHz spacing with RHC polarization.

Several generations of standards have been used by the Intelsat satellites. The older Intelsat IV birds provide 12 transponders at 40-MHz increments running from 3,725 to 4,225 MHz and use RHC polarization. The Intelsat IV-A satellites use the same frequency and polarization scheme, but increase the number of transponders to 20 by using 8 eastern hemispheric and 8 western hemispheric beams, combining both physical isolation and frequency reuse. The Intelsat IV-A satellites also use RHC polarization. The number of channels available per transponder are also doubled by using the "half transponder" transmission mode. In this mode each transponder can be used to carry two separate video feeds, with bandwidths of about 18-MHz each. The newest Intelsat satellite series, Intelsat V, increases the C-band video channel capacity to 37 feeds by using a combination of right-hand and left-hand circular polarization as well as east and west hemispheric and zone (spot) beams. In addition, the Intelsat V series satellites are also equipped with two linearly polarized spot beam transponders operating in the 10.95 to 11.7 GHz band. Fig. 12-6 clarifies the relationships among the various satellite systems.

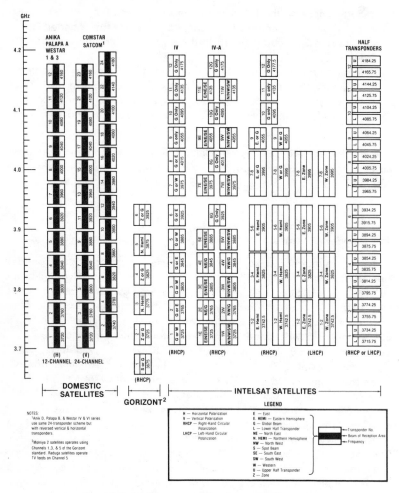

Fig. 12-6. Satellite transponder frequency schemes.

I-F Bandwidth Modification

Depending on the scheme in use, the conventional C-band TVRO receiver will need to be modified to optimize the satellite reception. First, the i-f bandwidth should be limited to match the bandwidth of the received signal. For half transponder schemes, this bandwidth reduction improves the signal-to-noise ratio and *real* overall system gain by 2 to 3 dB. When this modification is coupled with the circularly polarized feedhorn assembly retrofit, an overall system gain of 5 to 6 dB will be achieved. This is equivalent to reducing the effective required antenna diameter from 8 or 9 meters to 5 or 6 meters for a given system.

Energy Dispersal (ED) Signals

In addition to changing the receiver i-f bandwidth, the automatic frequency control (afc) circuits may also need to be modified to handle the energy dispersal (ED) signals which are different from those used by the North American satellites.

Most satellite systems use an energy dispersal technique designed to minimize interference with terrestrial microwave services sharing the same frequencies. The transponder carrier is slowly frequency shifted from its center frequency, requiring the receiver to "track" the carrier center frequency up and down the frequency band. North American TVRO receivers have built-in dispersion removal circuits, which are usually a part of the automatic frequency control system in the receiver. The usual energy dispersal convention uses a triangular waveform which is half the field frequency (25 or 30 Hz). The Russian scheme also uses a triangular ED waveform, but at 2 to 2.5 Hz (very slow), and with a deviation of 6- to 8-MHz peak-to-peak (very large). Feeding this signal into a North American C-band receiver will usually cause the video to be clamped on dispersal peaks producing sparklies every second or so. This problem can be solved by using an op-amp low-pass active filter inserted between the receiver afc line and the voltage-tuned oscillator (VTO) in the receiver rf stage. Several receiver manufacturers now make international-standard TVRO receivers which will handle the half transponder formats and the differing energy dispersal waveforms. One popular model which has been modified is the Avcom unit (see Appendix C for the address of the manufacturer in Virginia.)

NTSC vs PAL and SECAM

A satellite transponder is an electronic "mirror." When a signal is uplinked, it is translated and downlinked by the satellite without change. Thus, North American satellites relay NTSC 525-line video signals with 3.58 MHz color burst subcarriers and 60 fields per second. The German PAL (phase alteration by line) system is used by Britain and most other countries in Europe. PAL tv uses 625 lines of 50 fields per second with a color subcarrier at 4.43 MHz. Brazil operates a modified version (PAL-M) with a 525-line transmission and a 3.58 color subcarrier. Argentina has adopted yet another variation (PAL-N), with a 625-line transmission and 3.58 MHz subcarrier utilizing the PAL coding scheme.

A third international standard, the French SECAM system (for "sequence a memoire") operates with 625 lines at 50 fields per second (the European standard) but uses an fm modulation scheme to apply the color information onto 2 subcarriers at 4.406 and 4.250 MHz. The CCIR pre-emphasis characteristics vary between the two line standards, and require differing de-emphasis and filtering by the TVRO receiver.

Luckily, a standard NTSC black and white television set can also be used to receive PAL and SECAM pictures (in black and white) without modification. With minor adjustment of the vertical hold and linearity controls, the thin and elongated-looking people on the 625-line signal can be squashed down to a normal appearance. The best solution for the international satellite television viewer, however, is to purchase a good multistandard color monitor. These are readily available from the video mail order discount houses located in Miami, Florida, and London, England. One manufacturer, Barco (a Belgian company with sales offices in every major country), provides a number of 4-standard television sets that will correctly demodulate most of the world's terrestrial tv broadcast signals. In addition, these sets will also interface with certain ¾-inch video cassette recorders which produce NTSC 4.43-MHz subcarrier audio when playing NTSC 3.58-MHz encoded tapes.

A list of the major countries of the world and their television standards is presented in Table 12-2. When viewing an Intelsat feed, the satellite standard used will often (but not always) be the same one used by the terrestrial television service of that country. However, since each country maintains PAL-SECAM-NTSC conversion equipment at its Intelsat earth stations, this is not always true.

THE INTELSAT SATELLITE SYSTEM

The Intelsat international satellite consortium was founded in 1964 by the United States and ten other countries to oversee the operations of the multinational network of communications satellites, which Comsat (the US Communications Satellite Corporation) was constructing to operate. Intelsat today has over 106 members, with transmit and receive earth stations located in 155 countries. The United States holds the largest investment share (23%) through Comsat, whose stock is traded on the New York Stock Exchange.

There are at present 14 Intelsat satellites positioned over the three oceans providing 24-hour-per-day coverage (see Table 12-3). Television transmission has dramatically increased since the first months of 1965. In that year there were about 80 hours of tv delivered by Intelsat satellites worldwide. By 1981 this figure had increased to 36,658 hours of television traffic; in 1982 the figure had jumped to over 45,000 hours. By 1984 there may well be over 75,000 hours. Over two billion people are scheduled to see portions of the 1984 summer Olympics broadcast from Los Angeles via Intelsat satellites.

Intelsat has developed a series of satellite "generations" beginning with the first Intelsat I and extending to the new Intelsat VI (under development). The Intelsat V satellites are the newest workhorses providing both C-band and Ku-band transponders operating in global, hemispheric, and zone (or spot) beam configurations. Most Intelsat IV and IV-A satellites are also still functioning, providing a unified system of 6 main feeds: three for the Atlantic Ocean, two for the Indian Ocean, and

Table 12–2. Television Standards for Major Countries

Country	Standard	System
Algeria	625/50	PAL
Argentina	625/50	PAL-N
Australia	625/50	PAL
Brazil	525/60	PAL-M
Canada	525/60	NTSC
China	625/50	PAL
Colombia	525/60	NTSC
Cuba	525/60	NTSC
France	625/50	SECAM
Germany (FED)	625/50	PAL
India	625/50	PAL
Indonesia	625/50	PAL
Italy	625/50	PAL
Japan	525/60	NTSC
Luxembourg	625/50	PAL/SECAM
Malaysia	625/50	PAL
Mexico	525/60	NTSC
Morocco	625/50	SECAM
Netherlands	625/50	PAL
Niger	625/50	SECAM
Nigeria	625/50	PAL
Oman	625/50	PAL
Peru	525/60	NTSC
Philippines	525/60	NTSC
Portugal	625/50	PAL
Saudi Arabia	625/50	SECAM
Singapore	625/50	PAL
Spain	625/50	PAL
Sudan	625/50	PAL
Sweden	625/50	PAL
Switzerland	625/50	PAL
United Kingdom	625/50	PAL
U.S.A.	525/60	NTSC
USSR*	625/50	SECAM

*The Communist bloc countries of Eastern Europe use the USSR SECAM tv standard for all feeds.

one for the Pacific Ocean (Fig. 12–7). In addition, the in-orbit spares are actively maintained.

Intelsat earth stations use a standard antenna size. The "A" stations must provide a minimum G/T figure of merit of at least 40.7 dB per Kelvin (dB/K). This requires large antennas, in the range from 26 to 30 meters. The "B" standard antennas vary from 9 to 11 meters in size to provide a G/T figure of 31.7 dB/K, and the "C" standard T-R stations range from 19 to 20 meters to provide a minimum G/T of 39 dB/K. These enormous and expensive antennas are needed to receive the very low

Table 12—3. Intelsat Satellite System

Satellite	Location	Ocean	Use
IV—F1	174° E	Pacific	Primary
IV—F4	27.5° W	Atlantic	Spare
IV—F7	53° W	Atlantic	South American leased services
IV—F8 •	179° W	Pacific	Spare
IV-A—F1	18.5° W	Atlantic	Major Path 2
IV-A—F2	21.5° W	Atlantic	Major Path spare
IV-A—F3	60° E	Indian	Leased, Some tv (Spare)
IV-A—F4	34.5° W	Atlantic	Leased, Some tv (Spare)
IV-A—F6	63° E	Indian	Spare
V—F1	60° E	Indian	Major Path (Some leases)
V—F2	34.5° W	Atlantic	Major Path 1
V—F3	24.5° W	Atlantic	Primary Path (Some leases)
V—F4	27.5° W	Atlantic	Leased, Some tv (Spare)
V—F5	63° E	Indian	Primary Path

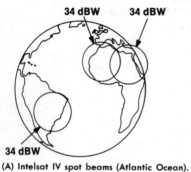

(A) Intelsat IV spot beams (Atlantic Ocean).

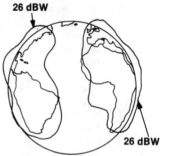

(B) Intelsat IV-A hemispheric beams (Atlantic Ocean).

(C) Intelsat IV-A hemispheric beams (Indian Ocean).

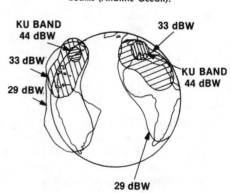

(D) Intelsat V beams (Atlantic Ocean).

Fig. 12-7. Typical Intelsat footprints for various satellites.

EIRP levels (22 dBW or less in some instances) of the satellite's, global beams. The global beams are designed to bridge the oceans, and cover 40% of the surface of the earth. Hemispheric beams, which illuminate only 20% of the earth, however, can be received by a 4- to 6-meter TVRO earth station. Zone or spot beams, which target 10% of the earth surface, can be picked up by a 3- to 4-meter dishes. This makes the viewing of most of the Intelsat television feeds possible for the satellite television enthusiast who may not wish to construct a 90-foot dish in his backyard!

The Intelsat satellites use a C-band frequency scheme similar to the domestic communications satellites, although the transponder nomenclature differs somewhat. Table 12-4 references the Intelsat transponder frequencies to the standard US transponder numbers for ease of conversion. Most of the Intelsat television transmissions are in the half-transponder format. In this mode, a television signal is fm modulated into an 18-MHz wide bandwidth instead of the conventional 36-MHz wide channel. By restricting the fm signal deviation, Intelsat can effectively double the capacity of each satellite transponder and allow two simultaneous programs to be transmitted, consuming the upper half and lower half of each transponder. A single television feed can also be assigned to one half of the 36-MHz bandwidth capacity; the other half can be used to transmit telephone message traffic using an FDM telephone carrier multiplexing scheme. Intelsat pays a price in increasing the effective signal "through put" of the birds in this manner. Delivering two simultaneous video pictures over the same physical transponder channel requires a minimum power reduction of 50%, or 3 dB, for each television feed. In reality, to minimize intermodulation interference, the transponder power is usually backed off an additional 2 or 3 dB. To successfully recapture the signal, it is important that a conventional 36-MHz bandwidth C-band TVRO receiver be modified to decrease its i-f bandwidth to 18 MHz; this will at least recapture 3-dB in signal-to-noise improvement, which would otherwise be lost.

By using left-hand and right-hand circular polarization, half-transponder modulation techniques, and physical beam isolation via eastward-oriented and westward-oriented hemispheric beams, the Intelsat V designers have raised the techniques of frequency reuse to those of a fine art. This four-fold spectrum reuse provides the equivalent of 37 standard transponders. When this capability is coupled with the Ku-band transponder package and the ability to "cross strap" C-band uplinks with Ku-band downlinks (and vice-versa), total capacity of each Intelsat V satellite increases to an equivalent of almost 58 standard "North American type" satellite television channels. Some of these flexibilities are found in the Intelsat IV satellites; specific configurations vary from bird to bird and month to month (see Fig. 12–8).

Table 12—4. Domestic Satellite vs Intelsat Transponders

US Transponder No.	Freq. MHz	Transponder No.	Intelsat*		
			IV	IV-A	V
1/2	3725	1	G or W	S or H	H and Z
3/4	3765	2	G or E	G or S	H and Z
5/6	3805	3	G or W	S or H	H and Z
7/8	3845	4	G or E	G or S	H and Z
9/10	3885	5	G or W	S or H	H and Z
11/12	3925	6	G or E	G	H and Z
13/14	3975	7	G or W	S or H	H and Z
15/16	4015	8	G or E	G	G,H, and Z
17/18	4055	9	G	S or H	H or G
19/20	4095	10	G	G	G
21/22	4135	11	G	S or H	G
23/24	4175	12	G	G	G

*G = global beam. H = Hemispheric beam (could be west or east). Z = Zone beam (west or east). E and W = east and west spot beams. S = spot beam (east or west).

Television Programming

Intelsat was originally planned to carry television programming on an occasional use basis; its first tariffs were designed accordingly. Over the years, the national television networks have leased television transmission time on a periodic and relatively consistent basis. Intelsat rents its transponder capacity to its member nation owners. It is the local telephone authorities such as British Telecommunications (BT) in London and Teleglobe Canada in Montreal which provide the actual interconnection with the television studios. In the United States, Comsat is the Intelsat carrier, and satellite feeds are coordinated through its L'Enfant Plaza control center in Washington, DC.

The North Atlantic segment, which primarily provides east-west feeds between North America and Europe, is the most heavily used communications path. All three of the commercial US television networks, United Press International, and other national tv networks operate daily feeds on a scheduled basis. Table 12—5 summarizes the principal feeds on the Primary, Major Path 1, and Major Path 2 Atlantic satellites. Intelsat reserves transponder 12 (transponder 23—24 on a North American TVRO receiver) configured in a half-transponder global beam format for occasional-use television news feeds. Similar transmission schedules are also maintained for the Pacific and Indian Ocean satellites with American, Australian, and Japanese television network traffic.

Within a short time, television transmissions through Intelsat have increased dramatically. Many countries have leased full-time 24-hour-per-day half-transponder television channels from Intelsat. Some of these television feeds are designed for truly global use such as the

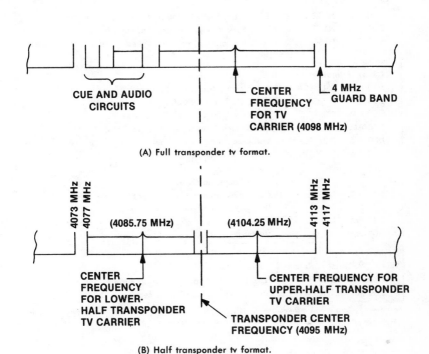

(A) Full transponder tv format.

(B) Half transponder tv format.

Fig. 12-8. Intelsat transponder formats (Transponder 10 illustrated).

American Armed Forces Radio and Television Service (AFRTS), which delivers a 24-hour-per-day programming stew of CBS, NBC, ABC, and CNN programs to armed forces installations and military bases worldwide. The service is first uplinked from New York via RCA on Satcom 2, transponder 20. This is received by the Comsat earth station at Andover, Maine for uplink to the Atlantic Intelsat IV-A (F2) satellite at 1° W longitude. The service uses Intelsat transponder 2 (US transponder 3) with a global beam, right-hand circular polarization, and program audio on a 6.8 MHz subcarrier. With a 19 to 22 dBW EIRP footprint from pattern edge to bore sight, a 6 to 7½-meter TVRO antenna is able to pick up this feed. The AFRTS service is also sent to Pacific military bases via the Intelsat V satellite at 181° W longitude. A separate NBC/CNN service feeding Austrialia also utilizes the Pacific Ocean bird at 181° W longitude, but operates in a "diplexing" mode using a full transponder

241

Table 12—5. Intelsat Occasional-Use North Atlantic TV Feeds

Time (EST)	Standard	Source and Content
Intelsat IV-A —F1		
18.5° W, Major Path 2		
6:00 AM	NTSC	Eurovision (Brussels) IntraEurope pool
2:30 PM	NTSC	ABC New York sports feeds to Europe (Sat. only)
5:30 PM	NTSC	ABC London—Bureau news feeds (World News Tonight)
Intelsat V —F3		
24.5° W, Primary Path		
5:30 AM	PAL/SECAM	Eurovision IntraEurope pool
6:30 AM	NTSC	ABC, NBC, CBS—Bureau news feeds
8:00 AM	PAL	UPITN/DSS/E News service to Americas
10:00 AM	SECAM	FR3 French Network feeds from Paris
11:00 AM	PAL/SECAM	Eurovision IntraEurope pool
1:00 PM	NTSC	Cable News Network bureau feeds
1:00 PM	PAL	Visnews to Johannesburg for South African Broadcasting Corp.
3:10 PM	NTSC	TVE News Service to South America
3:30 PM	PAL	RTP Tele-Journal news service to Brazil
4:45 PM	NTSC	Visnews Atlantic news service to North America
6:00 PM	NTSC	UPITN DSS/L news service to Europe
Intelsat IV-A —F4		
34.5° W, Major Path 1		
2:30 PM	NTSC	NBC New York sports feeds to Europe (Sat. only)
3:50 PM	NTSC	UPITN DSS/A London news service to U.S.

Note: Feeds are usually on Intelsat transponder 12 (4175 MHz) in half-transponder format, using global beams. UPITN also handles US network and tv global feeds feeds during their DSS service. UPITN may also be found in SECAM on Gorizont at 14° W, 3675 MHz, Gorizont TR No. 1.

to multiplex two simultaneous television feeds via the Thomson-CSF French video processing technology.

Many countries use full-time Intelsat leases for domestic service (Table 12-6). The Americas are particularly active. Argentina uses Intelsat V (F2) at 27.5° W longitude to carry Argentine Televisora Color (ATC) network to hundreds of low-power television stations scattered throughout that nation. A global beam pattern is used in half-transponder format carrying PAL-N television. An occasional use feed on Intelsat V (F3) at 24.5° W longitude (global beam) also carries feeds from Argentina to Spain and back.

Brazil uses Intelsat IV-A (F2) at 21.5° W longitude to operate three channels of national television programming. Rede Globo, the government television network, uses a global transponder to transmit PAL-M 525-line color feeds, a standard which is unique to Brazil. Boa Noite Brazil operates a hemispheric beam in half-transponder format for transmission to television stations throughout the country, often carrying Spanish-dubbed American programming such as Star Trek. This service

Table 12–6. Intelsat Full-Time Domestic-Lease TV Feeds

Country	Satellite	Transponder†	Bandwidth	Freq.	Beam	Standard	Program
Algeria	IV-A V 60° E	5 (9)	18	L	W. Hemi	625 PAL	RTA
Argentina	V 27.5° W	38 (24)	36	—	Global	625 PAL-N	ATC LS 82 Ca. 7
Australia	IV 174° E	3 (5)	18	L	W.Spot	625 PAL	ABC
Australia	IV 174° E	5 (6)	18	U	W.Spot	625 PAL	ABC
Australia	IV 174° E	5 (9)	18	L	Global	625 PAL	ABC
Brazil	IV-A 21.5° W	6 (11)	18	U	Global	525 PAL-M	Rede Globo
Brazil	IV-A 21.5° W	3 (5)	18	U	W.Hemi	525 PAL-M	Boa Noite Brazil
Colombia*	V 27.5° W	1 (1)	18	L	W.Hemi	525 NTSC	Cadena-1
France	IV-A 18.5° W	4 (7)	36	—	Global	625 SECAM	
Mexico	IV 53.0° W	1 (1)	18	L	W.Spot	525 NTSC	XEW
Mexico	IV 53.0° W	3 (5)	18	L	W.Spot	525 NTSC	TCW/TRM
Mexico	IV 53.0° W	5 (9)	18	L	W.Spot	525 NTSC	
Mexico	IV 53.0° W	7 (14)	36	—	W.Spot	525 NTSC	
Morocco	V 27.5° W	3 (5)	18	L	E.Hemi	625 PAL	RTM
Niger	IV-A 21.5° W	3 (5)	18	L	E.Hemi	625 SECAM	Tele-Sahel
Nigeria	IV-A/V 60° E	7 (14)	18	U	W.Hemi	625 PAL	NTV Ch.10 Lagos
Norway	IV-A 21.5° W	11 (21)	18	U	E.Hemi	625 PAL	
Oman	IV-A/V 60° E	36 (20)	18	L	Global	625 PAL	
Peru*	V 27.5° W	2 (4)	18	U	W.Hemi	525 NTSC	Enrad Peru
Saudi Arabia	IV-A 21.5° W	1 (1)	36	—	E.Hemi	625 SECAM	
Spain	IV-A 34.5° W	7 (14)	18	U	E.Hemi	625 PAL	TVE Cadena-1
Sudan	IV-A 21.5° W	7 (3)	18	L	E.Hemi	625 PAL	
Venezuela	V 27.5° W	3 (5)	36	—	Global	525 NTSC	
Zaire	IV-A 21.5°	9 (7)	18	L	E.Hemi	625 SECAM	Tele Zaire

*Audio on separate SCPC feed.
†Number in ( ) is corresponding US number.

can be received by 6-meter TVRO antennas throughout the satellite's useful footprint. Rede Globo operates a second feed for occasional news transmission use in the half-transponder format via a hemispheric beam.

Mexico is now a major user of Intelsat television service, consuming four transponders on Intelsat IV (F7) positioned at 53° W longitude. Transponders 1, 3, 5, and 7 (US transponders 1, 5, 9, and 14) are presently operating in half-transponder format using Western spot beams. No other services are currently using the other side of the transponders, allowing the Mexican programming to be seen with smaller antennas throughout much of the US, Central America, and South America. XEW-TV (Commercial Channel 2 from Mexico City) is located on transponder 1. XHITN (Channel 11, Mexico City) operated by the Telecomunicaciones de la Republica de Mexica (TRM) is a privately owned educational network carrying programming similar to PBS or BBC2. On transponder 7 may be found XETV (Mexico City cable channel 6) providing feeds from ABC, NBC, and CBS television affiliates picked up in Tijuana off-the-air from San Diego. This service is scheduled to be replaced by Mexican Educational Television, although since it is popular with Mexican SMATV systems, it may be continued. XHDF (Mexico City Channel 13) uses transponder 9 to deliver its commercial television programming. XHDF is owned and operated by the Mexican Ministry of Government Affairs.

Mexico intends to expand significantly its usage of Intelsat transponders during the interim period while it builds its own domestic satellites. Additional planned services include XHTB (Mexico City channel 5), a private, commercially sponsored station oriented to young Mexicans, and XHTN (Mexico City channel 8), a similar organization. With the launch of its Iluicahua satellites in 1985, the Mexican educational network XETV will be able to provide true *DBS* programming nationwide with 1-meter TVRO dishes. The Mexican satellites now being built by Hughes Aircraft will each be configured with 18 C-band and 6 Ku-band transponders, allowing mixed-mode message traffic and DBS operation.

Other neighbors using Intelsat domestically include Venezuela, Colombia, and Peru. Venezuela operates VeneVision, provided by Venezolana de Television on Intelsat V (F2) at 27.5° W longitude. This service uses a full transponder with hemispheric beam (transponder 3; US transponder 5) providing excellent reception via antennas as small as 12 feet in diameter. Colombia operates a half-transponder Spanish service using a hemispheric beam on transponder 1 of the same satellite, as does Peru, whose Lima Channel 7 RTP television station may be found on transponder 2 (US transponder 4). Six-meter dishes should be adequate to pick up both of these services in most parts of the satellite footprint. Spanish-dubbed American programming including Sesame Street may be found on most of these South American services. NTSC tv standards are used.

In Africa, Intelsat transponders carry domestic television feeds for Saudi Arabia, the Sudan, and Nigeria, which deliver some English-language programming, and by Morocco, Niger, and Algeria, which provide French programming as well as local language service. Oman also operates an Arabic programming service provided by Radio Television Oman in Muscat. See Table 12–6 for details on these services.

European countries have not needed to use Intelsat for domestic television transmission as they operate extensive terrestrial microwave networks which deliver high-quality television signals from border to border. Interconnection of European television networks via microwave has also minimized the need for intra-European satellite feeds except for daily news packages shared by the Eurovision consortium. Moreover, Europe is constructing its own series of new Ku-band DBS satellites, as well as using the experimental OTS-2 Eutelsat satellite for television transmission (covered later in this chapter).

In Austral-Asia, the Australians are principal users of Intelsat IV located at 174° E longitude, operating at least 3 half transponders using Western spot and global beams. These are contracted to the Australian Broadcasting Commission (ABC) for national television service while Australia constructs its own domestic satellites. The ASEAN South Pacific and Asian countries (Singapore, Malaysia, Indonesia, Thailand, and the Philippines) use a separate regional satellite system operated by Indonesia. These Palapa satellites were built by Hughes Aircraft, the designers of the original Anik and Westar birds. Consequently, Palapa operates on North American standards with 12 transponders per bird and horizontal polarization, providing footprints of approximately 34 dBW throughout the region. Palapa A1 and A2 are located at 83° and 77° E longitude, and are scheduled to be replaced with second generation Palapa Bs carrying 24 transponders each. Palapa B satellites are similar to the Westar IV and V series. They will deliver slightly higher EIRPs (in the region of 36 dBW or so) through the use of higher power TWT amplifiers. They will be stationed at 108° E and 113° E longitude. Television programming services of the ASEAN countries (with the exception of Singapore) are carried over the Palapa satellites. Palapa is not a part of the Intelsat consortium and is operated independently, although Intelsat theoretically maintains global monopoly rights to non-Soviet-bloc international satellite communications facilities. As more and more regional satellite networks like Palapa are born, Intelsat will be forced to lower its tariffs and international satellite tv transmission will skyrocket. The next regional consortium scheduled for launch is Arabsat in the mid-1980s.

INTERSPUTNIK SYSTEM

While Intelsat was busy creating the western international satellite network, the Russians were building the Soviet bloc's geosynchronous satellite network—Intersputnik. The Intersputnik satellites are stationed over the Indian and Atlantic Oceans where they can be directly controlled from TT&C (Telemetry Tracking and Command) earth stations inside the Soviet Union (see Table 12−7). A number of "statsionar" (stationary) slots were defined by the Russians for use in constructing their global satellite communications network. The current active locations are Statsionar 1−80° E; Statsionar 2−35° E; Statsionar 3−85° E; Statsionar 4−14° W; Statsionar 5−53° E; and Statsionar 6−90° E. Fig. 12−9 shows the footprints of Statsionar 4 and 5. Into these orbital slots, the Russians have launched several different generations of satellites ranging from the initial Raduga (rainbow) satellites to be followed (but not necessarily replaced) by the Gorizont (horizon) series. Raduga-10 (launched in October of 1981) carries 6 transponders operating on either global and hemispheric beams in the 3450 to 3900 MHz band. The Gorizont satellites also carry 6 C-band transponders with hemispheric and global beams, as. well as steerable spot beams which can deliver an EIRP in excess of 45 dBW at transponder bore site. Gorizont uses the 3.65 to 3.95-GHz band, which means that most but not all its transponders can be received by domestic C-band TVRO receivers. The Gorizont satellites also carry a military X-band transponder package for independent use by the Soviet military high command.

Within the Soviet Union, there are four national television networks, which to a varying degree cover all areas of the country. All originate from the Moscow television center, with breakaways to local and regional feeds in the Soviet states. Although program listings for all four Moscow tv channels are published daily in the State newspapers Izvestia and Pravda (the Sunday edition provides listings for a week in advance), only the two most popular services are carried via the Russian Intersputnik system. Video feeds are in the PAL standard, with audio delivered by way of subcarriers located at 6.5, 7.0, and 7.5 MHz, allowing independent radio feeds to be carried simultaneously with program audio. "Moskva" remote-controlled satellite-fed low-power tv and cable

Table 12−7. Intersputnik Satellite System

Satellite	Location	Ocean	Use
Gorizont 2—S4	14° W	Atlantic	I Programma
			II Programma
			Euro-Spot
			Cuba feeds
			Intersputnik news exchange
Raduga 9—S2	35° E	Indian	II Programma West
Gorizont 5—S5	53° E	Indian	Orbita III—Vostok II Programma
Raduga 10—S3	85° E	Indian	II Programma East
Gorizont 4—S6	90° E	Indian	African and Asian feeds

Transmissions are in Russian, Spanish, and English, as well as Soviet-bloc languages.

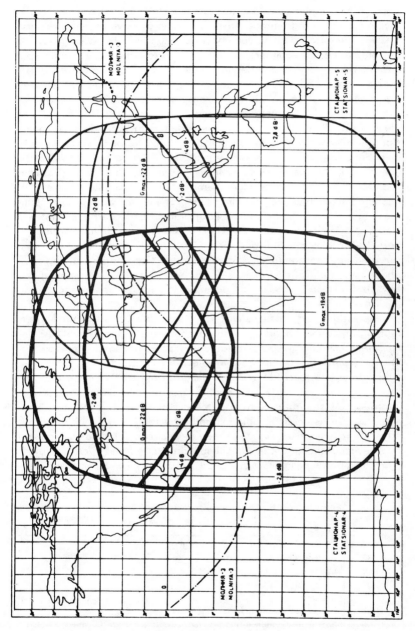

Fig. 12-9. Footprints of Gorizont Stationar 4 and Stationar 5 systems.
(Courtesy Intersputnik)

lines were inaugurated during the summer 1980 Olympic games held in Moscow. The I Programma and II Programma television network services were introduced to the country via Gorizont II at 14° W longitude. At present, three transponders on this bird carry video. The Europe spot beam, which provides a variety of I Programma feeds as well as UPITN London news and Cuban programming may be found on Gorizont transponder 1 at 3675 MHz. This transponder is below the tuning frequency of the North American C-band TVRO receivers. Modifications to the local oscillator tuning circuit are required to properly pick up the signal. Gorizont transponder 4 at 3825 MHz (US transponder 6) provides a global beam which occasionally provides II Programma feeds, but is only occasionally used. Gorizont transponder 5 at 3875 MHz (US transponder 9) operates a northern hemispheric beam which delivers Intersputnik news and programming feeds among the Soviet block countries. In addition, time-shifted I Programma feeds and some Intelsat traffic relaying occasionally may be seen on this transponder. Transponders 2, 3, and 6 at 3725, and 3925 MHz provide only occasional video service with transponder 3 (US transponder 3) primarily carrying telephone message traffic. Table 12-8 presents a recent programming day for the Gorizont 2 satellite as observed on the European spot beam (transponder 1).

The Russian Statsionar satellites transmit on a lower but partially overlapping frequency band, as compared to the Intelsat satellites. Intersputnik uses right-hand circular polarization, a massive energy dispersal waveform, and the SECAM television standard.

In addition to these differences, the program audio on the Eurospot beam of the Gorizont 14° W satellite uses a compander processor to improve the signal-to-noise ratio of the audio signal. This requires, however, that an associated audio expander be used at the TVRO to properly restore the audio signal. Although one can listen to the Russian audio without the expander, it will sound distorted and overmodulated. Many compander circuits have been published in the consumer elec-

Table 12-8. Gorizont Programming Schedule

Time (GMT)	Program
13:00	UPITN Euro-Spot Beam London (English)
13:30	Intersputnik news feeds
14:00	Czechoslovakia entertainment to Cuba (Spanish)
15:00	Hungarian entertainment
16:00	Test Patterns
17:00	I Programma—Moscow Nightly News
18:00	Sports, music, dance, movies from Eastern European Countries
21:00	Cuban news feeds to Eastern Europe. Also movies, cartoons taken "off the air" from US (Spanish and English)

tronics project construction magazines (*Radio Electronics, Computers and Electronics,* etc.), and consumer-quality hi-fi expanders are available in the marketplace.

MOLNIYA RUSSIAN DOMESTIC SATELLITE SYSTEM

The Molniya satellite system is the only one of its kind in the world. Using a nongeosynchronous polar orbit, the complete system consists of a series of high-power satellites which are placed into the same U-shaped orbit with apogees over Russia and Canada. The Molniya satellite in orbit swoops down from one apogee past its south pole perigee and swings back up again to its second apogee. As it approaches each apogee, each Molniya satellite appears to hover relatively motionless for several hours, and is visible for about six hours from a northern TVRO in the USSR (or North America). The mechanics of tracking the Molniya satellites have been covered in a previous section. The satellites use an unusual "sound in sync" audio program transmission scheme which has also been discussed previously. Each bird carries three transponders of 43 MHz each, spaced 66 MHz apart, at 3675, 3775, and 3875 MHz, roughly equal to transponders 1, 3, and 5 of the Raduga Intersputnik series. The transponders use right-hand circular polarization with a global coverage, putting out an EIRP of over 32 dBW in the primary footprint. The visibility limit for the satellites extends well down to South America and northern Africa, and can be readily tracked using the home TVRO earth station.

The original Molniya-1 satellites downlinked in the uhf television band and were spaced 120° apart in orbit. Three birds were launched to provide 24-hour coverage. Today, the Molniya-3 satellites use a four-bird configuration, spaced 90° apart, with each bird providing active on-air service for six hours of each Molniya orbit (slightly less than a 24-hour day).

At present, two separate Molniya-3 systems operating four satellites each are in operation, with military and message traffic assigned to a number of the transponders. In addition, a 4-bird Molniya-1 uhf system is also in orbit, providing a total of 12 Molniya satellites currently "on-line." The US-USSR "hotline" between the White House and the Kremlin uses a Molniya transmit-receive earth station located at Ft. Detrick, Maryland. The satellite system may also link a number of Soviet embassies and military outposts back to Moscow. One of the Molniya-3 series is used for civilian traffic, and transponder 5 (US 9) at 3.875 GHz is used to carry the "Orbita-I" programming service, which is produced by I Programma in Moscow. The program service follows the time zone at the extreme eastern edge of the USSR which is a full eight hours ahead of Moscow. Thus at midnight in the Kremlin, viewers watching Orbita-I are waking up to the Russian version of "Good Morning America."

DBS AROUND THE WORLD

Big plans are afoot in dozens of countries to launch television broadcast satellites. The World Administrative Radio Conference on Broadcast Satellites was held in 1977 to establish unified DBS standards for the 11.7 to 12.5 gigahertz band. WARC-BS 1977 reserved 40 television channels, each with several audio channels, in this unused frequency spectrum which is relatively free from terrestrial interference. The DBS satellites will be spaced at 6-degree intervals, and the TVRO receiver standards are designed to operate with antennas 90 centimeters (about 2½ feet) in diameter. Fig. 12—10A presents the estimated footprint coverages for the European DBS satellites, assuming the use of standard 90-centimeter TVRO dishes. A somewhat larger antenna—say 5 to 8 feet—will allow a DBS transmission from the satellite of any country in Europe to be seen in any other country in that region. Fig. 12—10B shows the estimated coverage of a 2-meter antenna. This potential to broadcast satellite television pictures across national borders has raised the concern of many European states. It will be one of the most exciting issues of international law in the late 1980s.

The European DBS projects have benefited from significant experience obtained through the operation of two experimental satellite systems. The Symphonie I and II satellites were launched in the late 1970s as a joint effort of France and Germany. They are now colocated at 11.5° W longitude. The joint satellite system transmits with both east and west hemispheric beams, and provides a 29 dBW EIRP footprint at the edge of effective coverage, which extends from northern Europe to central Africa. The frequency plan is given in Fig. 12—11. Two 4-gigahertz transponders are presently in operation, and France continues to use Symphonie to deliver national television network programming from Paris to its overseas departments in northern Africa and elsewhere. The birds are scheduled to be replaced by Telcom I, a new dual-band 4/12 gigahertz satellite scheduled for launch in 1984. To conserve the onboard fuel supplies of the aging satellites, the French ground technicians have allowed the birds to wander in an extended figure eight orbit around their nominal Clarke-orbit center. TVRO earth stations must therefore be equipped with automated tracking-correction mounts. Using this approach, 5-meter antennas can be used in the French overseas territories. Symphonie uses the SECAM system, and can readily be seen throughout Europe.

The second experimental European satellite is OTS (the Orbital Test Satellite), operating in the 11.4 gigahertz band. OTS was launched in the late 70s by the European Space Agency (ESA) to validate the DBS concept. OTS II carries 6 transponders, 4 of which can be used for commercial television transmission with 40 MHz or 120 MHz bandwidths. Two are narrow band transponders (5 MHz bandwidth) used for message traffic and data communications testing. The two 40-MHz transponders use the same 11.51 gigahertz frequency operating with vertical and horizontal polarization. The horizontal transponder has failed, and the remaining vertical transponder feeds an elliptical "Eurobeam," providing an EIRP of 35 dBW from northernmost Europe to northern Africa, including the offshore Canary Islands. This transponder is used for a variety of experimental services. To date, several European countries have conducted video conferences, news relays, and DBS tests, using all of the popular European television transmission standards. The 120-MHz transponders operate with a spot beam centered on Switzerland, and provide an EIRP of over 47 dBW at the bore sight. 42 to 43 dBW power levels are available throughout most of western Europe, allowing for the use of 1.8 to 3.7 meter TVRO antennas. In Great Britain, 5- to 6-foot dishes (Fig. 12–12) are quite adequate for reception of these 2 transponders, which share an 11.64 gigahertz frequency using vertical and horizontal polarization. In practice, the entire 120-MHz bandwidth is not used. Instead, the tv signals are restricted either to 27-MHz bandwidth (the DBS format), or a 40-MHz bandwidth (Eurovision format), improving the effective EIRP.

The French commercial television network Antenne-II uses the vertically polarized transponder to transmit its service to Tunisia in Northern Africa. Under pressure from other countries, the French have reluctantly scrambled the service by means of a line dicing technique. The audio signal is not scrambled. Most of the Antenne-II programming day is carried on the OTS II satellite, and can be seen throughout Europe with the help of a scrambling decoder. The French service operates in standard SECAM.

The horizontal transponder is presently used by a variety of organizations during the day, and in the evening is turned over to British Telecoms, which leases it to Satellite Television, Ltd., the European "Super Station" (brainstorm of ex-Thames tv producer Brian Haynes). Satellite Television transmits a PAL-TV signal scrambled using the Oak Orion system, in which the standard sync pulses are eliminated, and replaced by a "sync sine wave" located in the midband of the video channel. The audio program is also scrambled using a digitized "sound in sync" technique.

The Satellite Television service presents several hours per night of commercially supported entertainment to cable companies scattered throughout Europe, including Switzerland, Finland, Norway, and Malta (Fig. 12–13). In addition, apartment complexes and hotels in several countries are licensed to receive Super Station Europe, and the company is negotiating with a number of other countries, including Spain and Germany, to allow the service within their borders. PAL-TV is the transmission scheme used. OTS II is also coming to the end of its effective service, having been in operation since May 1978. (OTS I was never

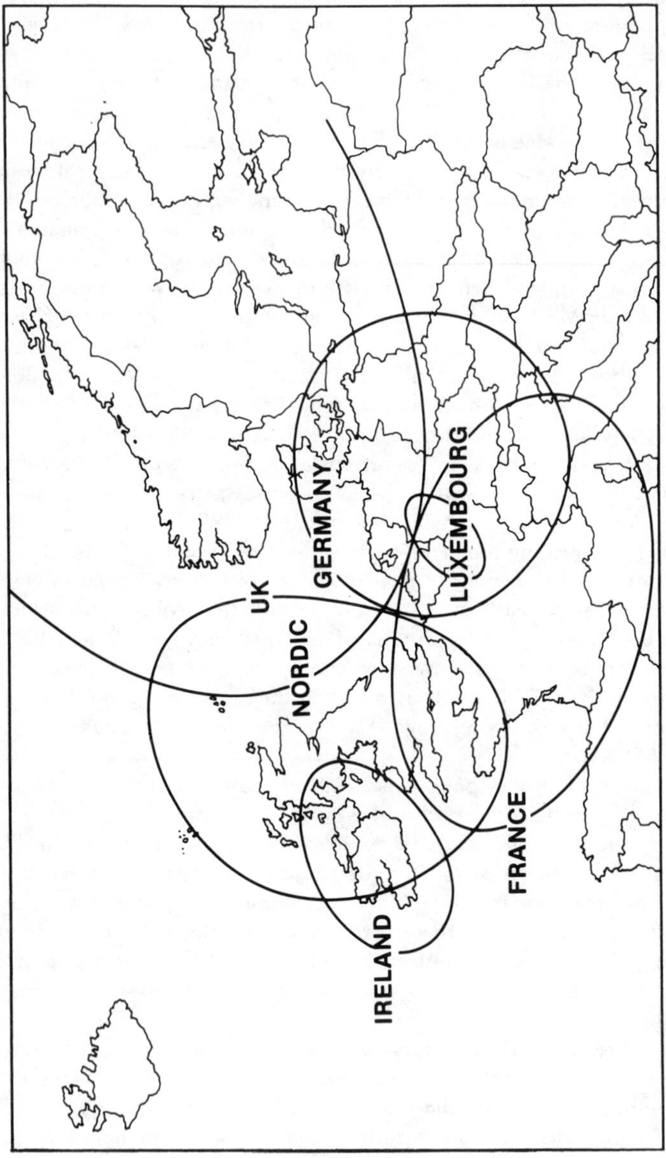

(A) Individual reception—0.9-meter antenna.

Fig. 12-10. European
(Courtesy

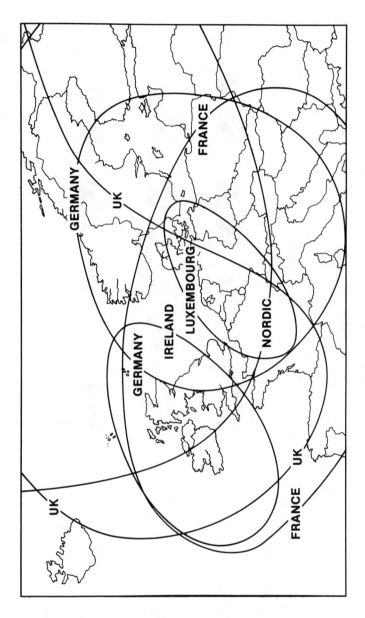

(B) Community reception—2.0-meter antenna.

DBS satellites.
WARC)

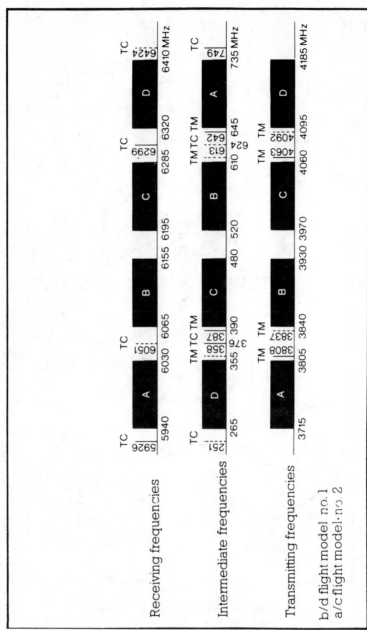

Fig. 12-11. Symphonie frequency plan.
(Courtesy European Space Agency)

Fig. 12-12. Small KU band antenna atop the London Headquarters of Satellite Television Ltd.
(Courtesy Satellite Television Ltd.)

successfully placed into orbit; it was destroyed in 1977 on launch when its Delta rocket exploded.) OTS II will be replaced by the new European Communications Satellite (ECS-1) which should be launched in early 1984. ECS uses a television format with 80-MHz channel bandwidth and 8-MHz guard bands. The ETS satellite (Fig. 12–14) will carry 12 20-watt transponders with orthogonal linear polarization (vertical and horizontal) for frequency reuse. A projected EIRP of 35 dBW for the Eurobeam pattern will provide effective coverage from Iceland to northern Africa using 2½- to 4-meter antennas. Three spot beams (Fig. 12–15) targeted on the Atlantic Ocean islands (Spot-Atlantic), western Europe (Spot-West), and eastern Europe (Spot-East), will provide an EIRP of over 42 dBW, allowing the use of 1½- to 3-meter antennas in television service, and 4.5-meter antennas in high speed digital TDMA message traffic. As many as seven transponders may be reserved for television use with the higher-frequency transponders assigned to cable television service.

ECS is the last of the interim DBS satellites operating in the 10.95 to 11.70 gigahertz band, which is primarily reserved for fixed satellite serv-

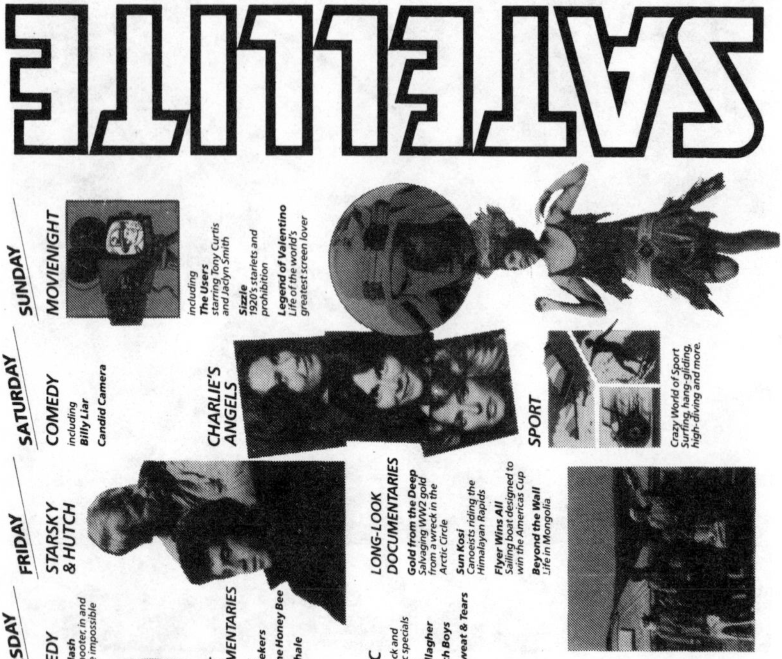

SATELLITE

(Times shown are U.K. Time)

MONDAY

19.00 SKYWAYS — Drama series set in a Pacific airport with tensions, disasters, intrigue

19.45 BIG LEAGUE SOCCER — Highlights from the week's best matches

TUESDAY

19.00 MUSIC

19.30 MOVING ON — Trucking across America

WEDNESDAY

19.00 COMEDY — Please Sir

ENTERTAINMENT — Wolfman Jack

SKYWAYS

THURSDAY

19.00 COMEDY — Monty Nash Troubleshooter: in and out of the impossible

SHORT DOCUMENTARIES — including Thrill Seekers, Life of the Honey Bee, Killer Whale

MUSIC — Best of rock and pop music specials including Rory Gallagher, The Beach Boys, Blood, Sweat & Tears

FRIDAY

STARSKY & HUTCH

LONG-LOOK DOCUMENTARIES
Gold from the Deep — Salvaging WW2 gold from a wreck in the Arctic Circle
Sun Kosi — Canoeists riding the Himalayan Rapids
Flyer Wins All — Sailing boat designed to win the America's Cup
Beyond the Wall — Life in Mongolia

SATURDAY

COMEDY — including Billy Liar, Candid Camera

CHARLIE'S ANGELS

SPORT — Crazy World of Sport Surfing, hang-gliding, high-diving and more.

SUNDAY

MOVIENIGHT — including
The Users starring Tony Curtis and Jaclyn Smith
Sizzle 1970's starlets and prohibition
Legend of Valentino Life of the world's greatest screen lover

20.30 NEWS/CURRENT AFFAIRS — U.P.I.T.N. A weekly look at current affairs

UPITN World-wide Television

Times: 19.00 · 19.30 · 19.45 · 20.00 · 20.30 · 21.00

Fig. 12-13. Program schedule for "Super Station Europe."
(Courtesy Satellite Television Ltd.)

256

Fig. 12-14. The ECS satellite.
(Courtesy Eutelsat)

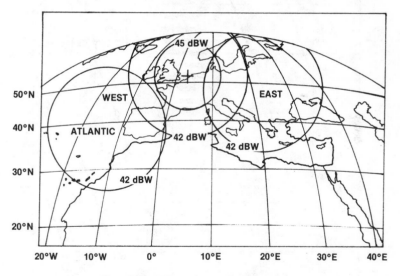

Fig. 12-15. ECS spot beams (footprint map).

ices (FSS) used in common carrier message traffic transmission. The true European DBS satellites will operate in the 11.7 to 12.5 gigahertz band. This is close enough to enable reception of both the FSS and DBS satellites, with minor retuning, on the same TVRO receiver.

257

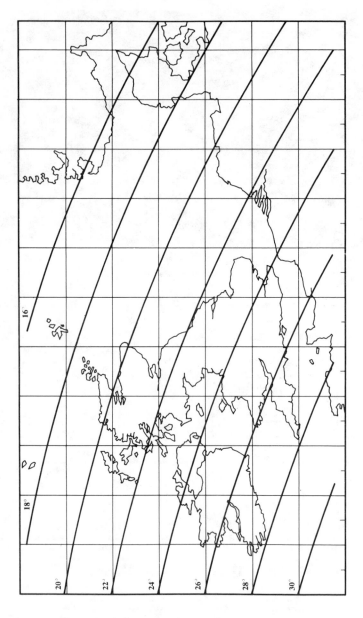

Fig. 12-16. Angle of elevation in the United Kingdom of a geostationary
satellite at 31° west.
(Courtesy The Hunt Report)

Table 12–9. DBS Orbital Positions, Channel Assignments and Polarizations for Countries of Western and Southern Europe

Orbital position	Lower half (11.7–12.1 GHz)		Upper half (12.1–12.5 GHz)	
	Left-hand circular polarization	Right-hand circular polarization	Left-hand circular polarization	Right-hand circular polarization
37° West	Andorra 4, 8, 12, 16, 20	San Marino 1, 5, 9, 13, 17 Lichtenstein 3, 7, 11, 15, 19		Monaco 21, 25, 29, 33, 37 Vatican 23, 27, 31, 35, 39
31° West	Portugal 3, 7, 11, 15, 19	Ireland 2, 6, 10, 14, 18 United Kingdom 4, 8, 12, 16, 20	Iceland 21, 25, 29, 33, 37 Spain 23, 27, 31, 35, 39	
19° West	West Germany 2, 6, 10, 14, 18 Austria 4, 8, 12, 16, 20	France 1, 5, 9, 13, 17 Luxembourg 3, 7, 11, 15, 19	Switzerland 22, 26, 30, 34, 38 Italy 24, 28, 32, 36, 40	Belgium 21, 25, 29, 33, 37 Netherlands 23, 27, 31, 35, 39
5° East	Finland 2, 6, 10 Norway 14, 18 Sweden 4, 8 Denmark 12, 16, 20	Turkey 1, 5, 9, 13, 17 Greece 3, 7, 11, 15, 19	Nordic* 22, 24, 26, 28, 30, 32, 36, 40 Sweden 34 Norway 38	Cyprus 21, 25, 29, 33, 37 Iceland† 23, 27, 31, 35, 39

*Eight channels in a wide beam covering the Nordic countries: these are assigned to Finland (22, 26), Sweden (30, 40), Denmark (24, 36) and Norway (28, 32).

†Beam covers Iceland, the Azores and part of Greenland. Channels 27 and 35 registered under Denmark.

Each European country has ambitious plans for its DBS satellite service. The British, assigned the orbital slot at 31° W longitude, have been allocated channels 4, 8, 12, 16, and 20, using right-hand circular polarization. Fig. 12–16 shows the angle of elevation in the United Kingdom for the satellite. Their "Hailey" satellite will carry 2 BBC channels providing a scrambled pay-tv service and a free "Best of Britain" channel for national and international broadcast. The Independent Broadcast Authority (IBA) and other British commercial programmers, including Satellite Television, have also announced plans to use the British DBS bird.

The Federal Republic of Germany has been allocated five channels on the DBS satellite to be positioned at 19° W longitude. Known as TV-SAT A3, this DBS service may be the first operational in Europe, with a tentative launch date in the summer of 1985. The French will follow with the TDF-1 satellite also to be located at 19° W longitude, and scheduled for launch in late 1985. Table 12–9 lists the European DBS channels assigned to the various countries.

By the end of the 1980s, over 30 DBS satellites should be in operation worldwide. They will broadcast on several hundred new national and international television channels from dozens of countries, and in almost a hundred languages. All this, directly broadcast to tiny rooftop antennas.

CHAPTER 13

Those Other TV Channels: STV, LPTV, and MDS

The 1980s are the decade of the tv revolution. In addition to the television satellites which can bring dozens of channels to one's backyard via a TVRO earth station, new types of earthbound television stations are also springing up throughout the land. Subscription tv broadcasting (STV) stations are turning on monthly. New low power, or LPTV, mini-broadcasters are springing up in the thousands. Multipoint distribution services (MDS) operate one or more private channels in every major American city. Four, eight, and twelve-channel "wireless cable" MDS systems are now being built.

Each of these services is unique, using different technologies and different television frequencies to deliver video signals. All use local transmitters whose range is limited to a 40-mile radius or less. Their transmitters operate with power levels significantly higher than the five-to eight-watt output of a satellite transponder to provide service to a community.

The cost of equipment, a direct function of the antenna size and the complexity of the system, is far lower for these terrestrial television services. Most of them operate in a pay-television scramble mode. In a number of states, laws have been passed to make it illegal to watch the scrambled television pictures on bootleg equipment without paying the monthly subscription fee.

Significant legal differences exist between watching satellite television programming direct from space and the unauthorized viewing of these private terrestrial tv channels. Appendix F discusses more fully the legal subtleties of the issues. SPACE, the Society of Private and Commercial Earth Stations is an active and powerful home TVRO satellite lobby with thousands of members who support the rapidly growing field of home satellite television. SPACE has taken an active stand against the viewing of local pay television programming on an unauthorized basis. But, how these other services work is of interest because of their differences and similarities to satellite television technology.

THE SUBSCRIPTION TELEVISION (STV) STATIONS

Commercial television was introduced in the United States in the 1940s. By the early 50s, pioneers like Ike Blonder were experimenting with scrambled over-the-air pay-television stations. The mid-50s saw several prototype subscription tv stations providing commercial-free tv shows and programs to narrowly targeted audiences in New York, Connecticut, and elsewhere. These first efforts were a disaster. A combination of poor programming, unreliable technology, and pay-per-view billing difficulties coupled with an effective "free-tv" lobby buried the STV idea for over two decades.

By the end of the 70s, HBO's satellite-delivered pay movie service had established a precedent in the cable television marketplace. STV soon followed, as ON-TV was fired up by Oak (the CATV equipment manufacturer) in Los Angeles. SelecTv™ followed a year later with a second Los Angeles STV station, Wometco began operations in New York, and other entrepreneurs jumped on the STV bandwagon. Over 100 STV stations are now either in operation or have applied for licenses nationwide. These stations are capable of reaching over 100 million people. The original ON-TV now has almost half a million subscribers in Southern California who pay over $20 per month to see first-run commercial-free movies, and SelecTv™ is following close on their heels.

The STV operator buys first-run movies from the major motion picture studios, and runs these films over a conventional uhf television broadcasting station. The station itself may or may not be owned by the STV operator. The programming is transmitted in an encrypted mode using video and audio scrambling techniques which are FCC-type approved. About a dozen such proprietary scrambling systems have been approved to date. Most suppress the vertical and/or horizontal synchronization information present in the video signal as shown in Fig. 13–1, preventing the ordinary television set from properly "locking onto" the video picture. The scrambling systems also use a "barker channel" to provide an audible announcement that invites the casual tv viewer tuning across the channel to phone in and subscribe to the movie service. The program audio track is hidden on a higher-frequency subcarrier which is ignored by the sound demodulator on the television set. When an STV subscriber orders the service, an STV decoder (Fig. 13–2) is installed between the antenna and television set that restores the video synchronization information, and shifts the hidden program audio channel back to the proper subcarrier frequency.

Subscription fees for STV services run between $15 and $25 per month; this includes both the monthly programming charge and the decoder unit. Many of the scrambling schemes are quite sophisticated, allowing multiple levels or tiers of programming to be seen at various times throughout the programming calendar. By using microprocessor-

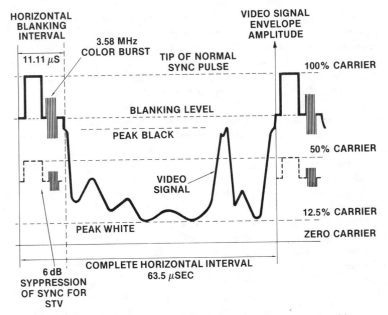

Fig. 13-1. Simple sync suppression used by the STV operators to scramble their over-the-air tv signals.

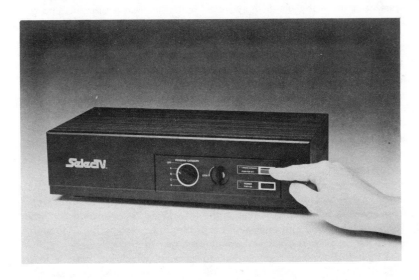

Fig. 13-2. A modern pay-tv decoder used by a Los Angeles STV system.
(Courtesy SelecTv)

equipped decoders with internal electronic serial numbers, the STV station can turn a subscriber box on and off from the broadcasting studio, depending upon the service ordered. This allows for the transmission of pay-per-view events like the boxing matches, as well as a second adult-movie tier broadcast in the early morning hours. Most STV operators charge from $5 to $10 more per month for the second tier, and from $10 to $25 for each special event presented.

STV stations are broadcasting facilities, not common carriers; therefore, the rules and regulations issued by the FCC Broadcast Bureau apply. Although these television pictures have been made secure, the schematics for the decoder boxes are not. A number of organizations provide construction plans and information on how to build decoder boxes for most of the scrambling schemes in the country today. Electronic hobbyist magazines regularly publish articles on how the systems work and on how to build decoders. Their classified advertisement sections are loaded with announcements on how to obtain schematics, constructions, plans and even complete operating "bootleg" decoder systems.

Because the STV technology is proprietary, these bootleg operations are illegal and their use inevitably ruins the market, raising the costs to the legitimate STV subscribers. STV stations now rent their decoder boxes on a monthly basis as part of the overall service. Future FCC regulations will probably allow home users to build their own decoder systems if they pay the STV operator a monthly programming charge. Ultimately, however, the battle will be won in the marketplace through the introduction of highly secure military-grade cryptographic equipment. One such STV box manufacturer, Telease of Los Angeles, produces a system which digitally encodes the audio program track, making it effectively impossible for the bootleg manufacturer to recover any audio sound. While this may not be of primary concern when watching a fight or adult movie through a pirate decoder, watching a first-run movie without its sound track should drive 90% of the potential bootleg buyers out of the marketplace.

LOW POWER TELEVISION (LPTV) STATIONS

The television station "sleeper" of the decade is LPTV. By the time that this service is fully implemented in the late 1980s, a revolution will have occurred in commercial television programming, and the Big 3 networks will have long since disappeared as we know them today.

LPTV is a spin off of the tv translator service, which has been in existence since the early 1950s. Television translators are extremely low power television receiver and transmitter combinations. Translators pick up distant city vhf television channels and rebroadcast them to local communities on uhf channels. They are found in rural and suburban communities throughout the United States and Canada, and are used, for example, to deliver a signal to the isolated valley behind the nearby mountain. Over 5000 translators are in operation in the United States, transmitting with power levels of 1 to 10 watts over areas of one to three miles in radius.

On October 17, 1980, the FCC announced the creation of a new low powered tv service. Final rules issued in the spring of 1982 allow the existing translators to immediately upgrade their status to LPTV operation, permitting a power output of up to 1000 watts. This effectively increases the range of an LPTV station to up to 20 miles or more. LPTV applicants can originate programming from their own studios, or receive programming from the television satellites or a distant television station and rebroadcast it.

The response to this new television service was overwhelming. By April 9, 1981, over 5000 applications had been received in Washington, and the FCC was forced to place a freeze on further applications. Every type of organization had applied. Nonprofit, educational, community service programming groups applied in mass. KUSK-TV, a Prescott, Arizona country and western affiliate of the Allstate Venture Capital Corporation (Sears & Roebuck) filed to own and operate 140 new television stations throughout the country. These LPTV stations would receive their programming via a TVRO satellite terminal from the new Arizona superstation, instantly establishing a new fourth national television network. Many applicants intend to operate STV pay television services, and several proposed networks plan to utilize a central computer facility to address and command the decoder boxes of up to several million STV subscribers scattered throughout the country.

LPTV is opening new vistas for satellite-based television networks and innovative programmers. By the mid-1980s, dozens of new specialty television networks will be distributing their satellite television feeds to LPTV transmitters in every town and village in America. Many of these new services will be commercially sponsored; some will require subscriptions to underwrite their operations. By 1986, STV, LPTV, and the new "minicable" MDS operations will create demand for at least 50 new television networks beaming their programming earthward via communications satellites.

MULTIPOINT DISTRIBUTION SERVICE (MDS)

Multipoint distribution service (MDS) is a local microwave television service created by the FCC in 1970. MDS utilizes a reserved piece of the microwave spectrum located at about 2 GHz. The MDS operator is authorized to use these frequencies to transmit data communications and television programming within the community. The MDS license also provides for a narrow-bandwidth "reverse channel" which can transmit information back from the subscriber's home to the central MDS transmission point.

MDS is a common carrier, not a broadcasting service. The two MDS channels (2150–2156 MHz and 2156–2162 MHz) are not considered public frequencies. The MDS signal is broadcast in standard NTSC vestigial-sideband video and fm audio format—but with a twist. The transmission passband is inverted, resulting in lower vestigial-sideband video with the program audio carrier located 4.5 MHz below the video carrier, which is in turn located 1.25 MHz below the upper band edge. This transmission scheme is a mirror-image reversal of the standard television broadcast, and the MDS audio and video can be readily inverted and restored to a normal signal by a small MDS down converter connected to a specially tuned MDS antenna at the home, or apartment (Fig. 13–3).

Another group of 31 MDS channels known as the Instructional Fixed Television Service (IFTS) provides a standard NTSC-compatible color transmission in the normal vhf/uhf tv broadcast format. These channels are located between 2.5 and 2.69 GHz, and have typically been used by schools and universities to deliver four or more simultaneous programming feeds, broadcast from a common transmit antenna, to classrooms and campuses geographically disbursed throughout a region. The IFTS service is being restructured, allowing the creation of dozens of new commercial MDS channels in each city overnight. This opens up the exciting possibility of "wireless cable": four to twelve channel local, scrambled, pay-TV systems which will deliver their feeds to the subscriber's home by radio rather than cable transmission. Major players like CBS and Microband Corp. have announced plans to build *their own* 5 to 8 channel nationwide wireless cable system, and an 8-channel service is now operating experimentally in Salt Lake City.

MDS was a narrowcast service designed to reach selected target audiences. Broadcast television is a mass communications service. Because of these original philosophical differences, MDS has historically been considered a common carrier service, but this view is rapidly changing at the FCC. By the late 1970s, MDS was used primarily as a subscription television programming service, bringing HBO movies inexpensively and efficiently to hotels, condominiums, and multiple-family dwellings. A single MDS transmitter located on top of the tallest building in the area could easily cover a 30-mile radius. The MDS programmers now use this system to deliver pay television directly to small parabolic and shotgun-like antennas for home reception.

MDS service operates line-of-sight, and only authorized subscribers are allowed to view its programming. MDS pay television fees are usually lower than a comparable STV channel's since the MDS transmitter output power ranges from 10 to 100 watts, compared to the hundreds of thousands of watts that a conventional STV uhf station requires. An

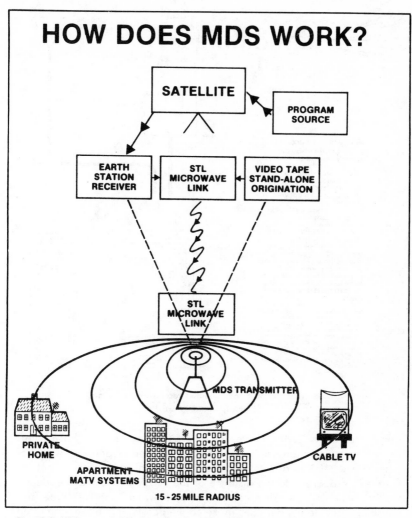

HOW DOES MDS WORK?

Fig. 13-3. MDS uses a 2-GHz frequency band to transmit pay-tv, movies, and entertainment throughout a region.
(Courtesy Microband Corp of America)

MDS transmitter costs between $25,000 and $100,000, a fraction of the equipment expense for a uhf television station.

These costs differences have made MDS an attractive vehicle for pay movie programmers. MDS transmission costs are not related to the value of a broadcasting station air time, which is defined by the amount of money a minute of commercial advertising will bring. MDS air time is available from the common carrier which owns and operates the MDS

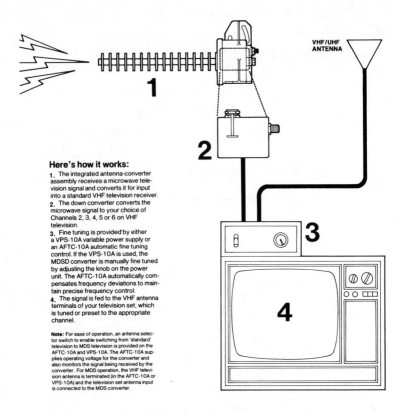

Here's how it works:

1. The integrated antenna-converter assembly receives a microwave television signal and converts it for input into a standard VHF television receiver.

2. The down converter converts the microwave signal to your choice of Channels 2, 3, 4, 5 or 6 on VHF television.

3. Fine tuning is provided by either a VPS-10A variable power supply or an AFTC-10A automatic fine tuning control. If the VPS-10A is used, the MDSD converter is manually fine tuned by adjusting the knob on the power unit. The AFTC-10A automatically compensates frequency deviations to maintain precise frequency control.

4. The signal is fed to the VHF antenna terminals of your television set, which is tuned or preset to the appropriate channel.

Note: For ease of operation, an antenna selector switch to enable switching from 'standard' television to MDS television is provided on the AFTC-10A and VPS-10A. The AFTC-10A supplies operating voltage for the converter and also monitors the signal being received by the converter. For MDS operation, the VHF television antenna is terminated (in the AFTC-10A or VPS-10A) and the television set antenna input is connected to the MDS converter.

Fig. 13-4. An MDS roof-mounted receive-only antenna.
(Courtesy Conifer Corp.)

transmitter under tariffed rates of only $4000 per month (base rate). Some tv stations charge this much for one minute of prime commercial time. Since MDS is a wireless medium, transmission costs do not increase in proportion to the distance of the subscriber to the "headend" (unlike CATV). Areas which cable companies would not likely wire (because low population densities make cabling costs prohibitive) can often be profitably served by MDS. The MDS marketing company (the organization which rents the MDS channel from the common carrier) sells the MDS pay tv service to the home subscriber. The MDS operator installs a small directional microwave antenna (Fig. 13—4) and down converter on the subscriber's premises. The cost for these receiver systems (Fig. 13—5) used to run several thousand dollars apiece, but like the satellite TVRO earth terminal costs, MDS antennas have collapsed in price over the past several years. It is now possible for the MDS operator to purchase a complete antenna/down-converter package for under $50.

Fig. 13-5. The MDS television system.
(Courtesy Standard Communications)

MDS has exploded. Dozens of MDS common carriers lease their transmission facilities to pay-television marketers and programmers throughout the country. The largest common carrier, Microband Corporation of America, owns MDS transmission facilities in over 50 top markets, and they only began operation in 1974. In fact, most MDS systems are under 5 years old. Many MDS systems now have between 10,000 and 30,000 subscribers, each paying $15 to $25 per month. Like the STV television service, dozens of entrepreneurial manufacturers have sprung up to produce illegal MDS receiving equipment that sells for between $150 and $500. With colorful names like "Bootleg TV," and "Pirate TV" their ads (Fig. 13—6) inundate the back pages of the popular hobbyist electronics publications. They have capitalized upon the strong American belief that any radio signal radiated (trespassing) onto the user's property is his to do with as he wishes. Thus, tens of thousands of people have purchased unauthorized MDS antenna units to pick up the unscrambled MDS signals. Because of recent court rulings at both the

Fig. 13-6. Typical ad from a bootleg microwave antenna sales company.
(Courtesy Pirate TV)

federal and state levels, however, it is usually unlawful to both manufacture MDS receiving systems and to use them even in the privacy of
one's own home. Watching MDS with a pirate antenna is not at all the
same as legally viewing home satellite television via a backyard TVRO
earth station.

CHAPTER **14**

DBS and You: Direct Broadcast Satellites and the Future of Satellite Television

THE SATELLITE TELEVISION CORPORATION

The new satellites are coming. On December 17, 1980, Comsat, through its subsidiary the Satellite Television Corporation, applied on FCC Broadcasting Application Form 301 to construct and operate a new television broadcasting station. What made this application unique, however, was that the transmitter would be located in the Clarke geosynchronous orbit. The new tv service would use high-power satellite transponders operating in the 12-GHz Ku band for direct transmission to every point in the continental United States. These tv signals would be picked up by special microwave receivers and miniature parabolic dish antennas to be installed by the Satellite Television Corporation on rooftops nationwide. Fig. 14—1 diagrams such a system.

Comsat had budgeted initial working capital of $625 million for this Direct Broadcasting Satellite System. By the time the first three-channel pay-TV scrambled-service DBS satellite begins operation in 1986, Comsat will have poured well over a billion dollars into the project.

STC plans to launch four satellites (Fig. 14—2) to each feed three channels of programming to Eastern, Central, Mountain, and Pacific time zones, and to eventually add service to Alaska and Hawaii.

STC retained a nationally known marketing analysis firm to conduct an in-depth survey of prospective viewers to determine their preferences and desires. The result of the survey indicates that a balanced combination of distinctive entertainment, educational, and informational programming on a diversified multichannel schedule was desired. STC plans to provide general and timely entertainment, plays, movies, concerts, night-club shows, dance, and opera performances. Children's programming for preschool-age tots as well as teenage and adult educational programming will be provided. Broadcasting time will be made

271

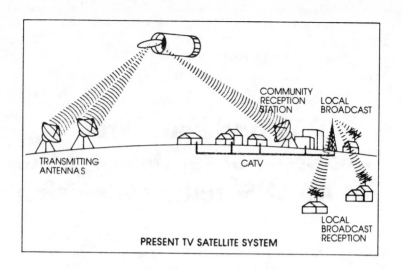

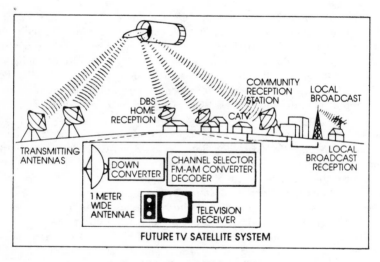

Fig. 14-1. The new DBS satellites.
(Courtesy Canadian Government Department of Communications)

available for public affairs activities as well as for special interest, minority, and religious groups.

Some programs will be sold on a pay-per-view basis; others will be provided on a per-service basis. In all cases, the DBS system will be supported directly by viewer subscriptions. The monthly subscriber will receive 24 hours per day of programming that will be unscrambled by an addressable tv set-top decoder. Through the use of an "intelligent"

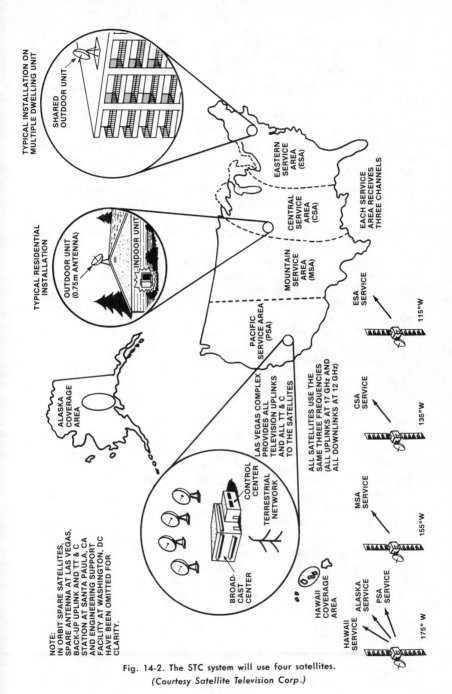

Fig. 14-2. The STC system will use four satellites.
(Courtesy Satellite Television Corp.)

273

decoder system, instant "narrowcast" markets can be created nation-wide. This will allow groups of viewers who are interested only in modern dance, or pet grooming, or medical instruction, or Japanese theater to cover the continuing production costs by subscribing to that specific subservice.

The initial DBS system will be divided into three channels. Channel A, "Superstar," will present movies, theatrical specials, concerts, and general family entertainment. Channel B, "Spectrum," will deliver film classics, variety shows, children's programming, performing arts, cultural, and public affairs events. Channel C, "Viewers Choice," will feed adult education, sports and instructional sports events, special interest programming, and pay-per-view theatrical productions and other special events.

Satellite Television Corporation has concluded that the first-run movie services have the greatest interest nationally. Approximately 82% of Channel A—roughly 138 hours per week—will consist of first-run, feature-length movies. Eight percent of Channel A's air time will present sports specials; 4% will provide "star specials," with both Broadway and other theatrical events filling up the remainder of the available air time.

Channel B's programming fare will vary from children's shows to film classics, from public affairs to performing arts. Channel B will target a wider viewing audience adding an occasional pay-per-service and pay-per-view event. During 45% of its programming "clock," Spectrum will carry classic films. Nineteen percent of the channel will be devoted to children's programming, both instructive and entertaining, with no gratuitous violence or advertising. Seventeen percent of the time the service will present performing arts and cultural events including the STC Journal for women and minorities. "Singled-out," a show discussing social problems, and "Transition," a semiweekly program on American lifestyles are among the planned STC services.

Channel C's viewers will have a significant choice of special programming including regularly scheduled lectures, discussions, regional theater presentations, social and science issues, and political debates. By counterprogramming Viewers Choice against Channels A and B, STC hopes to provide the greatest number of viewers with their first choice shows at any given time. Channel C will also carry an "On Air" magazine providing how-to information for the individual and family. Over one-third of the air time each week will be devoted to adult high school and continuing education/special interest courses. "Curtain Call," airing regional theater throughout the US and Canada, rounds out the Viewers Choice Channel.

The STC DBS service is an ambitious undertaking for Comsat, which until now has been operating solely as a common carrier, renting its transmission facilities for others to use. When fully operational, STC will be a new fifth network competing directly with ABC, NBC, CBS, and PBS. All together, STC plans to present 387 hours a week of programming; two-thirds of which will be general entertainment, 20% educational, 7% children's, and 6% public affairs programming.

The STC Direct Broadcast Satellite application produced immediate competition from other common carriers and programming companies. The FCC proceeded to issue Comsat a construction permit to build the DBS satellites, and also accepted the applications from nine other direct broadcasting satellite aspirants. These DBS entrepreneurs include: CBS, RCA, Western Union, US Broadcasting Corporation, Direct Broadcast Corp., Focus Broadcast Satellite Corp., Graphic Scanning Corp., Video Satellite Systems, and United Satellite.

Four other DBS applicants, the National Christian Network, Oak Industries/ON-TV, Satellite Syndicated Systems, and Satellite Development Trust have since applied to construct and operate US domestic Direct Broadcasting Satellite Systems in the 12.2- to 12.7-GHz band. The CBS proposal is particularly intriguing. The network plans to create a new high definition television (HDTV) DBS satellite service which will provide a picture quality equal to 35-mm film. It will require the purchase of new noncompatible wide-screen television sets to view the picture at home. Table 14—1 lists the various DBS services that have been proposed while Chart 14—1 gives the addresses of the principal companies.

The success of DBS technology depends on the players' ability to compete in the consumer marketplace with programming services which are equal or better than those the subscriber can now obtain through other distribution channels. Strangely, this rash of DBS applicants has appeared just at a time when the growth of C-band home satellite TVRO earth stations has taken off. With new "dish extender" technology bringing the size of parabolic antennas down from 10 to 8 to 6 to 4 feet, the effective arrival of DBS for the homeviewer has already occurred. The consumer will surely benefit most from the new DBS competition and the resultant increase in viewing options, but there will be a major shake-out as these industry giants battle for the best DBS orbital slots and DBS market share. This big bang of bankruptcies and consolidations should occur in the late 1980s.

HOW DBS OPERATES

Most of the DBS applications have presented proposals similar to the United Satellite (USTV) plans. Initially, USTV plans to use the Canadian Anik satellites to produce the coverage of Fig. 14—3. Then in 1984—1985, they will switch to the new G Star satellites providing coverage to 95% of the US as shown in Fig. 14—4. USTV will install a tiny roof-mounted parabolic antenna and associated electronics at the customer's

Table 14–1. Ku Band DBS Applicants in the US

Applicant	System Operation	Marketing Strategy	Programming
Satellite Television Corporation (STC)	4 satellites each with 3 channels @ 185 watts per transponder to cover 4 time zones	STC will construct and operate the satellites, and program and market the system	Pay-tv with pay-per-view and pay-per-service
Columbia Broadcasting System (CBS)	4 satellites, 1 for each time zone; three 400-watt transponders each	High-Definition Satellite Transmission Service (HDTV) to be owned and operated by CBS	Commercially supported tv plus pay-tv
RCA American Communications, Inc. (RCA)	4 satellites, 1 for each time zone; six 230-watt transponders each	To be leased by RCA on noncommon carrier basis. Will keep 1+ channel themselves	Leased science plus NBC ad-supported network
Western Union Telegraph Corporation (WU)	4 satellites, 1 for each time zone; four 100-watt transponders each	To be leased by WU on noncommon carrier basis. Use advanced Westar Satellite for first DBS	Leased service to other programmers
Graphic Scanning Corporation (GSC)	2 satellites each covering half of US, two 300-watt transponders each	To be operated and programmed by GSC itself	STV operation on 2 full-time channels
Direct Broadcast Satellite Corporation (DBSC)	3 satellites, each covering 1/3 of US. Six 200-watt transponders each	Common carrier service; will lease to others	Combination pay-tv and ad-supported programming
Video Satellite Systems (VSS)	2 satellites to cover 1/2 of US each. Two 150-watt transponders each	VSS will program and operate satellite itself	Pay-tv movie service
United States Satellite Broadcasting Corp. (USSB)	2 satellites to cover 1/2 of US each. Six 230-watt transponders each	USSB owned and operated by independent broadcasters for membership usage	Ad-supported tv

Applicant	System Operation	Marketing Strategy	Programming
United Satellite/General Instruments (USTV)	4 satellites, 1 for each time zone; three 300-watt transponders each	Owned and operated by United Satellite with programming and sales by the company	Pay-tv STV service
Focus Broadcasting Satellite Corp.	Will lease several transponders from WU on advanced Western Satellite	Regional service feeds to "Top 10" cities nationwide	Pay-tv during prime-time hours. Ad-supported otherwise

Chart 14-1. DBS Applicants With Addresses

Advance Incorporated
1835 K Street, N.W.
Suite 404
Washington, DC 20006

*CBS, Inc.
51 West 52 Street
New York, New York 10019

*Direct Broadcast Satellite Corporation
Suite 520E
7315 Wisconsin Avenue
Washington, DC 20014

*Focus Broadcast Satellite Company
Suite 825
One Commerce Plaza
Nashville, Tennessee 37239

*Graphic Scanning Corp.
99 West Sheffield Avenue
Englewood, New Jersey 07631

Home Broadcast TV Partners
5601 South 22nd Street
Milwaukee, Wisconsin 53221

National Christian Network
% Shrinsky, Weitzman and Eisen
1120 Connecticut Avenue
Washington, DC 20036

*RCA American Communications, Inc.
David Sarnoff Research Center
Princeton, New Jersey 08540

Satellite Development Trust
1106 North La Cienega Boulevard
Los Angeles, California 90069

*Satellite Television Corporation[1]
950 L'Enfant Plaza, S.W.
Washington, DC 20024

*United States Satellite
  Broadcasting Company, Inc.
(Hubbard Broadcasting)
3415 University Avenue
St. Paul, Minnesota 55114

Unitel Corporation
P.O. Box 51
Traverse City, MI 49684

*Video Satellite Systems, Inc.
29201 Telegraph Road, Suite L-8
Southfield, MI 48034

*Western Union Telegraph Company
1828 L Street, N.W.
Washington, DC 20036
or
One Lake Street
Upper Saddle River, NJ 07458

*Accepted for filing by the FCC.
[1]Construction permit to build the first DBS system has been issued.

Satellite Coverage

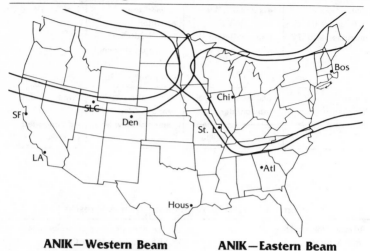

ANIK—Western Beam   **ANIK—Eastern Beam**

Fig. 14-3. Projected coverage of the USTV system using Anik satellites.
(Courtesy United Satellite)

Satellite Coverage

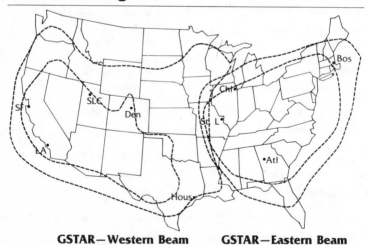

GSTAR—Western Beam   **GSTAR—Eastern Beam**

Fig. 14-4. Projected coverage of the USTV system using the GSTAR satellites.
(Courtesy United Satellite)

residence as part of the turnkey system installation (Fig. 14—5). The outdoor unit (ODU) consists of a 0.5- to 1.0-meter (2- to 3-foot) receiving antenna with microwave down-converter electronics. The indoor unit (IDU) will be connected to the ODU through up to 100 feet of miniature

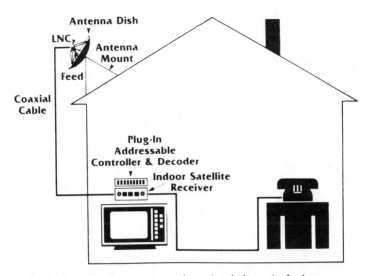

Fig. 14-5. TVRO antenna system and associated electronics for home use.
(Courtesy United Satellite)

coaxial cable. The IDU consists of a receiver and decoding electronics package with necessary cabling to connect the unit to the subscriber tv set (and optionally to a hi-fi system for reception of the stereo program audio). United Satellite (USTV) will have the first operational system using an addressable decoder electronics package which connects to the main computer via the home telephone.

In the STC system multiple dwelling units will have a single master ODU and special intermediate frequency distribution amplifiers installed to distribute programming to multiple IDUs located in each subscriber location (Fig. 14–6).

The STC business plan envisions the customer buying a basic monthly subscription television service for $15 to $20. This base fee will include access to all three DBS channels on the STC service. Additional tiers of programming will periodically be substituted in place of several of the regular programs, and these programming tiers will provide for either pay-per-program or pay-per-series billing. Specific features of the STC service include stereo sound, a second language track, closed captioning for the hearing impaired, and a teletext "newspaper of the air."

The home equipment rental for the satellite antenna and electronics will run between $10 and $15 per month, with a one-time charge of $100 for the installation of the antenna and electronics package. As an alternative, the customer will be able to purchase the hardware for a one-time cost of $500 to $1,000.

If the customer fails to continue the monthly programming payments,

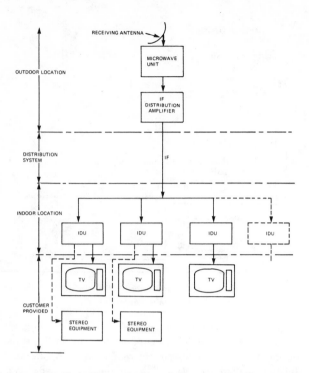

Fig. 14-6. A multifamily building using a modular approach with an i-f diatribution amplifier and a separate IDU for each home viewer.
(Courtesy Satellite Television Corp.)

the DBS computer can transmit a signal to the home terminal to deactivate the individual unit. To obtain service or equipment installation, the subscriber will probably call a toll-free 800 number, and a DBS installation service person will appear within the next 2 or 3 days. Comsat believes that everyone will eventually want to have their own DBS satellite terminal, and the company sees tv outlets being installed in each room in the same way that modularized telephone jacks are installed today. When a DBS subscriber moves from building to building or city to city, the customer can simply unplug the indoor electronics unit and take it along with the television set. The outdoor unit (antenna) will remain behind on the roof. When unpacking at the new home, the IDU can be reconnected to the new outlet for immediate restoral of television service. Unfortunately, however, the various DBS systems will most likely be incompatible with each other—at least as far as their IDU electronics.

Comsat was created by President John F. Kennedy and the US Congress in 1962 as a private corporation whose charter was to bring satellite communications to the world. The company has had an outstanding track record, and is betting over a billion dollars that this DBS enterprise is not science fiction. However, given the enormous competition of existing C-band satellite and conventional television services, as well as the new competition of other DBS operators, it will take a company with deep pockets to survive this cutthroat marketplace. If only one-half of the DBS applicants launch their own television satellites, the videophile of the late 80s will have to install seven incompatible parabolic antennas on his roof to pick up all of the competitive services. And, this does not include the eighth satellite dish—the original TVRO earth station that he installed in the early 80s to pick up the dozens of television satellite channels which are already being broadcast direct from space.

THE CANADIAN NORTHERN TERRITORIES EXPERIMENT

To see what the future might look like in the United States, one need simply look across the border to Canada. Since the late 1970s, Telesat Canada has been experimenting with a high powered, 12-GHz Anik satellite system. The Canadian Broadcasting Corporation (CBC) began using satellites to deliver low-cost network television programming directly to isolated homes in the Northern Provinces. Both French and English language programs are provided on this service, which is known as "CBC-North." Viewers use one meter (approximately 3-foot) dishes to pick up the DBS signals (Fig. 14-7).

CBC-North is a successful service. Thousands of individual homes and communities are now receiving television for the first time, watching perfect pictures coming from Montreal, Toronto, and Vancouver, eliminating what had been a profound isolation from the rest of the country.

The costs for this multichannel programming service are in the range of ordinary terrestrially based television networking. Experience shows that low cost DBS terminals can be easily installed in large quantities, and a reliable national service can be fed from a single satellite transmitting from 22,300 miles in space. Hundreds of local television broadcast translators and transmitters have been eliminated, saving a significant amount of energy. Some delivery of local news has been curtailed, however. The very nature of DBS tends to make it a national rather than a regional programming service.

Many other countries are now experimenting with direct broadcasting satellite services or are planning to construct their own DBS satellites by the late 1980s. The Japanese experimental broadcast satellite was launched in the mid-1970s. Several west European entrepreneurs are planning to deliver multination DBS television service in several languages across the political boundaries of the EEC nations. The social, cultural, and political ramifications of supranational DBS broadcasting have yet to be felt, but already a series of intensive meetings have been held among the European broadcasting authorities to attempt to come

Fig. 14-7. A one-meter TVRO dish to pick up CBC via the Anik satellite.
(Courtesy Telsat Canada)

to terms with this new and free-thinking technology. The large scale C-band domestic satellite deployment in the United States coupled with a unified political structure and common language which spans a continent, have worked to establish the 3.7-GHz satellite television service in the USA. No similar C-band domestic satellite systems exist in Europe today. European cable television systems, whose American counterparts provided the economic incentives for the existing US communications satellites, are rudimentary or nonexistent in most countries. Thus, the path is clear for the rapid development of many regional and national 12-GHz DBS satellite systems throughout Europe and the rest of the world.

The Canadian Anik Ku-band DBS transmissions can be picked up throughout Canada and the United States by many C-Band TVRO earth stations with the addition of a different low noise amplifier and down-converter combination. Several TVRO manufacturers produce stand-alone 12-GHz TVRO terminals which have been sold throughout Europe and Canada. Panasonic and NEC are now manufacturing Ku-band *consumer* TVRO systems. Anik DBS viewers using other equipment have reported excellent reception throughout Central America, Haiti, and the rest of the Caribbean, a success not envisioned by Telesat, since its service is for Canada only.

NOWHERE TO GO BUT UP

By the mid-1980s, over 50 communications satellites will be circling the globe, providing 600 or more channels of television programming. A dozen languages and as many cultures will be represented. RCA predicts that in the United States alone, more than 41 communications satellites will be operated by six or seven companies before 1990. Fig. 14–8 graphs the RCA projection. The "old line" satellite companies, Western Union, RCA, and Comsat, will be joined rapidly by AT&T, GTE, Hughes, SPC, and American Satellite. More and more networks will appear (and disappear) overnight.

As the satellites themselves get bigger, larger launching vehicles are needed. The space shuttle has arrived just in time to deliver these new monster "super satellites" into geosynchronous orbit (Fig. 14–9). By the

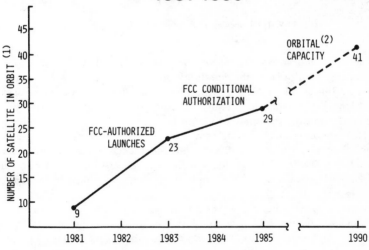

PROJECTED SATELLITES IN ORBIT
1981-1990

(1) HYBRID C/K-BAND SATELLITES COUNTED AS TWO

(2) ASSUME 3° SPACING FOR BOTH C-BAND AND K-BAND FOR ORBITAL CAPACITY

Fig. 14-8. Projected satellites in orbit—1981–1990.
(Courtesy RCA Corp.)

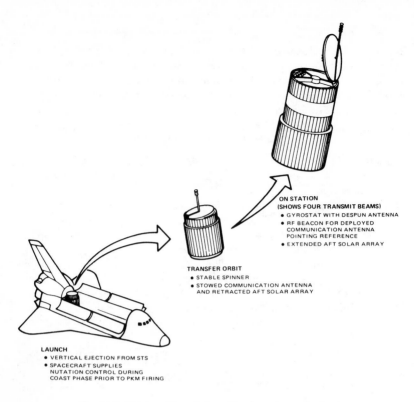

ON STATION
(SHOWS FOUR TRANSMIT BEAMS)
- GYROSTAT WITH DESPUN ANTENNA
- RF BEACON FOR DEPLOYED
  COMMUNICATION ANTENNA
  POINTING REFERENCE
- EXTENDED AFT SOLAR ARRAY

TRANSFER ORBIT
- STABLE SPINNER
- STOWED COMMUNICATION ANTENNA
  AND RETRACTED AFT SOLAR ARRAY

LAUNCH
- VERTICAL EJECTION FROM STS
- SPACECRAFT SUPPLIES
  NUTATION CONTROL DURING
  COAST PHASE PRIOR TO PKM FIRING

Fig. 14-9. The launching of a satellite from the space shuttle.

mid-1980s, the space shuttle will be operating with bi-weekly flights, delivering hundreds of thousands of pounds into space every several weeks. The shuttle vehicle can handle a satellite package which is 60 feet long by 15 feet in diameter, the equivalent of launching five Volkswagens into orbit bumper to bumper!

New rocket launching systems are also being developed by the European Space Organization (ESO) using the French Equatorial launching facility on Devils Island in the Pacific. The ESO Ariane high capacity rocket will provide a welcome and significant alternative to the space shuttle capabilities.

The ability to radiate a higher-power signal increases with the size of a satellite. As the power outputs of the satellite transponders jump, smaller and smaller TVRO antennas are needed. Satellites carrying hundreds of transponders are now on the drawing boards, some of which will be launched in pieces for final assembly in space. New television bandwidth reduction techniques will allow two or more television signals to be squeezed into one transponder, allowing hundreds of television pictures to be broadcast by a single "super-satellite." The reduction of satellite spacing in the Clarke Orbit from 4 degrees to 2 degrees, to be complete by the mid-1980s, will make room for at least 30 additional satellites which can be seen by North America.

Building bigger satellites with giant antenna arrays means that smaller uplink ground stations can also be contemplated, allowing tv pictures to be transmitted from mobile vans and moving vehicles from any city, creating an instant nationwide television relay (Fig. 14—10). Six- to ten-foot transmit antennas can beam regional sports activities and special events to independent and regionalized television networks. The modern electronic news gathering (ENG) video link will go global as instant news coverage from the actual site of an overseas story will become commonplace. Thousands of television stations and organizations will own uplink terminals, creating a new revolution in narrowcasting television programming. The Fortune-100 Companies will be running daily videoconferences between their headquarters and their regional offices.

As both the TVRO and uplink antennas shrink in size, other innovative and *ad hoc* uses of the communications satellites will emerge. Satellite technology can fulfill, for example, the promise of providing universal education to everyone. The libertarian dream of the ultimate free speech society where every culture is free to explore and to grow independently may well be achieved. Of course, the "Baskin Robbins" syndrome may surface. Given the overwhelming choice of hundreds of

Fig. 14-10. Satellite communications in the not too distant world of the future.
(Courtesy Audio Sound Magazine)

285

"flavors" of television, many people will resort to watching only one or two channels. This phenomenon may already be at work: most American television viewers watch only one or two channels for the majority of their television entertainment. CBS, ABC, and NBC will still retain their respective market shares by fragmenting into specialized "mininetworks" of sports, news, adventure, and situation comedy programming. The satellite channel capacity will certainly be there to support such an endeavor. The creation of hundreds of new specialty networks and programming services will ultimately provide the society with a giant electronic "magazine stand of the air" replacing the hardcopy Gutenberg printing press technology with a cornucopia of real time electronic images.

THE NEXT STEP

Well, you've read the whole book by now, and may be dazzled by the prospect of having your own satellite television terminal. The romance and adventure of the medium cannot be denied. Or, perhaps, you are an old hand at this sort of thing. You may have owned your own backyard dish for several years now. Possibly you have worked in the broadcasting or cable television industries since the beginning days of satellite tv way back in the early 70s. But, this field is still very young, and changing every month. Where do you go from here if you want further information about owning your own "backyard" TVRO system for fun and profit?

You might consider joining SPACE, the Society of Private and Commercial Earth Stations. The Washington, DC organization is an active supporter of the expansion of satellite television terminals for both the home user and private cable operator. SPACE's annual conference brings together satellite tv enthusiasts from around the world for several days of shop talk and basic all around fun.

Subscribe to Coop's Satellite Digest, the folksy magazine written by the industry guru, Bob Cooper, Jr. Not only does Cooper have his finger on the pulse of this industry, his fingerprints may be found virtually all over its body! Coop's Satellite Digest is (as it has been from the very first issue) a joy to read.

Orbit magazine, Channel Guide, and Channel Chart should all be subscribed to for their excellent articles and programming content. They are monthly, weekly, and bimonthly publications in respective order.

Satellite Television Technology, the organization originally founded by Robert Cooper, holds a series of conferences and trade shows throughout the country each year. You should definitely attend the next one, if at all possible.

The Satellite Center in San Francisco, California publishes "how to" manuals and management reports for private cable operators, TVRO dealers, and home satellite television enthusiasts. The Satellite Center also runs a series of seminars held periodically for people and firms who are interested in entering the satellite TVRO dealership business. Another set of seminars is held for the apartment building manager, condominium association, mobile home park owner, and private cable entrepreneur who plans to own and operate an SMATV system. These seminars are held throughout the country at various times. Further information on all of The Satellite Center services can be obtained directly from the office in San Francisco.

For further information on The Satellite Center, Coop's *Satellite Digest*, SPACE, and other organizations and magazines mentioned, simply turn to the Appendices following this chapter. Feel free to drop a note to the other pioneers in the industry and to get involved by calling the manufacturers and distributors directly. They are a friendly group of people. SPACE can help you track down some of its members; just about everybody in the industry seems to belong to it (I do). And feel free, of course, to contact me by way of The Satellite Center, PO Box 330045, San Francisco, CA 94133. I may not always be able to speak to you personally, but we always try to answer every inquiry promptly and with the answer to your question. Above all, have fun viewing the best of television programming—beamed direct from space to you!

Glossary*

Afc (Automatic Frequency Control)—A circuit used in television receivers which automatically locks onto the signal, preventing the receiver from becoming out of tune due to temperature changes or component variations.

Agc (Automatic Gain Control)—A circuit used in television receivers to automatically adjust the amplification of the receiver so that the signal will remain constant, despite atmospheric variations or signal levels which may be different from transponder to transponder.

Alignment—The process of "tweaking" an electronic circuit through the use of screwdriver-adjustable components to maximize the sensitivity and signal-receiving capability of the circuit.

Alternating current (ac)—The voltage found in one's house. Produced by a generator or "oscillator," alternating current and the corresponding voltage oscillates periodically between some maximum positive and negative value as determined by its "frequency." Ac can be used as signals which can be sent along wires or transmitted via radio over great distances.

AM (Amplitude Modulation)—The process of varying a power output of a transmitter in direct relationship to an incoming signal (either audio or video). Amplitude modulation techniques are more susceptible to impulse and natural noise (see fm modulation). In the North American television system, the sound channel is transmitted using fm, while the picture is transmitted using am. This is why during an electric storm the sound will remain static-free while the visual pictures will be broken up and distorted with each flash of lightning. Unlike terrestrial television, satellite tv systems use the frequency modulation scheme for both the sound and picture signals. The home satellite tv receiver must then convert these signals to the standard format required by the television set.

Analog—Continuously variable. Examples of signals which are inherently analog in nature include human speech, music, and television pictures. Satellite systems in use today primarily relay analog signals back to earth.

Anik—The Canadian domestic satellite system which transmits the Canadian Broadcasting Corporation (CBC) Network feeds throughout the country. Anik is also used heavily by the Canadian telephone companies to bring voice communications to major metropolitan centers and small hamlets throughout the country.

Anik 2 (retired) and Anik 3 utilize twelve transponders in the horizontal format, located at 114° west longtitude. Anik B uses twleve transponders in the horizontal format and four 12-GHz transponders especially developed for direct broadcast to homes in northern Canada; (located at 109° west longitude). Anik D with 24 transponders is located at 104.5° west longitude. All Anik satellites are operated by Telsat Canada, Ottawa.

---

*Reprinted with permission of The Satellite Center. Copyright © 1982, 1983, The Satellite Center.

Antenna—A metallic transponder connected to a transmitter which converts the high-frequency ac signals produced by the transmitter into radio and television signals which are propagated to the atmosphere or, conversely to convert the received radio and television signals into corresponding ac signals which the receiver can use. Depending on their use and operating frequency, antennas can take the form of a single piece of wire or a sophisticated parabolic-shaped dish.

AP (Associated Press)—A national news service which transmits both hourly audio news-wire feeds and photographs to newspapers and broadcast stations via satellite.

Apogee—The point in an elliptical satellite orbit which is farthest from the surface of the earth. Geosynchronous satellites which maintain circular orbits around the earth are first launched into highly elliptical orbits with apogees of 22,300 miles. When the communication satellite reaches its initial apogee, small on-board thruster rockets are fired to nudge the satellite into its permanent circular orbit of 22,300 miles. (See *Geosynchronous*.)

Attenuation—The process of lowering the signal level of an incoming signal to prevent a receiver from being overloaded. Attenuators are passive devices which usually are placed between the coaxial cable and the satellite receiver. They are also heavily used in the CATV business to compensate for excessively high level signals for those homes nearest the CATV distribution amplifier. (See *Distribution Amplifier*.)

Azimuth—The angle of rotation (left to right) that a ground-based parabolic antenna must be rotated through to point to a specific satellite in a geosynchronous orbit. (See *elevation*.) The azimuth angle for any particular satellite can be determined for any point on the surface of the earth given the latitude and longitude of that point.

Baseband—The basic direct 6-MHz output signal from a television camera, satellite television receiver, or video tape recorder. Baseband signals can be viewed only on studio monitors. To display the baseband signal on a conventional television set, a "remodulator" is required to convert the baseband signal to one of the vhf or uhf television channels which the tv set can be tuned to receive. (See *NTSC*.)

Beta Format—One of the two popular home consumer video tape recorder formats, the Beta system was developed by Sony Corporation. It is incompatible with the VHS system. (See *VHS*.)

Bird—The actual communications satellite itself, a typical "bird" weighs several thousand pounds, has an average lifetime of seven years, and is "parked" into a circular orbit at an altitude of 22,300 miles above the earth. The satellite acts as an electronic mirror, retransmitting telephone, video, and data signals fed up to it by network control centers located in the United States, and relaying these signals earth bound to other geographically scattered receive-only satellite terminals. (See *Geosynchronous*, and *TVRO*.)

Blanking (retrace)—An ordinary television signal consists of 30 separate still pictures or "frames" sent every second. They occur so rapidly, the human eye blurs them together to form an illusion of moving pictures. This is the basis for television and motion picture systems. The blanking interval is that portion of the television signal which occurs after one picture frame is sent and before the next one is transmitted. During this period of time special data signals can be sent which will not be picked up on an ordinary television receiver. (See *Teletext*.)

BNC Connector—A standard miniature twist-lock coaxial cable connector used throughout the television industry and by many satellite tv receivers. Other popular connectors include the larger PL-259 screw-in type and the smaller RCA type phono plug connectors.

Broadcast Bureau—The department of the Federal Communications Commission which issues television and radio station licenses. (See *Federal Communications Commission*.)

Broadcasting—The process of transmitting a radio or television signal via an antenna to multiple receivers which can simultaneously pick up the signal. Different from Cable Television. (See *Narrowcasting*, and *CATV*.)

C-Band—The 3.7 to 4.2-GHz frequency band used by the domestic and international communications satellites to transmit their signals down to the earth. (See *Microwave*.)

C/N (Carrier to Noise) Ratio—The ratio of the power in a satellite signal to the received noise as measured in decibels. The larger the ratio, the better the television picture looks. If C/N is below 7 dB, the picture quality is usually terrible; above 11 dB, the picture quality is excellent. (See *Threshold Extension*, and *dB*.)

Carrier—The basic radio or television transmitter center of frequency signal. The carrier is modulated by manipulating its amplitude (making it louder or softer) or its frequency (shifting it up or down) in relation to the video signal coming to the television, or audio signal originating from a microphone. Satellite carriers are frequency modulated. (See *Frequency Modulation*.)

Carrier Frequency—The main frequency on which a radio station, television station, or microwave transmitter operates. Am radio stations operate in the frequency band from 535,000 cycles per second (535 kilohertz) to 1600 kilohertz. Fm radio stations operate in the frequency band from 88 megahertz to 108 megahertz. Terrestrial television station transmitters operate in the frequency band whose channels range from 54 megahertz (channel 2) to 890 megahertz. Microwave and satellite communications transmitters operate in the band from 1 to 14 gigahertz (a gigahertz is one billion cycles per second). (See *Modulation*.)

Cassegrain Antenna—A popular antenna used for the reception of satellite television signals, the folded construction of the antenna eliminates bulky feed supports while retaining the advantages of long focal length and higher gain.

CATA—The Community Antenna TeleVision Association, a nonprofit organization composed of several thousand small cable television systems, some of who have as few as a dozen or so subscribers.

CATV—Originally meant "Community Antenna TeleVision" where independent "mom and pop" companies in rural communities would build a large television receiving antenna on a nearby mountain to pick up the weak tv signals form the distant metropolis. These signals were amplified, modulated onto television channels, and sent along a coaxial cable strung from house to house. Now standing for "Cable Television," most independent CATV companies have long since been purchased by national organizations which own multiple cable systems in rural and urban areas. (See *Multiple Service Operator*.)

CATV Convertor—A small box which sits atop the television set and is connected to the incoming CATV cable. The convertor replaces the function of the television set tuner, allowing the subscriber to select the various television channels sent over the cable.

CATV Decoder—Similar to a CATV convertor, the tv-set decoder includes an additional module which descrambles pay-television premium channels on the cable.

CCIR (Consultative Committee for International Radio)—One of the two major arms of the International Telecommunications Union, a United Nations agency located in Geneva. The CCIR developed international standards for radio, television, and satellite television transmissions. (See *CCIT*.)

CCITT (Consultative Committee for International Telephone and Telegraph)—Second major arm of the International Telecommunications Union, the CCITT developed international standards for telephone and telegraph communications including certain engineering specifications for satellite transmissions. (See *CCIR*.)

Channel—A frequency band in which a specific broadcast signal is transmitted. Channel frequencies are specified in the United States by the Federal Communications Commission. Television signals require a 6 megahertz frequency band to carry all the necessary picture detail. (Compare this with a typical high fidelity audio channel of 15,000 hertz.)

Christian Broadcast Network (CBN)—The most highly watched satellite program network, CBN is carried on the RCA Satcom 3 satellite. Now known as the Continental Broadcast Network, CBN is expanding its entertainment programming to reach more households. CBN is received by 10 million households on over 2,000 cable systems.

Circular Polarization—Unlike domestic North American satellites which utilize vertical or horizontal polarization, the international Intelsat satellites transmit their signals in a rotating corkscrew-like pattern as they are down-linked to earth. On some satellites, both right-hand rotating and left-hand rotating signals can be transmitted simultaneously on the same frequency; thereby doubling the capacity of the satellite to carry communications channels.

Clarke Orbit—That circular orbit in space 22,300 miles from the surface of the earth at which geosynchronous satellites are placed. This orbit was first postulated by the science fiction writer Arthur C. Clarke in *Wireless World* magazine in the late 1940s. Satellites placed in these orbits, although traveling around the earth at thousands of miles an hour, appear to be stationary when viewed from a point on the earth, since the earth is rotating upon its axis at the same angular rate that the satellite is traveling around the earth.

Coaxial Cable—A transmission line in which an inner conductor is surrounded by an outer conductor or shield and separated by a nonconductive dielectric, typically a foam. Coax cables have the capacity to carry enormously high frequency signals in the television range, and thus are used by catv companies for their signal distribution. Coax cables also connect the backyard satellite tv antenna to the tv set-top satellite television receiver.

Common Carrier—Any organization which operates communications circuits used by other people. Common carriers include the telephone companies as well as the owners of the communications satellites, RCA Americom Inc., Western Union Telegraph Corporation, Comsat General Corporation (which leases its Comstar satellites to AT&T), and AT&T itself.)

Common Carrier Bureau—The division of the Federal Communications Commission which regulates all activities of common carriers in the United States, including the service price "tariffs" issued by the common carriers themselves.

Communications Act of 1934—The federal law which established the Federal Communications Commission and which regulates all interstate wire and broadcast communications in the United States.

Comsat—The Communications Satellite Corporation, a New York Stock Exchange company established an Act of Congress in 1962. Comsat launches and operates the international satellites for the Intelsat consortium of countries, and through its Comsat General subsidiary, owns the Marisat series of marine communication satellites and the three 24-transponder Comstar domestic satellites leased to AT&T for domestic telephone, data, and television use.

Comstar—Three 24-transponder horizontal and vertical-polarized domestic satellites designated Comstar 1 & 2 (co-located), Comstar 3, and Comstar 4, located at 95.0° west, 86.9° west, and 127° west, respectively.

Copyright Tribunal—Established by the Copyright Act of 1976, the Copyright Tribunal collects fees from the cable television operators for their use of off-the-air broadcasting station signals. The copyright office collects these fees through a mandatory copyright license, and distributes the money to the motion picture and television producers and copyright owners.

dB (Decibel)—A mathematical ratio developed originally by Alexander Graham Bell to express logarithmically the relationship between two different levels or signals. Typically used in the measurement of the power of a signal in a channel compared with the power of the noise in that channel, a 3 dB increase in signal is equivalent to a signal which is twice as strong. A hi-fi stereo set requires a signal-to-noise ratio of at least 60 dB to be considered acceptable by the average listener. A telephone channel can have a signal-to-noise ratio of only 30 dB to be considered acceptable. The signal-to-noise ratio of a good quality television picture should be at least 45 dB.

dBm—The ratio in dB of the power of a signal as compared with a one millawatt reference power.

DBS (Direct Broadcast Satellite)—A new generation of satellites which are designed to operate in the 12-GHz range and up. Still experimental, the Canadian Anik B satellite has four 12-GHz transponders which are used for direct reception by three to five foot antennas in northernmost Canada. Higher frequency DBS satellites have the advantage of allowing smaller receive antennas to be used, but numerous difficulties need be overcome before the service can be introduced in any significant fashion. In the United States, Comsat has proposed the creation of a new DBS satellite system to be constructed and operated by its Satellite Television Corporation subsidiary in the mid to late

1980s. Although not originally intended for this use, the existing C-band communications satellites essentially function as direct broadcast satellites when their signals are picked up by an eight foot or larger backyard home satellite tv antenna.

Decibel—See *dB*.

Demodulator—A satellite receiver circuit which extracts or "demodulates" the video and audio signals from the received carrier.

Detector—A demodulation circuit used in a satellite television receiver to recover the audio and video signals from the carrier. (See *Demodulator*.)

Digital—Discreetly variable. A digital signal has, typically, two "states" corresponding to two different signal conditions. For example, slow speed computer terminals which operate over ordinary telephone lines use a "modem" (modulator and demodulator unit) to shift an audio tone between two frequencies to correspond to the "one" and "zero" computer logic states. Some satellite systems carry special digital signals for news services, etc. as well as the analog television signals. To decode these secret digital signals, a special digital demodulator must be used.

Dipole—A type of antenna. The typical rabbit ears television antenna is a simple dipole. Other dipoles are sometimes used in the feedhorn assembly of a satellite television antenna.

Direct Current (dc) —Like the current (and corresponding voltage) produced by a battery, a dc signal flows continuously in one direction on a wire, and because dc voltages do not produce sinusoidal magnetic fields, they cannot be transmitted through space via an antenna.

Dish-Stretching Technology—Two types—post receiver video processing, and i-f filtering, designed to reduce the visible "sparklie" noise present in the picture for a given antenna size, allowing a smaller antenna to be used.

Distribution Amplifier—Wide-band amplifier operating at the vhf television channel frequencies used by the CATV companies to periodically strengthen the weakened signals as they are transmitted down the CATV company's cable network. Distribution amplifiers are also used in apartment and other master antenna television (MATV) installations. Some models operate on baseband signals as well.

Distribution Center—The central point from which the television signals are distributed from a television programming organization (such as HBO) to its network stations or receivers. The distribution center for a satellite television programmer will usually consist of a bank of video tape machines, studio facilities, and ten-meter uplink (transmit) antenna for transmission of the signal directly to the desired satellite.

Dithering—The process of shifting the six-megahertz satellite-tv signal up and down the 36-MHz satellite transponder spectrum at a rate of 30 times per second (30 hertz).

The satellite signal is "dithered" to spread the transmission energy out over a band of frequencies far wider then a terrestrial common carrier microwave circuit operates within, thereby minimizing the potential interference that any one single terrestrial microwave transmitter could possibly cause to the satellite transmission.

Domsat—A domestic satellite. In the United States, there are three major domestic satellite systems: the AT&T Comstar series, the RCA Americom Satcom series, and the Western Union Westar series. Among the eight satellites, there are a total of 136 potential satellite transponders, each one of which can carry a television signal. There are over 60 transponders presently carrying video signals of one kind or other, including Picturephone Meeting Service teleconferences, network prefeeds, distribution feeds, and pay television programming.

Down Convertor—That portion of the satellite television receiver which converts the signals from the 3.7 GHz microwave range to (typically) the more readily used 70 MHz range. Down converters used to be located physically at the "front end" of the receiver, requiring bulky and expensive coaxial cable feeds from the antenna to the receiver. Newer designs have seen the down converter placed physically at the antenna itself, often in combination with the LNA, allowing miniature CATV-like coaxial cable to bring the satellite tv segment into the house.

Down Link—The frequency channel utilized by the satellite to retransmit the television signal down to earth for reception by the TVRO ground stations. (See *TVRO*.)

Earth Station—The term used to describe the combination of satellite antenna, LNA, down converter, and receiver electronics used to successfully pick up a signal transmitted by a satellite in space. Earth stations vary from 8-foot miniaturized arrangements to gigantic 90-foot Intelsat systems.

EIRP (Effective Isotropic Radiated Power)—A measure of the approximate power of a satellite television signal received on earth. EIRP is expressed in dBW or the ratio of the power of the signal as compared to a one watt reference signal. Satellite common carriers calculate EIRP signal strengths when the satellites are tested before launch. These studies are used to produce "footprint" maps showing the varying satellite signal strength expected to be received throughout the United States. At the "boresight" of a typical Domsat, usually aimed at the continental United States or "CONUS" center-country point, the satellite EIRP may be as high as 37 dBW or higher. In areas such as Southern California or Florida, the EIRP can drop to 30 dBW. There is a direct relationship between EIRP as received on Earth and required antenna size. In areas with signal strengths of 34 dBW or above, satellite television reception can be accomplished with an 8-foot parabolic antenna and a 120° LNA, a rather inexpensive combination.

Elevation—The upward tilt to a satellite antenna measured in degrees required to aim the antenna at the communications satellite. When aimed at the horizon, the elevation angle is zero. If tilted to point directly overhead, the satellite antenna would have an elevation of 90°.

FCC (Federal Communications Commission)—The federal agency established by the Communications Act of 1934 to regulate all interstate telecommunication services in the United States located at 1919 M Street, N.W., in Washington, DC. (See *Broadcast Bureau* and *Common Carrier Bureau*.)

Feed—A slang term which describes the transmission of video programming from a distribution center.

Feedhorn—A satellite tv receive antenna component which collects the weak signals from the main surface reflector and focuses these signals into the LNA detector circuits.

Feedline—The transmission line, typically coaxial cable, from the satellite antenna to the receiver.

Field Strength Meter—A test instrument used in the cable television and broadcast industries to measure the power of a signal in a transmission line or from an antenna.

FM (Frequency Modulation)—A process of varying the frequency of a sinusoidal carrier wave in relation to an incoming signal thereby modulating the carrier with the signal. When the modulated carrier is transmitted by an antenna, an fm radio or television signal results. (See *Am Modulation*.)

FM Threshold—That point at which the input signal power is just strong enough to enable the receiver demodulator circuitry to successfully detect and recover a perfect television picture from the incoming video carrier. Using threshold extension techniques, a typical satellite tv receiver will successfully provide good pictures with an incoming carrier noise ratio of 7 dB.

Footprint—A map of the signal strength showing the EIRP contours of equal signals as they cover the surface of the earth. Different satellite transponders on the same satellite will often have different footprints of signal strength. Footprint maps are available from the common carriers, and from both STT and CATA. (See Appendix E.)

Frequency—The property of an alternating current signal measured in cycles per second or hertz. In general, the higher the frequency of a signal, the smaller the required antenna, and the more susceptible the signal is to absorption by the atmosphere and physical structures. At microwave frequencies, radio signals take on a line-of-sight characteristic, and require highly directional and focused antennas to successfully see them.

Frequency-Agile—The ability of a satellite tv receiver to select or tune all 12 or 24 channels (transponders) from a satellite. Receivers not frequency-agile are dedicated to a single channel, and are most often used in the CATV industry.

Frequency Coordination—A computerized service utilizing an extensive data base to analyze potential microwave interference problems which arise between organizations using the same 4-GHz microwave band. The same C-Band frequency spectrum is used by the communications satellites and the terrestrial toll telephone network; therefore a CATV company contemplating the installation of a TVRO earth station will often obtain a frequency coordination study to determine if any potential interference might be received at the desired earth station location.

Frequency Modulation—(See *FM*.)

Frequency Reuse—A technique to expand the capacity of a given set of frequencies or channels by separating the signals either geographically or through the use of polarization techniques. Two approaches are used in the domestic communications satellites. First, the satellites are spaced approximately 4° apart in their orbits. Since the beam width of a typical antenna operating at the 3- to 4-GHz range is only about 2°, a TVRO earth station pointed at satellite one will not detect any signal from satellite two 4° away, even though the two satellites are operating in the same frequency band. Secondly, 24-transponder satellites use vertical and horizontal polarization techniques to reuse the same frequencies on the same satellite, allowing 24 television signals to be squeezed into the space normally capable of carrying only 12. (See *Polarization*.)

Gain—The level or strength of a signal, usually measured in decibels. A unit of amplitude.

Geostationary—(See Geosynchronous.)

Geosynchronous—The Clarke circular orbit above the equator. For a planet the size and mass of the earth, this point is 22,300 miles above the surface (See *Clarke Orbit*.)

Ghost—An unwanted, weaker reflected copy of the picture which interferes with the main picture itself. In vhf and uhf receptions, ghosts are caused by the receiver antenna detecting both the direct line-of-sight signal from the transmitter antenna and one or more weaker signals bounced off nearby buildings, mountains, etc.

Gigahertz (GHz)—One billion cycles per second. Signals operating above 1 Gigahertz are known as microwaves, and begin to take on the characteristics of visible light.

Global Beam—An antenna down-link pattern used by the Intelsat satellites, which effectively covers one-third of the globe. Global beams are aimed at the center of the Atlantic. Pacific, and Indian Oceans by the respective Intelsat satellites, enabling all nations on each side of the ocean to receive the signal. Because they transmit to such a wide area, global beam transponders have significantly lower EIRP outputs at the surface at the Earth as compared to a US domestic satellite system which covers just the continental United States. Therefore, earth stations receiving global beam signals need antennas much larger in size (typically 30 feet and up).

G/T (Gain over noise temperature)—A relationship, in dB, between the gain of a TVRO antenna system and the surrounding ambient noise. As this number increases, the picture quality increases. G/T can be raised by using an LNA with a lower noise temperature or increasing the size of the receive antenna.

Guard Channel—Television channels are separated in the frequency spectrum by spacing them several megahertz apart. This unused space serves to prevent the adjacent television channels from interfering with each other.

Headend—The master distribution center of a CATV system in which the incoming television signals from space and distant broadcast stations are received, amplified, and remodulated onto television channels for transmission down the CATV coaxial cable.

Heliax—A special type of thick coaxial cable often used to connect TVRO antennas to commercial satellite receivers. The cable has a low-loss transmission characteristics at microwave frequencies.

Hertz—Cycles per second. An ac signal or carrier completes a sinusoidal geometric pattern periodically in time. This phenomena is known as its frequency, and its repetition rate is measured in hertz.

Home Box Office—The original premium pay-TV cable television programmer owned by Time-Life, Inc. HBO transmits first-run movies and special shows over several domestic satellites daily.

Integrated Circuit (IC)—A miniaturized solid-state device which takes the place of thousands of discrete-component circuit elements. ICs can be fabricated to function as microcomputer circuits, stereophonic amplifiers, and television receivers.

Intelsat—The international satellite consortium which owns and operates the global satellite communications system. Intelsat is headquartered in Washington, DC. Its US member, Comsat, operates the international satellite system under management contract. Intelsat satellites carry telephone and television transmission to all parts of the world.

International Telecommunications Union (ITU)—The United Nations organization, headquartered in Geneva, responsible for all international telecommunications coordination. The ITU predates the UN by almost a century, having been founded by Napoleon in the midnineteenth century. (See *CCITT* and *CCIR*.)

Intersputnik—The Soviet Union rapidly expanding international satellite network of stationary satellites which are used to carry television and telephone communication between Communist countries.

Kelvin (K)—The temperature measurement scale used in the scientific community. Zero K represents absolute zero, and corresponds to minus 459 degrees Farenheit. Thermal noise characteristics of an LNAs are measured in Kelvins. NOTE: Unlike other temperative measurements, the word "degree" or symbol ° is not used when expressing temperatures in Kelvins.

Ku Band—The band of frequencies ranging from 11 to 14 gigahertz for use by common carriers and DBS operators.

Latitude—An angular measurement of a point on the earth above or below the equator. The equator represents zero degrees, the North Pole +90 degrees, and the South Pole −90 degrees.

License—An operating permit issued by the Federal Communications Commission to run a radio, television, or satellite transmitter. Voluntary licenses can also be obtained from the FCC by mini-CATV companies for their 15-foot or larger TVRO earth stations. While not legally required, the issuance of a TVRO license by the FCC guarantees that no future microwave transmitter can be constructed in the path of the TVRO antenna, thereby minimizing the possibility that future terrestrial interference will occur.

LNA (Low Noise Amplifier)—A very sensitive preamplifier used at the feedhorn of the TVRO satellite antenna to strengthen the very weak satellite signal. The most important parameter of the LNA is its noise figure as described in Kelvins. In general, the lower the noise figure, the better the signal quality will be. Typical LNAs have noise figures of 120 K. There is a limited but useful trade off between the noise figure of the LNA used and the size of the satellite-receive antenna. The larger the noise figure, the bigger an antenna required.

LNC (Low Noise Convertor)—A combination Low Noise Amplifier and down convertor built into one antenna-mounted package.

Longitude—An angular measurement of a point on the surface on the earth in relation to the meridian of Greenwich (London). The earth is divided into 360 degrees of longitude, beginning at the Greenwich mean. As one travels westerly around the globe, the longitude increases.

Low-Power TV (LPTV)—A new television service established by the Federal Communications Commission in October of 1980. LPTV broadcasting stations typically radiate between 100 and 1000 watts of power, covering a geographic radius of 10 to 15 miles. Upwards of 5000 LPTV stations will ultimately be licensed for operation, many of these will obtain their programming from new satellite television networks now being formed.

Master Antenna Television (MATV)—The master distribution coaxial cable television antenna system found in a modern apartment building, condominium complex, or hotel. Really a mini-CATV system in itself, an MATV system can easily be modified, with the addition of a TVRO earth station, to pick up and distribute first-run movies and other satellite programming direct to the MATV users.

MDS (Multipoint Distribution System)—A common carrier licensed by the FCC to operate a broadcast-like omnidirectional microwave transmission facility within a given city.

MDS carriers often pick up satellite pay-tv programming and distribute it via their local MDS transmitter to specially installed antennas and receivers in hotels, apartment buildings, and individual dwellings throughout the area.

MDS Pirate—Although the MDS system is similar to a broadcasting station, it has been licensed by the FCC Common Carrier Bureau. Therefore, the signal cannot legally be received and divulged without the permission of the MDS operator. A number of enterprising electronics firms have sprung up nationwide to manufacture unauthorized MDS receivers, prompting the use of the term "pirate" for the owner of such a receiver. MDS should not be confused with satellite television, which is a totally different service, falling under different regulations, and using different frequencies and technologies.

Microwave—A radio or television signal whose carrier is oscillating at a frequency of one gigahertz or higher. (See *Gigahertz*.)

Microwave Interference—Interference which occurs when a TVRO earth station aimed at a distant satellite picks up a second, often stronger signal, from a local telephone terrestrial microwave relay transmitter. Microwave interference can also be produced by nearby radar transmitters as well as the sun itself. Relocating the TVRO antenna by only several feet will often completely eliminate the microwave interference.

Modulation—The process of manipulating the frequency or amplitude of a radio or television carrier in relation to an incoming video or audio signal. (See *Carrier*.)

Modulator—A device which modulates a carrier. Modulators are found as components in broadcasting transmitters and in satellite transponders. Modulators are also used by CATV companies to place a baseband video television signal onto a desired vhf or uhf channel. Home video tape recorders also have built-in modulators which enable the recorded video information to be played back using a television receiver tuned to vhf channel 3 or 4.

Multiple Service Operator (MSO)—A major cable television owner. The top fifty MSOs own cable television franchises throughout the United States. Among the big ten are Teleprompter Corporation, Time-Life/ATC, Warner-Amex Cable Communications, Times-Mirror Cable Television (L.A. Times), Viacom International, and Westinghouse Broadcasting.

Narrowcasting—The concept of delivering a television program to a very small audience market share. By carrying twenty or more channels of television, CATV companies have begun to threaten the established television networks with their narrowcasting capabilities. LPTV stations threaten to erode the market share of the major television networks as dozens of new and alternative satellite-based television services begin in the eighties.

National Association of Broadcasters (NAB)—A trade association headquartered in Washington, DC whose members are the major radio and television stations in the United States. The NAB holds an annual convention and trade show which attracts visitors from around the world to see the latest technology of the broadcasting industry.

National Cable Television Association (NCTA)—A trade association whose members are most of the cable television companies in the United States. Headquartered in Washington, DC. Like the NAB, the NCTA also sponsors its own annual trade show and convention.

National Television Standards Committee (NTSC)—The television system in use in the United States today, in which 30 separate still pictures are transmitted per second using an electron beam to scan 525 horizontal lines per picture on the face of the picture tube. (See *PAL* and *SECAM*.)

Network—Television programming and transmission arrangement consisting of a program creation and distribution center, telecommunications channels (either terrestrial, microwave, or satellite video circuits), and viewer delivery mechanisms (typically broadcasting station affiliates or CATV companies). There are over 30 television networks in operation today whose programming can be picked up by satellite. (See Appendix B.)

Noise Temperature—(See *Kelvin*, and *LNA*.)

Occasional-Use Transponder—A satellite transponder used for television transmission on a periodic basis, typically rented by the satellite common carrier to the programmer for an hourly fee.

Offset—The C-Band satellite transponder operates in the same frequency band that the terrestrial common carrier microwave service uses. To minimize potential interference from nearby microwave transmitters, each satellite channel is "offset" in frequency from its corresponding terrestrial microwave channel.

PAL (Phase Alternation System)—A European color television system incompatible with the US television system. US and Canadian satellites utilize the NTSC color television transmission system. Intelsat satellites often use the PAL system making them incompatible with US receivers.

Parabolic Antenna—The most frequently found satellite tv antenna, it takes its name from the shape of the dish described mathematically as a parabola. The function of the parabolic shape is to focus the weak microwave signal hitting the surface of the dish into a single focal point in front of the dish. It is at this point that the feedhorn is usually located.

Pay TV—The concept made popular by Home Box Office and the CATV companies in which a special decoder box is installed adjacent to the television set to enable the scrambled pay tv channel to be watched. The viewer pays a per-month or per-movie charge for the rental of the box. Pay tv services are also provided by MDS operators and STV television stations which transmit their pay tv signals over-the-air in a scrambled mode.

Perigee—The point in an elliptical satellite orbit which is closest to the surface of the earth. (See *Apogee*.)

Picturephone®—The brand name for a videoconferencing facility operated by Bell Telephone in dozens of US cities. Comstar satellites are often used to link together the various Picturephone Meeting Centers.

Polarization—A technique used by the satellite designer to increase the capacity of the satellite transmission channels by cleverly reusing the satellite transponder frequencies. In linear polarization schemes, half of the transponders beam their signals to earth in a vertically polarized mode; the other half horizontally polarize their down links. Although the two sets of frequencies overlap, they are 90° out of phase, and will not interfere with each other. To successfully receive and decode these signals on earth, the TVRO earth station must be outfitted with a properly polarized feedhorn to select the vertical or horizontally polarized signals as desired. In deluxe TVRO installations, the feedhorn has the capability of receiving the vertical and horizontal transponder signals simultaneously, and routing them into separate LNAs for delivery to two or more satellite television receivers. Unlike the US and Canadian domestic satellites, the Intelsat series use a technique known as left-hand and right-hand circular polarization. (See *Frequency Reuse*.)

Power—The strength of a signal as measured in watts. The output power of a typical satellite transponder is only five watts, equivalent to shining a small flashlight at the surface of the earth from a point 22,300 miles in space.

Private Terminal—A television receive-only antenna system used by anyone other than a CATV system.

Program Control Tones—Special musical tones which are transmitted by a Network Control Center before and after each program. Program Control Tones are used by the CATV companies to automatically switch earth station receivers and associated equipment at their feed ends.

Programmer—A supplier of television programs transmitted via satellite. Programmers include Home Box Office, Cable News Network, Christian Broadcasting Network, and the dozens of other users of satellite transponders who provide daily video feeds to CATV companies. (See Appendix B.)

Protected-Use Transponder—A satellite transponder provided by the common carrier to a programmer with a built-in insurance policy. If the protected-use transponder fails, the common carrier guarantees the programmer that it will switch over to another transponder, often preempting some other nonprotected programmer from the other transponder.

Public Service Satellite Consortium (PSSC)—The Washington-based nonprofit organization which provides public-interest programming and video teleconferencing for interested parties nationwide.

Radio Frequency Spectrum—Those sets of frequencies in the electromagnetic spectrum which range from several hundred thousand cycles per second (very low frequency) to several billion cycles per second (microwave frequencies).

Receiver—An electronic device which detects and decodes a radio or television signal. The typical home television set is a receiver designed to detect uhf and vhf television signals and display them on a built-in television screen. The typical satellite television receiver detects signals in the microwave range which are picked up by a satellite antenna and amplified by an LNA. The satellite television receiver provides its output as a baseband video signal to a tv monitor or as a vhf television channel which can be fed into an ordinary television set.

Registered TVRO—A satellite television receive-only earth station for which a nonmandatory license has been issued by the FCC. (See *License*.)

Reuters—The British-based news service similar to the Associated Press and United Press International. Reuters operates a high-speed global stock exchange and electronic news wire system called Monitor, and transmits this signal to thousands of users throughout the United States using an entire satellite transponder on the Satcom 3 satellite.

RF Adaptor—An add-on modulator which interconnects the output of the satellite television receiver to the input (antenna terminals) of the user's television set. The rf adaptor converts the baseband video signal coming from the satellite receiver to a radio frequency (rf) signal which can be tuned in by the television set on vhf channel 3 or 4.

Satcom—The series of 24-transponder domestic satellites owned and operated by RCA American Communications Inc. Satcom 3 (known as the cable bird) is located at 131° west longtitude. Satcom 2 is positioned at 119° west longtitude. The newest, Satcom 4 is at 83° west longitude. Satcom birds operate using 12 vertically and 12 horizontally polarized transponders.

Satellite—A smaller body revolving in orbit around a larger body in space. The moon is a natural satellite of the earth. Man-made satellites travel in a variety of orbits around the earth, as the earth itself revolves on its axis. All major communications satellites are placed into geosynchronous orbits around the earth. (See *Geosynchronous*.)

Satellite Receiver—A wide-band fm receiver operating in the microwave range, converting the incoming C-Band (or Ku Band) rf signals to a standard baseband video signal.

Satellite Television Technology (STT)—The publishing and seminar organization founded by Bob Cooper, Jr., the father of the home satellite tv business. STT specializes in home satellite television information. Located in Oklahoma, STT holds home satellite tv conferences and trade shows several times a year throughout the United States.

Satellite Terminal—A receive-only satellite earth station consisting of an antenna (typically parabolic in shape), a feedhorn, a low-noise amplifier (LNA), a down converter, and a satellite receiver.

Secam—A color television system developed by the French and used in the Francophile countries and the USSR. Secam operates with 625 lines per picture frame, but is incompatible in operation with the European PAL system.

Secrecy Act—That section of the Communications Act of 1934 which protects the privacy of communications transmitted over private common carrier communications circuits, such as microwave and satellite systems. The Secrecy Act prevents an unauthorized person from receiving and divulging communications not intended for that individual. However, simple reception of the signal does not appear to violate the Secrecy Act, nor does the Secrecy Act clause apply to broadcast signals.

Signal—Information sent through electronic means from a transmitter to a receiver, either by an electric circuit or radio transmission technique.

Signal-to-Noise Ratio (S/N)—The measure in decibels (dB) of the power or strength of a signal as measured in watts compared to the power of the noise surrounding the signal.

Just as signal-to-noise ratios are important in high-fidelity stereophonic systems, so a high signal-to-noise ratio means a better, clearer, snow-free television picture.

Single-Channel-Per-Carrier (SCPC)—A special high power audio service offered by the satellite common carriers used to transmit several dozen audio signals over a single satellite transponder. Although far fewer audio channels are handled (a transponder has the capacity to carry 2000 voice conversations normally), each audio signal is transmitted with far more power, thereby allowing much smaller receive-only antennas to be used at the earth stations. Using SCPC techniques, the American news wire services are transmitting audio news feeds to radio stations nationwide equipped with antennas only three to four feet in diameter.

Slot—That longitudinal angular position in the geosynchronous orbit into which a communications satellite is "parked." Above the United States, communications satellites are, typically positioned in slots which are based at three- to four-degree intervals.

Snow—A form of noise picked up by a television receiver caused by a weak signal. Snow is characterized by alternate dark and white dots randomly appearing on the picture tube. To eliminate snow, a more sensitive receive antenna must be used or better amplification must be provided in the receiver (or both).

Society of Private and Commercial Earth Station Users (SPACE)—The Washington, DC nonprofit association of individuals and mini-CATV organizations who own satellite terminals. SPACE has been active in encouraging the backyard development of home satellite television reception.

Space Shuttle—The reusable United States space transportation system which will carry most major satellite packages into near-earth orbit over the next 20 years. Since the space shuttle only reaches an orbit of several hundred miles, an additional booster rocket system will be carried in the space shuttle hold along with the communications satellite. Upon its release from the space shuttle hold, the booster rocket will "kick" the communications satellite into its 22,300 mile geosynchronous orbit. During each space shuttle video mission, live video from the shuttle is down-linked to NASA and retransmitted via RCA Satcom Satellites for commercial news distribution.

Sparklies—A form of satellite television "snow" caused by a weak signal. Unlike terrestrial vhf and uhf television snow which appears to have a softer texture, sparklies are sharper and more angular noise "blips." As with terrestrial reception, to eliminate sparklies, either the satellite antenna must be increased in size, or the low noise amplifier must be replaced with one which has a lower noise temperature.

Spherical Antenna—The second major form of satellite television earth station antenna, the spherical antenna (unlike the parabolic antenna) has the ability to simultaneously see, and thus receive, several satellites in orbit. Because of this feature, spherical antennas are becoming increasingly popular for the home satellite tv user who may wish to switch from satellite to satellite easily and conveniently. (See *Parabolic Antenna*.)

Spot Beam—A focussed transponder pattern used by the Intelsat satellites to a limited geographical area. Spot beams are also used by the US domestic satellites to deliver certain transponder signals to Hawaii, Alaska, and Puerto Rico.

Subscription Television (STV)—A broadcasting television station transmitting pay television (usually first-run movies) in a scrambled mode. Over 50 STV stations are presently in operation or planned for cities throughout the United States.

Subcarrier—A second signal "piggybacked" onto a main signal to carry additional information. In satellite television transmission, the video picture is transmitted over the main carrier. The corresponding audio is sent via an fm subcarrier. Some satellite transponders carry as many as four special audio or data subcarriers whose signals may or may not be related to the main programming.

Superstation—A term originally used to describe Ted Turner's WTBS Channel 17 uhf station in Atlanta, Georgia. With the addition of several other major independent television stations whose signals are also carried by satellite, the term superstation has come to mean any regional television station whose signal is picked up and retransmitted by satellite to cable companies nationwide.

Synchronization (Sync)—The process of orienting the transmitter and receiver circuits so that information which is sent in relation to a precise instant of time by the transmitter will be perfectly related to that same instant by the receiver. Home television sets are synchronized by an incoming sync signal with the television cameras in the studios 60 times per second. The horizontal and vertical hold controls on the television set are used to set the receiver circuits to the approximate sync frequencies of incoming television picture and the sync pulses in the signal then fine-tune the circuits to the exact frequency and phase.

Teleconference—An electronic multilocation, multiperson conference using audio, computer or slow-scan video systems. (See *Video Conference*.)

Teletext—An "electronic newspaper of the air" consisting of several hundred "pages" of 24 character by 20 line English text displays. Teletext signals can be transmitted simultaneously over a satellite transponder using a data subcarrier. Several satellite programmers are now transmitting teletext signals, and the national television networks have begun to implement their own teletext systems. A special teletext decoder is required to capture and display the teletext pages as they are transmitted. Three world systems now exist: the British Prestel (the original), the French Antiope, and the Canadian Telidon services.

Telstar—A new generation of AT&T owned-and-operated 24-transponder satellites replacing the older Comstar satellites.

Terrestrial TV—Ordinary vhf (very high frequency) and uhf (ultrahigh frequency) television transmission limited to an effective range of 100 miles or less. Terrestrial tv transmitters operate at frequencies between 54 megahertz and 890 megahertz, far lower than the 3.7 billion hertz (gigahertz) microwave frequencies used by satellite transponders.

Threshold Extension—A technique used by satellite television receivers to improve the signal-to-noise ratio of the receiver by approximately 3 dB (50%). When using small receive-only antennas (10 to 12 feet), a receiver equipped with the threshold extension feature can make the difference between obtaining a decent picture or no picture at all.

Translator—An unattended television repeater usually operating on a uhf channel. A translator picks up a broadcast tv station from a distant city and retransmits the picture locally on another channel. (See *Low Power Television*.)

Transmitter—An electronic device consisting of oscillator, modulator, and other circuits which produce a radio or television electromagnetic wave signal for radiation into the atmosphere by an antenna.

Transponder—A combination receiver, transmitter, and antenna package, physically part of a communications satellite. Transponders have a typical output of five to 8½ watts, operate over a frequency band with a 36 megahertz bandwidth in the four to six gigahertz microwave spectrum. Communications satellites typically have between 12 and 24 on-board transponders.

TT&C (Telemetry Tracking and Command) Stations—The master earth station operated by the common carrier which monitors the on-board operation of the satellite and directs the satellite electronics and rocketry equipment. Each satellite is controlled by its own TT&C station using a separate set of telemetry and command frequencies and transponders, unrelated to the communications satellite function.

Tuner—That portion of a receiver which can variably select under user control a desired signal from a group of signals in a frequency band. By setting the tuner control to the various vhf channels, the viewer is selecting one of 12 different incoming channel frequencies which range from 54 megahertz (channel 2) to 216 megahertz (channel 13).

Turnkey—An installation of a satellite earth station in which all aspects are performed by the manufacturer or dealer.

TV Receiver—A sophisticated electronic device consisting of a vhf and uhf tuner, audio and video amplifiers, and television display monitor. The "everything in one" tv receiver will probably go the way of the hi-fi console of the 1950s, as more and more modular video equipment such as video tape recorders, home computers, projector television sets, and satellite receivers become popular.

300

TVRO (Television Receive Only Terminal)—(See *Satellite Terminal.*)

Tweeking—The process of adjusting an electronic receiver circuit to optimize its performance.

Twin Lead—The old fashioned flat antenna wire used in CATV systems in the 1950s. Twin lead has since been largely replaced with coaxial cable, capable of carrying television signals with less distortion.

Ultrahigh Frequency (UHF)—Officially the band of frequencies ranging from 300 to 3000 MHz. In television use, refers to the set of frequencies starting at 470 MHz which are designated as channels 14 through 70.

Unprotected Use Transponder—A transponder rented to a programmer on an as-is basis by the common carrier. If the transponder fails, the programmer has no recourse to demand replacement or restoral of service by the common carrier. (See *Protected-Use Transponder.*)

United Press International (UPI)—A US news agency similar to the Associated Press which operates both audio and data news wire feeds via satellite transponder subcarriers.

Uplink—That set of frequencies used by the satellite distribution center earth station to transmit the video signal up to the satellite.

Video Cassette Recorder (VCR)—Originally developed for commercial use, VCR systems have been purchased by almost four million people in the United States. There are two mutually incompatible schemes: the Beta (Sony) system, and the VHS (Japan Victor Corp.) system.

Videoconference—A teleconference using conventional B&W or color-tv systems. Often carried via satellite to multiple locations.

Vertical Blanking—(See *Blanking Signal.*)

Very High Frequency (VHF)—Officially the band of frequencies from 30 to 300 MHz. In television use, refers to that frequency band reserved for television channels 2 through 13, beginning at 54 MHz and extending up to 216 MHz.

VHS Format—One of the two most popular home video tape recorder systems, VHS is supported by over a dozen manufacturers, and has obtained approximately 75% of the market share in North America. (See *Beta Format.*)

Video Monitor—A television set without the tuner. A video monitor only accepts a baseband video signal, not a vhf or uhf television broadcast signal. Video monitors usually have higher resolution and display quality than conventional television sets.

Waveguide—A metallic microwave conductor, typically rectangular in shape, used to carry microwave signals into and out of microwave antennas.

Westar—The US domestic satellite system owned by Western Union. There are 5 Westar birds. The three original ones are 12-transponder birds, horizontally polarized. They are located at 91° West longitude (Westar 3), and 79° West longitude (Westar 1 & 2 colocated). Two new 24-transponder Westar satellites are located at 99° West longitude (Westar 4) and at 123° West longitude (Westar 5) and have become the major alternative to the RCA Satcom birds for video/CATV programming.

Wind Loading—The pressure placed upon a satellite TVRO antenna caused by the wind. Well designed parabolic antennas should be able to operate in 40-mph winds without noticeable picture degradation, and be able to withstand winds of well over 100 mph.

World Administrative Radio Conference (WARC)—An international meeting coordinated by the ITU in which the countries of the world determine which frequencies will be allocated for what services and to whom. The use of domestic communications satellites located in orbits above the United States is determined by the WARC.

The Satellite TV Networks: A Satellite-by-Satellite Listing of What's on the Birds

This Appendix is divided into three parts. In the first part, each satellite is listed along with its location (west longitude), its polarization, and a transponder-by-transponder description of each programming service.

The second part is the alphabetical listing of the major satellite programming services, including their addresses, and the principal satellites with the transponders where their feeds may be found.

The third part has been provided for the SMATV or Mini-CATV operator who wishes to obtain programming services. This section provides information on those satellite-based services that will currently sell their program products to the SMATV market, as well as the cost (when known) to buy these services.

This information was provided, in part, by *Channel Guide* and *The Channel Chart*, and was used with permission from those two organizations. The listings are correct as of August, 1983, but, of course, are subject to change literally overnight. It is a good idea to subscribe to a weekly or monthly programming channel guide. Subscriptions to both of the above publications are highly recommended. Their addresses may be found in Appendix G.

A number of different satellite transponder designation conventions are in use by the various satellite common carriers. Also, it can be somewhat confusing to figure out which transponder on which satellite is polarized vertically or horizontally, etc. Table B–1 has been provided to cross reference the US and Canadian satellites to the corresponding "Channel Numbers" which usually appear on the tuning knob of a satellite TVRO receiver. It will be useful to refer to this table as the various programming services are reviewed.

There are over 16 North American geostationary satellites carrying some kind of television programming. Some satellites, like the Comstar series, are primarily used to carry occasional video and network television feeds. Others, like Satcom 3 and Galaxy 1 may have all 24 of their transponders feeding television programming 18 to 24 hours per day.

Table B–2 has been provided to present a "snapshot" picture, in a glance, for eight of the major television satellites and the programming to be found on each transponder. Over 110 channels of satellite television are listed.

PROGRAM LISTINGS BY SATELLITE

A transponder-by-transponder description of the programming for all the major North American television satellites is given in Table B-3. This table is organized by satellite as they appear in orbit. For each satellite, the location and polarization, as well as a listing of the transponders actively carrying video along with the programming on each transponder is included. The frequency for the accompanying audio subcarrier for the video is given in parentheses ( ) with the video listing. In addition, other audio programming services, and their audio subcarrier frequencies, are included in the table.

THE MAJOR PROGRAMMERS

An alphabetical list of each of the major satellite programmers and networks using the US and Canadian domestic satellite systems is given in Table B–4. The programming service name, most frequently-used transponders, and the address (and phone number) of the programming ming organization is included. In those instances where more than one transponder is operated by the programmer, the major-feed transponder is listed first.

SATELLITE FEEDS AVAILABLE FOR SMATV SYSTEMS

An alphabetical list of those satellite programmers and services which are available for use by the stand-alone SMATV system is given in Table B–5. In many cases, no service charge is levied; for others there is a nominal per subscriber fee. The movie programmers supply their first-run movie services at wholesale rates comparable to those paid by the STV operators and cable tv systems. Refer to Table B–4 for the addresses of the program services listed here. Note that for an SMATV operator to legally obtain any satellite programming, it must sign an affiliation agreement with the satellite programmer, and pay the required fee (if any). It is quite illegal for an SMATV operator to simply take the signals off of the birds without compensating the program owner. Additional satellite programmers are extending their feeds for legal use by SMATV systems every month; although this list is accurate as of March, 1983, the SMATV operator should periodically contact each of the program suppliers listed in Table B–4 for the latest status of their services. Most of the audio and newswire programming services on the cable-oriented satellites will also supply their feeds to SMATV operators at prices ranging from 1 to 5 cents per subscriber per month.

Table B–1. North American C-Band Satellite
Frequency/Transponder Conversions

Channel or Dial Number	Uplink Frequency (MHz)	Downlink Frequency (MHz)	Transponder Designation and Polarization*				
			Satcom 1,2,3, 4 & 5	Comstar 1,2,3, & 4	Anik D	Westar 4 & 5	Westar 1, 2 & 3 Anik 3 & B
1	5945	3720	1 (V)	1V (V)	1A (H)	1D (H)	1 (H)
2	5965	3740	2 (H)	1H (H)	1B (V)	1X (V)	
3	5985	3760	3 (V)	2V (V)	2A (H)	2D (H)	2 (H)
4	6005	3780	4 (H)	2H (H)	2B (V)	2X (V)	
5	6025	3800	5 (V)	3V (V)	3A (H)	3D (H)	3 (H)
6	6045	3820	6 (H)	3H (H)	3B (V)	3X (V)	
7	6065	3840	7 (V)	4V (V)	4A (H)	4D (H)	4 (H)
8	6085	3860	8 (H)	4H (H)	4B (V)	4X (V)	
9	6105	3880	9 (V)	5V (V)	5A (H)	5D (H)	5 (H)
10	6125	3900	10 (H)	5H (H)	5B (V)	5X (V)	
11	6145	3920	11 (V)	6V (V)	6A (H)	6D (H)	6 (H)
12	6165	3940	12 (H)	6H (H)	6B (V)	6X (V)	
13	6185	3960	13 (V)	7V (V)	7A (H)	7D (H)	7 (H)
14	6205	3980	14 (H)	7H (H)	7B (V)	7X (V)	
15	6225	4000	15 (V)	8V (V)	8A (H)	8D (H)	8 (H)
16	6245	4020	16 (H)	8H (H)	8B (V)	8X (V)	
17	6265	4040	17 (V)	9V (V)	9A (H)	9D (H)	9 (H)
18	6285	4060	18 (H)	9H (H)	9B (V)	9X (V)	
19	6305	4080	19 (V)	10V (V)	10A (H)	10D (H)	10 (H)
20	6325	4100	20 (H)	10H (H)	10B (V)	10X (V)	
21	6345	4120	21 (V)	11V (V)	11A (H)	11D (H)	11 (H)
22	6365	4140	22 (H)	11H (H)	11B (V)	11X (V)	
23	6385	4160	23 (V)	12V (V)	12A (H)	12D (H)	12 (H)
24	6405	4180	24 (H)	12H (H)	12B (V)	12X (V)	

*Polarization for each transponder denoted in parentheses.
NOTE: Westar and Anik satellites use a polarization convention which is opposite to the Satcom and Comstar series. Thus, the Westar/Anik odd number transponders are horizontally polarized, while the odd numbered transponders are Satcom and Comstar satellites are vertically polarized.

Table B–2. Programming on the Eight Major Television Satellites

Transponder	Satcom 3[2] (131° W)	Satcom 4[2] (83°W)	Galaxy 1 (135° W)	Westar 4[3] (99° W)	Westar 5[3] (123° W)	Comstar 3[2] (87° W)	Anik D[3] (104.5° W)	Anik B[1] (109° W)
1	ARTS/Nick	SIN	HBO	—	Hughes TV/MSG	NBC-TV	—	—
2	PTL	FNN/Bravo	Westinghse	Hughes TV	CBS-TV	—	—	—
3	WGN-TV	SPN	HBO	—	WOR-TV	—	—	—
4	Spotlight	HSE/Dallas	(Reserved)	Hughes TV	—	Transglobal	CANCOM-KOMO-TV	—
5	Movie Ch.	ABC-TV	Times-Mir.	Bonneville	SelecTv	—	—	—
6	WTBS-TV	ESPN	SIN	XEW-TV	—	—	—	—
7	ESPN	NCN[5]	Turner	—	CBS-TV	—	—	ov
8	CBN	—	Westinghse	—	SNC (Reg)	ABC-TV	CANCOM-CHCH	—
9	USA Net	—	(Reserved)	Wold	—	—	CANCOM-WDIV-TV	—
10	Showr'm(W)	—	Times-Mir	ov	Disney (W)	CBS-TV	—	—
11	Music TV	HSE/Houston	(Reserved)	Wold/CTNA	SNC(Nat'L)	—	—	CBC(N)
12	Showr'm(E)	Playboy	Westinghse	—	Disney(E)	—	—	—
13	HBO(W)	—	(Reserved)	—	—	ABC-TV	—	ov
14	CNN	—[7]	Viacom	Hughes TV	SNC(Reg)	—	CANCOM-TCTV	—

cont. on next page

Table B-2-cont. Programming on the Eight Major Television Satellites

Transponder	Satcom 3[2] (131° W)	Satcom 4[2] (83°W)	Galaxy 1 (135° W)	Westar 4[3] (99° W)	Westar 5[3] (123° W)	Comstar 3[2] (87° W)	Anik D[3] (104.5° W)	Anik B[1] (109° W)
15	CNN2	Bisnet/TVSN	(Reserved)	PBS-A	SNC2(Ntl)	—	—	CBC(French)
16	HTN/ACSN/NJT	—	Viacom	CNN	SNC(Reg)	—	Parlim't (French)/ CBC(N)	—
17	CHN	TBN	HBO	PBS	Nash'le Network	CBS	—	CBC(ov)
18	EWTN/Reuters	TIME: Teletext	Turner	Wold	SNC(Reg)	—	CANCOM-CITV	—
19	C-SPAN	ov	HBO	Wold	—	—	—	CBC(N)
20	Cinemax (E)	—	SIN	ABC-TV	American Net	—	—	—
21	Weather Channel	RCTV	HBO	PBS	Spotlight (W)	—	CANCOM-WTVS-TV	—
22	MSN/D'Tme/ov	ABC-TV	Westinghse	ov	—	—	CANCOM-CTV	—
23	Cinemax (W)	Galavision	HBO	PBS/ Bonneville	ARTS/Daytime	—	CANCOM-WJBK	—
24	HBO(E)	NBC-TV	(Reserved)	Bonneville	BET	—	Parlim't (English)	—

[1] Twelve channel satellite, horizontal polarization.
[2] Odd-numbered transponders, vertical polarization; even-numbered transponders, horizontal polarization.
[3] Odd-numbered transponders, horizontal polarization; even-numbered transponders, vertical polarization.
[4] CBC service scheduled to transfer to Anik D in late 1983.
[5] In addition, has multiple audio subcarrier service.
[6] In addition, has audio subcarrier service for Radio America.
[7] Audio subcarrier service for CNN Radio Network.
ov = occasional video

Table B–3. Satellite-by-Satellite Programming[1]

Satellite	Trans-ponder	V/A[2]	Programming[6]
Alascom Aurora[3] (143° W)	19	V	Occasional transmissions:sporting events, news, network feeds (6.8)
	20	V	Learn/Alaska Television Network—Educational (6.8)
	21	V	Occasional transmissions:sporting events, news, network feeds (6.8)
	24	V	Alaska Satellite Television Project—Various network & independent programming (5.8)
Satcom 1[3] (139° W)	8	V	Landsat—Internal network channel (6.8) International Television—Domsat/Intelsat link channel (6.8)
	12	A	NBC Radio Network—(Pacific) (4.2) NBC Radio Network—(Central/Mountain) (4.6) NBC Radio Network—(Eastern) (5.0) NBC Radio Network—The Source (5.8/5.4) NBC Radio Network—Talk Net (5.8) NBC Radio Network—Talk Net (Spare) (6.3)
	13	V	NASA Contract Channel—Live NASA mission & mission related coverage during ongoing missions (6.8)
	20	V	Armed Forces Satellite Network—various network & independent programming (6.8)
	22	V	Hi-Net Communications Network—end-to-end videoconferencing
	23	A	ABC Radio Network—various formats (digital audio)
Galaxy 1[4] (135° W)	1	V	Home Box Office
	2	V	Group W Broadcasting Co
	3	V	Home Box Office
	5	V	Times-Mirror
	6	V	Spanish International Network (SIN)
	7	V	Turner Broadcasting System
	8	V	Group W Broadcasting Co.
	10	V	Times-Mirror

Table B-3-cont. Satellite-by-Satellite Programming[1]

Satellite	Trans-ponder	V/A[2]	Programming[6]
Galaxy 1[4] (135° W) cont	12	V	Group W Broadcasting Co.
	14	V	Viacom International
	16	V	Viacom International
	17	V	Home Box Office
	18	V	Turner Broadcasting System
	19	V	Home Box Office
	20	V	Spanish International Network (SIN)
	21	V	Home Box Office
	22	V	Group W Broadcasting Co
	23	V	Home Box Office
Satcom 3[3] (131° W)	1	V	Nickelodeon—Premium children's programming (6.8) ARTS (Alpha Repertory Television Service)—Performing and cultural arts programming (6.8)
	2	V	PTL (People That Love)—Religious programming (6.8)
		A	Satellite Radio Network—Contemporary/religious format (6.2)
	3	V	WGN-TV—Chicago independent tv station (6.8)
		A	Moody Broadcasting Network—Religious (5.47/7.92 stereo) Satellite Music Network/Starstation—Adult contemporary format (5.58/5.76 stereo) Satellite Music Network/Country—Modern country format (5.94/6.12 stereo) WFMT-FM—Chicago fine arts/classical radio (6.3/6.48 stereo) Bonneville Beautiful Music (7.38/7.56 stereo) Seeburg Lifestyle—Music (7.695) Satellite Music Network/Stardust—Traditional MOR music format (8.055/ 8.145 stereo)
	4	V	Spotlight—24-hr/day first-run movies (5.8/6.2 stereo, 6.8 mono)
	5	V	The Movie Channel—24 hr/day first-run movies (5.8/6.8 matrix stereo)
		V	WTBS-TV—Atlanta GA independent tv station (6.8)

Table B—3-cont. Satellite-by-Satellite Programming[1]

Satellite	Trans-ponder	V/A[2]	Programming[6]
Satcom 3[3] (131° W) cont	6	A	Music in the Air—Cowboy/Western (5.4/5.94 stereo) Music in the Air—Broadway/Hollywood (5.58/5.76 stereo) Music in the Air—50s/60s (6.435) Music in the Air—Comedy Specials (7.695) Music in the Air—Big Band (7.785)
	7	V	ESPN (Entertainment & Sports Programming Network)—24 hr/day sports & business news (6.8)
		A	ESPN Affiliate Information Network—ESPN program schedule information (6.2)
	8	V	CBN Cable Network—Religious/general family entertainment (6.8)
		A	Cable Jazz Network—Jazz music format (5.94/6.12 stereo)
	9	V	USA Network—Professional sporting events, Calliope, and Ovation (6.8)
	10	V	Showtime (West)—First-run movies & entertainment specials (6.8)
	11	V	MTV (Music Television)—Pop/Rock Video (5.8/6.62 matrix stereo)
	12	V	Showtime (East)—First-run movies & entertainment specials (6.8)
	13	V	HBO (Home Box Office) (West)—First-run movies, sports & entertainment specials (6.8)
	14	V	CNN (Cable News Network)—24 hr/day news (6.8)
		A	CNN Radio Network—All news radio feed (6.3)
	15	V	CNN Headline News (CNN2)—CNN news brief headline service (6.8)

Table B—3-cont. Satellite-by-Satellite Programming[1]

Satellite	Trans-ponder	V/A[2]	Programming[6]
Satcom 3[3] (131° W) cont	16	V	ACSN (The Learning Channel)—Educational (6.8) HTN Plus (Home Theatre Network)—First run G and PG movies (6.2/6.8) NJT (National Jewish Television)—religious (6.8)
	17	V	CHN (Cable Health Network)—24 hr/day health/fitness-related programming (6.8)
	18	V	Reuters Monitor Service—Commodity/stock market information (digital video) EWTN (Eternal Word TV Network)—religious (6.8) Alternate View Network—religious (6.8)
	19	V	C-SPAN—Live coverage of The House of Representatives (6.8)
	20	V	Home Box Office Cinemax (East)—Time structured HBO (6.8)
	21	V	The Weather Channel—24 hr/day national & regional weather/environment reporting (6.8)
	22	V	MSN (Modern Satellite Network) The Information Channel—General information entertainment (6.8) Daytime—Programming for women (6.8) USA Blackout Network—Substitute sports programming for regional blackout applications (6.8) HBO Promo Channel (6.8) Hi-Net Communications Network—end-to-end video conferencing (6.8) Don King Sports and Entertainment Network—Championship boxing special events (6.8) Occasional transmissions—Sporting events, news, and network feeds (6.8)
	23	V	HBO Cinemax (West)—Time structured HBO (6.8)
	24	V	HBO (East)—First-run movies, sports & entertainment specials (6.8)

Table B—3-cont. Satellite-by-Satellite Programming[1]

Satellite	Trans-ponder	V/A[2]	Programming[6]
Comstar 4[3] (127° W)	4H(8)		Oak/Telstar ON Satellite Television—First-run movies, sports, & entertainment specials (fully encrypted Oak/ Orion)
	6V(11)	V	Oak/Telstar ON Satellite Television—First-run movies, sports, & entertainment specials (fully encrypted Oak/ Orion)
	7V(13)	V	ESPN (Entertainment & Sports Programming Network)—24 hr/day sports & business news (Oak/Orion feed) (6.8)
	8H(16)	V	Occasional transmissions—sporting events, news, & network feeds (6.8)
	9H(18)	V	CMTV (Country Music Television)—Video country music (5.58 & 5.76 discrete stereo/6.8 mono)
Westar 5[4] (123° W)	1D(1)	V	Hughes Television Network—Sports events feeds (6.2/6.8) The Bluemax Theatre Channel—"x"-rated adult movies (fully encrypted/Videonet Orion) Madison Square Garden Cable Network—Live sports events from Madison Square Garden (6.8) Occasional transmissions—sporting events, news, & network feeds (6.2/6.8)
	1X(2)	V	CBS Television Network Occasional Transmissions—sporting events, news, & network feeds (6.2/6.8)
	2D(3)	V	WOR-TV—New York independent tv station (6.8)
	3D(5)	V	SelecTv—STV/SMATV feed—First-run movies, concerts, etc. (6.8)
	4D(7)	V	CBS Television Network Occasional transmissions: sporting events, news, & network feeds (6.2/6.8)
	4X(8)	V	Group W Satellite News Channel—Regional service (6.8)

Table B—3-cont. Satellite-by-Satellite Programming[1]

Satellite	Trans-ponder	V/A[2]	Programming[6]
Westar 5[4] (123° W)	5X(10)	V	The Disney Channel (West)—Premium family entertainment from the Disney studios (5.8/6.8 matrix stereo)
	6D(11)	V	Group W Satellite News Channel-1—National Service (6.8)
	6X(12)	V	The Disney Channel (East)—Premium family entertainment from the Disney studios (5.8/6.2 matrix stereo)
	7X(14)	V	Group W Satellite News Channel—Regional services (6.8)
	8D(15)	V	Group W Satellite News Channel 2—Nat'l Service (6.8) (Begins Jan. 1, 1984) TVSC (Television Syndication Center)—Vidsat/Synsat distribution (6.2/6.8)
	8X(16)	V	Group W Satellite News Channel—Backup/incoming news feeds (6.2/6.8)
	9D(17)	V	The Nashville Network—Premium country music/entertainment (5.58/5.76 stereo)
	9X(18)	V	Group W Satellite News Channel—Regional service (6.8)
	10X(20)	V	The American Network—First-run movies & medical programming (6.2)
	11D(21)	V	Spotlight (West)—24-hr/day first-run movies (5.8 & 6.2 stereo/6.8 mono)
	12D(23)	V	ARTS (Alpha Repertory Television Service—Performing & cultural arts programming (5.58 & 5.76 discrete stereo/6.8 mono) Daytime—Programming for women (6.8)
	12X(24)	V	BET (Black Entertainment Network (5.58 & 5.76 discrete stereo/6.8 mono)
Satcom 2[3] (119° W)	12	V	NBC Network Contract Channel—Live/taped network feeds (6.8)

Satellite	Trans-ponder	V/A[2]	Programming[6]
Satcom 2 cont	23	V	Time Video Information Services—Teletext testing (primary uplink feed) (6.8)
Anik 3[5] (114° W)			No Video—Telephone Message Traffic
Anik B[5] (109° W)	4(7)	V	Occasional transmissions:Sporting events, news, & network feeds (6.8)
	6(11)	V	CBC North—CBC network programming (Pacific time zone feed (6.8)
	7(13)	V	Occasional transmissions:sports events, news, & network feeds (6.8)
	8(15)	V	CBC (French Channel)—French language CBC programming (6.8)
	9(17)	V	CBC Occasional transmissions (6.8)
	10(19)	V	CBC North—CBC network programming (Atlantic time zone feed) (6.8)
Anik D[4] (104.5° W)	2B(4)	V	KOMO-TV, Seattle—ABC Network affiliate (fully encrypted/Oak Orion)
	4B(8)	V	CHCH-TV—Hamilton Ontario independent tv station (fully encrypted/Oak Orion)
		A	CKO-FM—Toronto Ontario all news & information radio (6.17 mono)
	5A(9)	V	WDIV-TV, Detroit—NBC Network affiliate (fully encrypted/Oak Orion)
	7B(14)	V	TCTV (Telemedia Communications Television)—TVA network (French) from CHLT, Sherbrooke/CFTM, Montreal (fully encrypted/Oak Orion)
		A	CKAC-AM—Montreal, Quebec—French language MOR/contemporary music (5.41 mono) CITE-FM—Montreal, Quebec—French language traditional MOR music (6.17)
	8B(16)	V	CBC Parliamentary Network (French)—Daily live coverage of the Canadian House of Commons from Ottawa (6.8)

Satellite	Trans-ponder	V/A[2]	Programming[6]
Anik D[4] (104.5° W) cont			CBC North—CBC network programming (Pacific time zone feed) (6.8)
	9B(18)	V	CITV-TV—Edmonton, Alberta independent tv station (fully encrypted/Oak Orion)
		A	CKRW-AM, Whitehorse, Yukon—"Voice of the North" MOR Music (5.41)
			CIRK-FM—(K-97) Edmonton Alberta progressive rock (6.17 mpx stereo)
	11A(21)	V	WTVS-TV, Detroit—PBS Network affiliate (fully encrypted/Oak Orion)
	11B(22)	V	BCTV (British Columbia Television)—Vancouver, B.C. British Columbia CTV network station (fully encrypted/Oak Orion)
		A	CFMI-FM—New Westminister, B.C. adult rock (6.17 mono)
	12A(23)	V	WJBK, Detroit—CBS Network affiliate (fully encrypted/Oak Orion)
	12B(24)	V	CBC Parliamentory Network (English)—Daily live coverage of the Canadian House of Commons from Ottawa (6.8) CBC (French Channel)—French language CBC programming (6.8)
		A	CFQM-FM, Moncton, New Brunswick—Uptown Country music (5.41) VOCM, St. John's Newfoundland—Adult contemporary music (6.17)
Westar 4[4] (99° W)	1X(2)	V	Hughes Television Network—Sports events feeds (6.2/6.8)
	2X(4)	V	Hughes Television Network—Sports events feeds (6.2/6.8)
	3D(5)	V	Bonneville Satellite Corporation—Occasional transmissions: Sporting events, news, & network feeds (6.2/6.8)
	3X(6)	V	XEW-TV—Mexico City network station (6.2)

Table B—3-cont. Satellite-by-Satellite Programming[1]

Satellite	Trans- ponder	V/A[2]	Programming[6]
Westar 4[4] (99° W) cont	5D(9)	V	Robert Wold Communications—Occasional transmissions: Sporting events, news, & network feeds (6.2/6.8)
	5X(10)	V	Occasional Transmissions: sporting events, news, & network feeds (6.2/6.8)
	6D(11)	V	Robert Wold Communications—Occasional transmissions: Sports events, news, & network feeds (6.2/6.8) CTNA (Catholic Telecommunications Network of America)—Religious (6.2/6.8)
	7X(14)	V	Hughes Television Network—Sporting events feeds (6.2/6.8)
	8D(15)	V	PBS (Public Broadcasting Service)— Schedule A programming (6.8)
	8X(16)	V	CNN (Cable News Network) Contract channel—Incoming regional bureau news feeds (6.2/6.8)
	9D(17)	V	PBS (Public Broadcasting Service)— Schedule B programming/occasional video conferencing (6.8)
	9X(18)	V	Robert Wold Communications—Occasional transmissions, sporting events, news & network feeds. (6.2/6.8)
	10D(19)	V	Robert Wold Communications—Occasional transmissions: sporting events, news, & network feeds (6.2/6.8) Wold Satellite Network—Syndicated programming feeds (6.2/6.8)
	10X(20)	V	ABC Network Contract Channel—Live/ taped network feeds (6.2/6.8)
	11D(21)	V	PBS (Public Broadcasting Service)— Schedule C programming (6.8)
	11X(22)	V	TSI Limited—Not in use Occasional Transmissions: Sporting events, news, & network feeds (6.2/6.8)

Table B—3-cont. Satellite-by-Satellite Programming[1]

Satellite	Trans-ponder	V/A[2]	Programming[6]
Westar 4[4] (99° W)	12D(23)	V	PBS (Public Broadcasting Service)— Schedule D programming (6.8) PBS occasional feeds (6.8) Bonneville Satellite Corporation— Occasional transmissions: Sporting events, news, & network feeds (6.28/6.8)
	12X(24)	V	Bonneville Satellite Corporation— Occasional transmissions: Sporting events, news, & network feeds (6.2/6.8)
Comstar 1/2[3] (95° W)	6H(12)	V	Occasional Transmissions—Picturephone/ teleconferencing, sporting events & net-work feeds (5.8)
	8V(15)	V	ABC Television Network—International news feeds; scheduling information (5.8)
	11H(22)	V	Occasional transmissions—Picturephone/ teleconferencing, sporting events & net-work feeds (5.8)
	12H(24)	V	Occasional transmissions—Picturephone/ teleconferencing, sporting events & net-work feeds (5.8)
Westar 3[5] (91° W)	3(5)	V	CNN (Cable News Network) Contract Channel—Incoming regional bureau news feeds (6.2/6.8)
	10(19)	V	Hughes Television Network—Sports events feeds (6.2/6.8)
	11(21)	V	Occasional transmissions: Sports events, news, & network feeds (6.2/6.8) Independent Network News (6.2) CBS Television Network Program Feeds (6.2)
	12(23)	V	Occasional transmissions: Sporting events, news, & network feeds (6.2/6.8) HSC-TV/The BAMC Hour (Brooke Army Med-ical Center—Academy of Health Science)—Medical training (6.8)
Comstar 3[3] (87° W)	1V(1)	V	Satellite Television Service—NBC regularly scheduled network programming (central time zone feed) (5.8)

Table B—3-cont. Satellite-by-Satellite Programming[1]

Satellite	Trans-ponder	V/A[2]	Programming[6]
Comstar 3[3] (87° W) cont	1V(1)	V	NBC Network Contract Channel—Live/taped network feeds (5.8)
	2H(4)	V	Transglobal Galactica—Occasional transmissions
	4H(8)	V	ABC Television Network (5.8)
	5H(10)	V	CBS Television Network (5.8)
	7V(13)	V	Satellite Television Service—ABC regularly scheduled network programming (central time zone feed) (5.8)
	9V(17)	V	Satellite Television Service—CBS regularly scheduled network programming (central time zone feed) (5.8)
Satcom 4[3] (83° W)	1	V	SIN (Spanish International Network) (6.8)
	2	V	FNN (Financial News Network)—Financial/business reporting with stock market readings (6.8) Bravo-Performing and cultural arts programming (6.8)
	3	V	SPN (Satellite Program Network)—Variety entertainment (6.8)
		A	Nationality Broadcasting Network—Multilingual music, news, and sports (6.435) Georgia State Radio Network (7.695) Rock-A-Robics—Upbeat top-40 rock for dancing and exercising (7.38/7.56 discrete stereo) Rhythm & Blues-Contemporary jazz/soul music (5.4 & 6.3 discrete stereo)
	4	V	HSE/Dallas (Home Sports Entertainment Network)—Regional cable sports network (6.8)
	5	V	ABC Television Network Remote Feeds (6.2/6.8)
	6	V	ESPN (Entertainment & Sports Network)—Remote sports events feeds (6.8)

Table B—3-cont. Satellite-by-Satellite Programming[1]

Satellite	Trans-ponder	V/A[2]	Programming[6]
Satcom 4[3] (83° W) cont	7	V	NCN (National Christian Network)—Religious (6.8)
		A	Joy Radio—Adult contemporary praise music (5.4)
			Family Radio Network (East) (5.58/5.76 stereo)
			Family Radio Network (West) (5.94/6.12 stereo)
			Astro Radio Network—Religious multidenominational talk/instructional format (6.3)
			SBN (Sheridan Broadcasting Network) (7.38/7.56 stereo)
			Blue Suede Radio Network—50s/60s classic music (7.74)
			People's Satellite Network—Ethnic programming format (8.055)
			Gold Mine Radio Network—Country & Western classics music (7.92)
	8	V	The Entertainment Channel—Not in Use
	11	V	HSE/Houston (Home Sports Entertainment Network)—Regional sports network (6.8)
			Netcom International—End-to-end video conferencing/occasional transmissions (6.8)
	12	V	The Playboy Channel—Adult-oriented entertainment & sexually oriented R-rated movies (6.8)
	15	V	Biznet (American Business Network)—Business videoconferencing (6.8)
			TVSN (Television Syndication Center)—Vidset/Synset distribution (6.2/6.8)
	17	V	TBN (Trinity Broadcasting Network)—Religious (6.8)
		A	Satellite Jazz Network—Contemporary/traditional jazz music from KKGO-FM, Los Angeles (5.58/5.76 discrete stereo).
	18	V	Time Video Information Services—Video teletext testing (turnaround uplink telechannel) (6.8)

Table B—3-cont. Satellite-by-Satellite Programming[1]

Satellite	Trans-ponder	V/A[2]	Programming[6]
Satcom 4[3] (83° W) cont	19	V	Occasional transmissions: Sporting events, news & network feeds (6.8)
	21	V	RCTV (Rockefeller Center Television)—Not in Use Occasional transmissions: Sporting events, news, & network feeds (6.2/6.8)
	22	V	ABC Television Network—Regularly scheduled network programming (Pacific time zone feeds) (5.8)
	23	V	Galavision—Premium Spanish-oriented entertainment programming (6.2/6.8)
	24	V	NBC Television Network Remote Feeds (6.8)
Westar 2[5] (79° W)	2(3)	V	Occasional transmissions—Sporting events, news, & network feeds (6.2/6.8)
	6(11)	V	Occasional transmissions—Sporting events, news, & network feeds (6.2/6.8)
	10(19)	V	Occasional transmissions—Sporting events, news, & network feeds (6.2/6.8)

[1]Courtesy "The Satellite Channel Chart," © 1983 Westsat Communications. Used with permission of the publisher.

[2]V = Video Programming; A = Audio Subcarrier Service.

[3]Odd transponders, vertical polarization; even transponders, horizontal polarization.

[4]Odd transponders, horizontal polarization; even transponders, vertical polarization.

[5]All horizontal polarization.

[6]The number in parentheses is the audio subcarrier frequency in megahertz, where two frequencies are separated by a /, stereo subcarriers are indicated.

Table B-4. Major Satellite Programmers and Networks

Name	Company	Address	Satellite/Transponder[1]
ABC	American Broadcasting Co.	1330 Avenue of the Americas, New York, NY 10019 (212) 887-7777	W4(20), F4 (5,22), D3 (8)
ACSN	Appalachian Community Service Network (The Learning Channel)	1200 New Hampshire NW, Suite 240 Washington, DC 20036 (202) 331-8100	F3 (16)
ARTS	Alpha Repertory TV Service	1211 Ave. of the Americas New York, NY 10036 (212) 944-4250	F3 (1)
BET	Black Entertainment TV	3222 N. St., NW, Suite 300 Prospect Place Washington, DC 20007 (202) 337-5260	W5 (24)
BIZNET	The American Business	US Chamber of Commerce 1615 H. St., NW Washington, DC 20062 (202) 463-5810	F4 (15)
BLUE	The Bluemax Theater Channel	3 First National Plaza, 14th Floor Chicago, IL 60602 (312) 781-9607	W4 (9)
BRAVO	Bravo	Rainbow Prog Service 100 Crossway Park West Woodbury, NY 11797 (516) 364-2222	F4 (2)
BSC	Bonneville Satellite Corp.	130 Social Hall Ave. Salt Lake City, UT 84111 (801) 237-2597	W4 (5, 23, 24)

Table B—4-cont. Major Satellite Programmers and Networks

Name	Company	Address	Satellite/Transponder[1]
CANCOM	CTV: Feeds of Canadian Stations	Canadian Television Corp. 42 Charles East Toronto, Ont, Canada (416) 928-6000	Anik D (4,8,9,13,18,21,22,23)
CBC	Canadian Broadcasting Corp.	354 Jarvis Toronto, Ont, Canada (416) 935-3311	Anik B (11, 15, 17, 19), Anik D (16)
CBN	Christian Broadcasting Network	Continental Broadcasting Network, Inc. Pembroke Four Virginia Beach, VA 23463 (804) 424-7777	F3 (8), G1 (7,18)
CBS	Columbia Broadcasting System	1211 Avenue of the Americas New York, New York 10036 (212) 719-7230	F4 (12), D3 (10), W5 (2, 9)
CINEMAX	Cinemax (HBO)	Rockefeller Center New York, NY 10020 (212) 484-1241	F3 (20) (east) F3 (23) (west)
CNN	Cable News Network	1050 Techwood Drive Atlanta, GA 30318 (404) 898-3850	F3 (14)
CNN2	Cable News Network 2	1050 Techwood Drive Atlanta, GA 30318 (404) 898-8500	F3 (15)
CNS	Commodity News Service	2100 W. 89th Street Kansas City, KS 66206 (913) 642-7373	

Table B—4-cont. Major Satellite Programmers and Networks

Name	Company	Address	Satellite/Transponder[1]
C-SPAN	Cable Satellite Public Affairs Network	400 North Capital St., Suite 155 Washington, DC 20001 (202) 737-3220	F3 (19)
CTNA	Catholic Telecommunications Network	95 Madison Ave. New York, NY (212) 696-0800	W4 (11)
DAYTIME	Daytime	ABC Video/Hearst Company 959 Eighth Avenue New York, NY 10019 (212) 262-3315	F3 (22)
DISNEY	The Disney Channel	6767 Forest Lawn Drive Suite 11 Los Angeles, CA 90068 (213) 840-1000	W5 (10, 12)
DJNS	Dow Jones Cable News	Box 300 Princeton, NJ 08450 (609) 452-2000	
ESPN	Entertainment & Sports Programming Network	ESPN Plaza Bristol, CN 06010 (203) 584-8477	F3 (7), F4(6)
EWTN	Eternal Word TV Network	5817 Old Leeds Rd. Birmingham, AL 35210 (205) 956-9535	F3 (18)
FNN	Financial News Network	2525 Ocean Park Santa Monica, CA 90405 (213) 450-2412	F4 (2)

Table B—4-cont. Major Satellite Programmers and Networks

Name	Company	Address	Satellite/Transponder[1]
GALA	Galavision	250 Park Avenue New York, NY 10020 (212) 563-8900	F4 (23)
HBO	Home Box Office	Time/Life Bldg. Rockefeller New York, NY 10020 (212) 484-1241	F3 (13) (west) F3 (24) (east) G1 (1,3,17,19,21,23)
HTN	Home Theater Network	465 Congress Street Portland, ME 04101 (207) 774-0300	F3 (16)
HUGHES	Hughes Television Network	4 Pennsylvania Plaza New York, NY 10001 (212) 563-8900	W3 (19) W4,(2,4,14) W5 (1)
MSG	Madison Square Garden	4 Pennsylvania Plaza New York, NY 10001 (212) 563-8900	W5 (1)
TMC	The Movie Channel	Warner-Amex Corporation 75 Rockefeller Plaza New York, NY 10019 (212) 484-6826	F3 (5)
MSN	Modern Satellite Network	45 Rockefeller Plaza New York, NY 10111 (212) 765-3100	F3 (22)
MTV	Music Television	Warner-Amex Satellite Entertainment Co. 1211 Avenue of the Americas New York, NY 10036 (212) 944-4250	F3 (11)

Table B—4-cont. Major Satellite Programmers and Networks

Name	Company	Address	Satellite/Transponder[1]
N-VILLE	Nashville Network	Group W Satellite Communication 41 Harbor Plaza Drive Stamford, CT 06904 (800) 243-9141	W5 (17)
NBC	National Broadcasting Co.	30 Rockefeller Plaza New York, NY (212) 664-4000	F4 (24) D3 (1)
NICKEL	Nickelodeon	Warner-Amex Satellite Entertainment Co. 1211 Avenue of the Americas 15th Floor New York, NY 10036 (212) 944-4255	F3 (1)
NJT	National Jewish Television	Jason Films 2621 Palisades Avenue Riverdale, NY 10463 (212) 549-4160	F3 (16)
ON-TV	On-TV	National Oak Media 1139 Central Ave. Glendale, CA 91201 (213) 507-6500	D4 (8, 11)
PLAYBOY	Playboy Channel	Rainbow Programming Service 100 Crossway Park West Woodbury, NY 11797 (516) 364-2222	F4 (12)
PBS	Public Broadcasting Service	475 L'Enfant Plaza Washington, DC 20024 (202) 488-5000	W4 (15,17,21,23)

Table B—4-cont. Major Satellite Programmers and Networks

Name	Company	Address	Satellite/Transponder[1]
PTL	People That Love TV Network	Charlotte, NC 29279 (704) 542-6000	F3 (2)
RCTV	Rockefeller Center Television	30 Rockefeller Plaza New York, NY 10112 (212) 930-4900	F4 (21)
REUTERS	Reuters, Ltd.	1700 Broadway New York, NY 10036 (212) 730-2740	F3 (18)
SELECT	SelecTv™ (Pay Movie Channel)	4755 Alla Road Marina Del Rey, CA 90291 (213) 827-4400	W5 (5)
SHOWTIME	Showtime Entertainment, Inc.	1211 Avenue of the Americas New York, NY 10036 (212) 575-5175	F3 (10) (west) F3 (12) (east)
SIN	Spanish International Network	250 Park Avenue New York, NY 10017 (212) 953-7500	F4 (1) G1 (6, 20)
SNC	Satellite News Channel	Group W Satellite Comm. 41 Harbor Plaza Drive Stamford, CT 06904 (800) 234-9141	W5 (8, 11, 14, 16, 18) G1 (2,8,12,22)
SPOT	Spotlight (Pay-Movie Channel)	Spotlight/Times-Mirror 2951 28th St., Suite 2000 Santa Monica, CA 90405	F3 (4) W5 (21) G1 (5,10)
SPN	Satellite Programming Network	P.O. Box 45684 Tulsa, OK 74145 (918) 481-0881	F4 (3)

Table B—4-cont. Major Satellite Programmers and Networks

Name	Company	Address	Satellite/Transponder[1]
TAN	The American Network	735 North Water Street Milwaukee, WI 53202 (414) 276-2277	W5 (20)
TBN	Trinity Broadcasting Network	P.O. Box A Santa Ana, CA 92711 (714) 832-2950 (212) 930-4900	F4 (17)
USA NET	The USA Network	208 Harristown Road Glenrock, NY 07452 (201) 445-8550	F3 (9)
WEATHER	The Weather Channel	Landmark Communications 2840 Mt. Wilkinson Pkwy. Atlanta, GA 30339 (404) 434-6800	F3 (21)
WGN-TV	WGN TV Channel 9, Chicago	United Video 520 S. Harvard, Suite 215 Tulsa, OK 74135 (800) 331-4806	F3 (3)
WOR-TV	WOR-TV Channel 9, New York City	Eastern Microwave P.O. Box 4872 Syracuse, NY 13221	W5 (3)
WTBS-TV	WTBS-TV Channel 17, Atlanta, GA	1050 Techwood Drive NW Atlanta, GA 30318 (404) 898-8500	F3 (6)

Table B—4-cont. Major Satellite Programmers and Networks

Name	Company	Address	Satellite/Transponder[1]
WOLD	Robert Wold Communications	10880 Wilshire Blvd. Suite 2204 Los Angeles, CA 90024 (213) 474-3500	W4 (9, 11, 19)
XEW-TV	Feed to SIN from Mexico	Mexico City, Mexico (905) 518-1220	W4 (6)

[1] D = Comstar, F = Satcom, G = Galaxy, W = Westar

Table B—5. Satellite Feeds Available to SMATV Systems

Service Name	Fee
ACSN (The Learning Channel)	$.025 to $.05 per month
BIZNET	Varies, low cost
BLUE	N/A
CBN	Free
CNN	$.20 (stand alone) per month
	$.15 (with WTBS) per month
CNN2 (Headline Service)	Free with CNN
C-SPAN	$.05 per month
ESPN	$.04 per month
EWTN	Free
FNN	Free
Galavision	Varies between $3.50 and $5.50 per month
HTN	Varies between $3.00 and $6.00 per month
MSN	Free
NCN	Free
NJT	Free
ON-TV	Varies. About $8.00 per month in package deal
PTL	Free
Reuters	Varies. Business service.
SelecTv™	About $7.45 per month
Showtime	Various. Limited to service in nonfranchised areas.
SIN	Free
SPN	Free
TAN (The American Network)	Varies. Between $3.00 and $6.00 per month.
TBN	Free
TMC[1]	Varies. About $7.00 per month.
WGN-TV[2]	$.15 per month per subscriber
WOR-TV[2]	$.15 per month per subscriber
WTBS-TV[2]	$.10 per month

[1]Service available through PATMAR Technologies, Inc. at P.O. Box 769, Bernardsville, NJ 07924 (201/766-4408).
[2]TV "superstations" available to SMATV operators licensed by FCC and copyright tribunal as CATV systems.

Manufacturers, National Distributors, and Dealers of Satellite Television and SMATV Equipment

MANUFACTURERS DIRECTORY

The following lists include both major manufacturers and national distributors of TVRO and CATV equipment for the home satellite tv and the commercial SMATV user. In addition, those Pioneer Members of SPACE are included and noted with an asterisk. This directory is further classified according to the six major categories:

1. National distributor of complete home TVRO satellite systems.
2. SMATV and private cable commercial TVRO and CATV equipment manufacturers and distributors.
3. Manufacturers of satellite TVRO antennas
4. Manufacturers of satellite TVRO receivers
5. Manufacturer of TVRO low noise amplifiers (LNAs) and/or low noise converters (LNCs)
6. Manufacturers of miscellaneous equipment (including feedhorns, antenna mounts and drives, towers, and LPTV/MDS transmitters/ antennas).

In general, commercial manufacturers of CATV equipment tend to charge more for their TVRO systems because their equipment is usually built to more rugged specifications. In many cases, the SMATV or private cable operator can readily use equipment intended for the home TVRO market. Some of the most sensitive and best-performing satellite receivers are made by the smallest companies.

It should be noted that a manufacturer or supplier eligible for classification in more than one particular category may not appear everywhere. In these cases, the most dominant classification will be listed. One should always contact the firm directly for the latest information on product availability. This list is not all-inclusive and a major TVRO supplier or manufacturer may have been unintentionally excluded.

Following the directory by category, an alphabetical listing of the complete names of the companies, including addresses, telephone numbers, and a brief description of their primary products and/or service is provided.

Category 1: National Distributors of Complete Systems

Allsat
American Microwave Technology
Birdview Satellite Communications
Boman Industries
Delta Satellite Center
Discom Satellite Systems
Earth Stations
Echosphere
Galaxscan
H&R Communications
High Frontier
Hoosier Electronics
Intersat Corporation
J V Electronics
Long's Electronics
Muntz Electronics
National Microtech
Paraclipse of California
Satellite America Marketing, Inc.
Satellite TV Specialists
Wespercom

Category 2: SMATV and Private Cable TVRO and CATV Equipment Manufacturers and Distributors

Allsat
Andrew
Anixter-Mark
Antenna Technology
Avcom
Blonder-Tongue
California Microwave
Channel One
Comtech
Fort Worth Tower
Gardiner Communications
Harris Corp.
Hero Communications
Hughes Corp.
Jerrold Electronics
KLM Electronics
M/A-Com Video Satellite
Magnavox
Microdyne
Newton Electronics
North Supply Corp.
Oak Communications
Paradigm Manufacturing
Prodelin
Rockwell International
Satellite Supplies
Satellite Systems
Satellite Technology Service
Scientific-Atlanta
Sylvania
Tocom
Toner Cable Equipment

Category 3: Manufacturers of Satellite TVRO Antennas

ADM
Avantek
Birdview
Comtech
Fort Worth Tower
Harrells
Hastings Antenna
Hero Communications
International Video Communications
Intersat
Janeil
KLM Electronics
Luly Telecommunications
Mac/Line
McCullough Satellite
Microwave General

330

Paradigm Manufacturing
Paraframe
Prodelin

United States Tower
Wilson Microwave

Category 4: Manufacturers of Satellite TVRO Receivers

Amplica
Arunta Engineering
Automation Techniques
Avcom
Blonder-Tongue
Channel Master
Dexcel
Earth Terminals
Electrophone
Gillaspie
M/A Com

Intersat
International Crystal
KLM
Lowrance Electronics
M/A-Com
Microwave General
R.L. Drake
Sat-Tec Systems
Satellite Technology
Telecom Industries

Category 5: Manufacturers of TVRO Low Noise Amplifiers and/or Low Noise Converters

Amplica
Avantek

Dexcel
M/A-Com

Category 6: Manufacturers of Miscellaneous Equipment

Advanced Design
Arvin
Bogner
Chaparral
Conifer
Emcee
Gillaspie
Hero Communications
Home Cable
Houston Satellite

Howard Engineering
KLM
Modulation Associates
Northwest Satlabs
Real-World Systems
Sigma International
Spaceage Electronics
Telcom Industries
Tel-Vi Communications
United States Tower

Alphabetical Directory of Distributors and Manufacturers

ADM (Antenna Development & Mfg., Inc.)
2745 Bedell Avenue
Poplar Bluff, MO 63901
(314) 785-5988
Popular 11-foot parabolic antenna manufacturer

Amplica, Inc.*
950 Lawrence
Newbury Park, CA 91320
(805) 498-9671
Major LNA supplier

Allsat Inc.*
7451 Switzer, Suite 107
Shawnee Mission, KS 66203
(913) 236-9692
Space Pioneer Member

Andrew Corporation
10500 West 153rd Street
Orland Park, IL 60462
(312) 349-3300
3- to 10-meter CATV-oriented antenna

American Microwave Technology
Box 824
Fairfield, IA 52566
(800) 247-5005
Regional TVRO distributor for Midwest and
Canada

Advanced Design Engineering Corporation
11684 Liburn Park
St. Louis, MO 63141
(800) 325-4058
Disk Motor activator and controller

Anixter-Mark, Inc.
2180 S. Wolf Road
Des Plaines, IL 60018
(312) 298-9420
Stocking distributor and manufacturer of
CATV-oriented equipment

Antenna Technology Corp.
895 Central Florida Parkway
Orlando, FL 32809
(305) 851-1112
Manufacturers *Simulsat*, a commercial
stretched-parabolic multisatellites TVRO
antenna

Arunta Satellite Television Systems
Department CSP, Box 15082
Phoenix, AZ 85060
(602) 956-7042
Major home TVRO receiver supplier

Arvin (Applied Technology Group)
4490 Old Columbus Road, NW
Carroll, OH 43112
(614) 756-9211
Builds the Sat-Weather tv Receiver used by
tv stations nationwide

Automation Techniques, Inc.*
1846 N. 106 E. Avenue
Tulsa, OK 74116
(918) 836-2584
One of the "Big 5" TVRO receiver manufacturers

Avantek, Inc.
3175 Bowers Avenue
Santa Clara, CA 95051
Major supplier of LNAs

Avcom of Virginia, Inc.*
10139 Apache Road
Richmond, VA 23235
(804) 794-2500
Top-notch TVRO receivers

Birdview Satellite Communications, Inc.*
Post Office Box 963
Chanute, KS 66720
(316) 431-0400
Major national distributor

Blonder-Tongue Laboratories, Inc.
One Jake Brown Road
Old Bridge, NJ 08857
(201) 679-4010
Old-line CATV equipment manufacturer

Bogner Broadcast Equipment Corp.
401 Railroad Avenue
Westbury, NY 11590
(516) 997-7800
LPTV equipment

Boman Industries*
9300 Hall Road
Downey, CA 90241
(213) 869-4041
Old-line auto radio importer/distributor now
in TVRO sales.

California Microwave, Inc.
455 W. Maude Avenue
Sunnyvale, CA 94086
(408) 732-4000
Major commercial TVRO manufacturer

Channel Master, Div. of Avnet, Inc.*
Ellenville, NY 12428
(914) 647-5000
Major TVRO receiver manufacturer

Chaparral Communications, Inc.*
103 Bonaventura Drive
San Jose, CA 95134
(408) 262-2536
Makes the Chaparral circular feedhorn and
electronic Polarotor V/H polarizer

Channel One, Inc.
Wallarch Road
Lincoln, NE 01773
(617) 899-1025
One of the original national dealers in the
home TVRO business.

Comtech Antenna Corp.
3100 Communications Road
St. Cloud, FL 32769
(305) 892-6111
Commercially oriented TVRO receivers and
antennas.

Conifer Corporation
1000 N. Roosevelt
Burlington, Iowa 52601
(319) 752-3607
MDS Antennas

Delta Satellite Center
1003 Washington Street
Grafton, WI 53024
(800) 558-5582
Regional midwest TVRO distributor

Dexcel, Inc.*
2285 Martin Avenue
Santa Clara, CA 95050
(408) 727-9833
Major LNA and receiver manufacturer. Introduced concept of LNC to industry.

Discom Satellite Systems, Inc.*
4201 Courtney, Box 8699
Independence, MO 64054
(816) 836-2828
Space Pioneer Member

Earth Stations, Inc.*
957 Washington Road
Grosse Pointe, MI 48230
(313) 885-9181
Space Pioneer Member

Earth Terminals, Inc.*
Post Office Box 636
Fairport, NY 14450
(716) 223-7457
Builds top quality "videofile" TVRO receivers

Echosphere Co.
5315 S. Broadway
Littleton, CO 80120
(303) 797-3231
National distributor of home TVRO products

Electrohome Electronics*
Kitchener, Ontario, Canada N2G 4J6
(519) 744-7111
Major TVRO receiver manufacturer

Emcee Broadcast Products Div.
Electronics Missiles and Communications, Inc.
Post Office Box 68
White Haven, PA 18661
(800) 233-6193
LPTV/MDS transmitters

Fort Worth Tower Co., Inc.
1901 East Loop 820 South
Post Office Box 8597
Ft. Worth, TX 96112
(817) 457-3060
Command broadcast towers and TVRO antennas

Galaxscan, Inc.*
3606 N. Harold
North Little Rock, AR 72118
(800) 643-8737
Space Pioneer Member

Gardiner Communications Corp.
1980 South Post Oak Road, Suite 2040
Houston, TX 77056
(713) 961-7348
Major CATV manufacturer of receivers, etc.

Gillaspie & Associates, Inc.*
950 Benicia Avenue
Sunnyvale, CA 94086
(408) 730-2500
Major receiver manufacturers; builds portable receivers and signal strength/dish alignment meters.

Harrell's Southside Welding*
Old Highway 7 North
Route 2, Box 46
Grenada, MS 38901
(601) 226-4081
Space Pioneer Member

H&R Communications, Inc. Starview Systems Div.
Route 3, Box 103G
Pocahontas, AR 72455
(501) 647-2001
National Distributor

Harris Corp., Satellite Communications Division
Post Office Box 7700
Melbourne, FL 32901
(305) 725-2070
Commercial TR and TVRO systems.

Hastings Antenna Co., Inc.*
948 West 1st
Hastings, NE 68901
(402) 463-3598
Space Pioneer Member; major antenna supplier.

Hero Communications, Div. Behar Enterprises, Inc.*
1783 West 32nd Place
Hialeah, FL 33012
(305) 887-3203
Space Pioneer Member who has led the way for international TVRO systems; supplies modified Intelsat receivers

High Frontier Corp.
2230 East Indian School Road
Phoenix, AZ 85016
(602) 954-6008
Regional TVRO Distributor for Southwest

Home Cable, Inc.
2123 Lewis Street
Salina, Kansas 67401
(913) 825-7939
"Crank eliminator" and digital satellite-locator control motorized drive

Hoosier Electronics, Inc.
Post Office Box 3300
Terre Haute, IN 47803
(812) 238-1456
Amateur-radio distributor now into TVRO systems

Houston Satellite Systems, Inc., Tracker Systems Div.
8000 Harwin, Suite 397
Houston, TX 77036
(713) 784-8953
"Tracker III" motorized mount

Howard Engineering
Post Office Box 48
San Andreas, CA 95249
Developer of advanced receiver technology; circuit boards

Hughes Corp., Microwave Communications Products Div.
Post Office Box 2999
Torrance, CA 90509
(213) 534-2146
Major CATV manufacturer of 4.5–6 meter disks, receivers

International Crystal Manufacturing Co., Inc.*
10 North Lee
Oklahoma City, OK 73012
(405) 236-3741
Major "old line" home TVRO receiver manufacturer

International Video Communications, Inc.
4005 Landski Drive
North Little Rock, AR 72118
(501) 771-2800
Dish manufacturer.

Intersat Corp.*
2 Hood Drive
St. Peters, MO 63376
(314) 278-2178
Major national manufacturer and distributor of home TVRO systems.

Janeil Corp.*
6860 Canby, #113
Reseda, CA 91335
(213) 881-4155
Major baked-enamel mesh parabolic antenna manufacturer

Jerrold Electronics, Inc.
Hatboro, PA 19040
(215) 674-4800
CATV decoder, line amps, etc. equipment manufacturer

J. V. Electronics
41 Canal Street
Landing, NJ 07850
(201) 347-3206
Regional TVRO distributor for Northeast

KLM Electronics, Inc.*
Post Office Box 816
16890 Church Street
Morgan Hill, CA 95037
(408) 779-7363
One of the "Big 5" TVRO receiver manufacturers. Also makes autotrack dish

Long's Electronics
P.O. Box 11347
Birmingham, AL 35202
(800) 633-6461
Major distributor of many lines of TVRO equipment

Lowrence Electronics, Inc.
12000 E. Skelly Drive
Tulsa, Oklahoma 74128
(918) 437-6881
Receiver "System 7" with stereo

Luly Telecommunications, Inc.
Post Office Box 2311
San Bernardino, CA 92405
(714) 888-7525
Unique fold-up, lightweight, portable 4–12 foot TVRO antenna

M/A-Com Video Satellite, Inc.
Home TVRO Div.
63 Third Avenue
Burlington, MA 01803
(617) 272-3000
One of the divisions of the giant Microwave Associates Corporation

Mac/Line, Inc.
W. 709 Seltice Way
Post Falls, ID 83854
(208) 773-7900
One-piece 8–12 foot parabolic antennas

Magnavox CATV Systems, Inc.
133 W. Seneca Street
Monduis, NY 13104
(315) 682-9105
CATV converter/decoders, line amps

McCullough Satellite Systems, Inc.
Post Office Box 57
Salem, AR 72576
(501) 895-3167
8–12 foot low-cost spherical multisatellite antennas and kits

Microdyne Corp.
Post Office Box 7213
Ocala, FL 32672
(904) 687-4633
CATV receivers, antennas. Major commercial supplier

Microwave General
2680 Bayshore Frontage Road
Mountain View, CA 90019
Deluxe home TVRO systems with motorized mounts

Modulation Associates, Inc.
897 Independence Avenue
Mountain View, CA 94043
(415) 962-8000
Satellite audio and radio receivers

Muntz Electronics, Inc.
7700 Densmore
Van Nuys, CA 91406
(213) 782-7511
Originator of Muntz TV, now in the TVRO business

National Microtech, Inc.*
Post Office Box 417
Grenada, MS 38901
(601) 226-8432
Original TVRO national distributor with major share of market

Newton Electronics, Inc.*
2218 Old Middlefield Way
Mountain View, CA 94043
(415) 967-1473
Space Pioneer Member

North Supply Corp.
10951 Lakeview Avenue
Lenexa, KS 66219
(913) 888-9800
CATV products of all types. Complete SMATV systems.

Northwest Satlabs
806 NW 4th
Corvallis, OR 97330
(503) 754-1136
"Tweaker" electronic metering device plugs into down convertor for alignment

Oak Communications, Inc., CATV Div.
Crystal Lake, IL 60014
(815) 459-5000
Decoders and Converters; catv equipment

Paraclipse of California
26732 Oak Ave., Suite G
Canyon County, CA 91351
(805) 252-0432
Western distributor; also sell Solar-powered TVRO Systems

Paradigm Manufacturing, Inc.*
2962 Cascade Blvd.
Redding, CA 96003
(916) 244-9300
Major manufacturer of parabolic mesh antennas

Paraframe, Inc.
Box 423
Monee, IL 60449
Parabolic antenna kits and antennas. Original wood-frame TVRO dish manufacturer

Prodelin, Inc., Div., Ma/Com
Box 131
Hightstown, NJ 08520
(609) 448-2800
6–15 foot parabolic antennas; major manufacturer

R. L. Drake Co.*
540 Richard Street
Miamisburg, OH 45342
(513) 866-2421
One of the "Big 5" TVRO receiver manufacturers

Real-World Systems
128 Cross House Road
Greenside, Sheffield, England 530 3RX
Electronics and receiver boards, kits.

Rockwell International, Collins Transmission
Systems Div
Post Office Box 10462
Dallas, TX 75207
(214) 996-5340
Commercial TVRO receivers.

Satellite America Marketing, Inc.*
Post Office Box 552
Grenada, MS 38901
(601) 227-1820
Major National TVRO Distributor

Sat-Tec Systems*
Div. of Ramsey Electronics, Inc.
2575 Baird Road
Penfield, NY 14526
(716) 586-3950
Space Pioneer Member responsible for in-
troducing low-cost TVRO receivers

Satellite Supplies, Inc.*
164 B Gilman Avenue
Campbell, CA 95008
(408) 370-1515
Space Pioneer Member

Satellite Systems, Unlimited*
Post Office Box 43
Conway, AR 72032
(501) 326-6501
Space Pioneer Member and dealer

Satellite Technology Services, Inc.
11684 Lilburn Park Road
St. Louis, MO 63141
(314) 569-3720
Space Pioneer Member

Satellite TV Specialists, Inc.
5665 South State
Salt Lake City, UT 84107
(801) 262-8813
National TVRO Distributor

Scientific-Atlanta, Inc.
3845 Pleasantdale Road
Atlanta, GA 30340
(404) 449-2000
Granddaddy of all TR and TVRO commercial manufacturers; CATV equipment

Sigma International, Inc.
617 N. Scottsdale Road
Scottsdale, AZ 85257
(602) 994-3435
TVRO receiver modules and kits

Spaceage Electronics
Post Office Box 15730
New Orleans, LA 70175
(504) 891-7210
"Space-View II" Satellite locator and
retro-fit motor drive

Sylvania CATV Operations, GTE Sylvania
Inc.
10841 Pellicano Drive
El Paso, TX 79935
(915) 591-3555
CATV decoders, convertors, line amps, etc.

Telcom Industries Corporation
27 Bonadventura Drive
San Jose, CA 95134
(408) 262-3100
Home TVRO receivers, motorized disk
mount controllers

Tel-Vi Communications, Div. Linear Drive
Systems, Inc.
1307 West Lark Industrial Blvd.
St. Louis, MO 63026
(314) 343-9977
Linear drive motorized mount systems

Tocom, Inc.
3301 Royalty Row
Box 47066
Dallas, TX 75247
(214) 438-7691
CATV equipment manufacturer

Toner Cable Equipment, Inc.
969 Horsham Road
Horsham, PA 19044
(800) 523-5947
CATV-oriented TVRO, converter, decoder,
& equipment distributor

United States Tower Co.
P.O. Drawer S
Afton, OK 74331
(918) 256-4257
Spherical TVRO antennas for CATV market.

Wespercom, Inc.
Post Office Box 7226
Bend, OR 97701
(503) 389-0996
National distributor of many kinds of TVRO
equipment.

Wilson Microwave Systems, Inc.
4286 S. Polario Avenue
Las Vegas, NV 89103
(800) 634-6898
Builds Parabolic dishes.

---

*Member of SPACE (Society of Private and Commercial Earth Stations)

TVRO DEALER DIRECTORY

The following directory lists installing dealers who are members of the Society of Private and Commercial Earth Stations (SPACE), the national trade association as of January 1, 1983. To obtain an updated directory, or for further information concerning dealer, SMATV, or personal membership in SPACE, the Association may be contacted at 1920 N. Street, NW, Suite 510, Washington, DC 20036, phone (202) 887-0605.

A-1 Communications, Inc.
% George R. Summers
P.O. Box 11972
Aspen, CO 81611

A-1 TV & Appliance
% Jim Elrod
1003 First Avenue
Rock Falls, IL 61071

AZC-Tech Distributing
% Tim Olin
Box 7
Arnold, NE 69120

Action Earth Satellite Corporation
% Robert Spurgeon
1333 Sulphur Spring Road
Baltimore, MD 21227

Advanced Earthstations
% Ralph E. Cheek
4526 Cloudmount
Houston, TX 77084

Advanced Satellite Systems
% James E. Allen
655 Lunt Avenueq
Schaumburg, IL 60193

Agri-Marketing Group
5957 W. 21st Street
Topeka, KS 66614

Agricultural Enterprises
  d/b/a/Friendly Fire
% Bill Eckert
1802 Laport Avenue
Fort Collins, CO 80521

Ajak Industries, Inc.
% Mr. James L. Berry
200 E. Trail
Dodge City, KS 67801

Alamosa TV Shop
6893 Trinchenn Lane
Alamosa, CO 81101

Alaska Cable Systems
% Martha J. Sumner
P.O. Box 73680
Fairbanks, AK 99707

Albin Radio & TV Service, Inc.
% Wayne Albin, President
7 Forest
P.O. Box 1174
Lamar, CO 81052

Alexander & Rays TV & Appliance
% Al Alexander
211 E. Santa Fe
Olathe, KS 66061

Allen Jackson
319 S. Madison Avenue
El Dorado, AR 71730

American Microwave Technology
% Stacey Peterson
1601 N. 4th, Box 824
Fairfield, IA 52556

American Television Systems, Inc.
% Stephen R. Young, Pres.
779 East Aurora Road
Macedonia, OH 44056

337

Antenna Service Co.
% R.A. Schatz
P.O. Box 32402
Cleveland, OH 44132

Antennas Unlimited, Ltd.
% Denver Webber, President
361 Saline Road
Fenton, MO 63026

Antennavision, Inc.
% Paul K. Hakenicht
14810 Canterrel Road
Little Rock, AR 72212

Applied Space Technology
% Lance Friedman
6535 Seneca Street
Elmar, NY 14059

Aquarian Video
% Karas Burrows
1780 E. Colorado Blvd.
Pasadena, CA 91106

Arcand Electronics
Box 1065
Big Springs, TX 79720

Arnn's TV
% Gene G. Arnn
Hwy. 67 at I-10, Rt. 2
Biloxi, MS 39532

Arnold & Morgan
% Larry Morgan
510 S. Garland Avenue
Garland, TX 75040

Arnold Pool Company
% Claude & Carol Arnold
RR #2, Box 144
Ogallala, NE 69153

Art's Electronics
% Charles Schmidt
313 Main Street
Augusta, AR 72006

Audio Concepts
% David Stormoen
916 Broadway
Alexandria, MN 56308

Audio Video
% Russ Swallow
485 W. Broadway, P.O. Box 1198
Jackson, WY 83001

Audio Video Systems
% Bob Pechan
2390 Peoria, Suite 306
Aurora, CO 80010

Aurora Communications
% Don Cunning
P.O. Box 5412
Ketchikan, AK 99901

Automation Controls
3650 James Street
Syracuse, NY 13206

Balentines' TV
% T.J. Balentines
1301 E. Tuxedo
Bartlesville, OK 74003

Bernie's TV & Appliance
% Neil Braddy
915 Main Street
Canon City, CO 81212

Bertz (TV) Sales & Rentals
% Bertz Longhi
1920 Delmar
Granite City, IL 62040

Bill Baer Satellite Antenna
% Willard A. Baer
1330 Minnesota Avenue
Winter Park, FL 32789

Bill's Radio & Television
% Bill Simmons
501 E. 1st
O'Fallon, IL 60461

Birdsells TV
% Charlie Birdsell
836 West Morton Road
Jacksonville, IL 62650

Blocker Electronics
% Alvin Blocker
411 Winans Avenue
Hot Springs, AR 71901

Boatright's Electronics
% Mr. Charles Boatright
804 Tower Square
Marion, IL 62959

Bob Grannemann Radio & TV
% Bob Grannemann
554 E. 5th Street
Washington, MO 63090

Bob Johnson Appliance Center
% Mr. Robert V. Johnson
210 S. Independence
Harrisonville, MO 64701

Bone & Co., Inc.
% Will Bone
421 S.W. 6th
Portland, OR 97204

Bonsall TV, Inc.
% Richard Bonsall
618 Iowa Ave.
Dunlap, IA 51529

Boone TV & Electronics
% Paul Steelhammer
Rt. 2, Box 60
Hallsville, MO 65255

BoSal Satellite Systems, Inc.
200 Ft. Mead Road
Laurel, MD 20707

Brancio Enterprises
% Gary Brancio
7171 Warren
Denver, CO 80221

Brazos Record & TV Center, Inc.
Brazos Shopping Center
Mineral Wells, TX 76067

Braodway Electronics
% Terry Bradly
1010 South Broadway
Pillsbury, KS 66763

Bruce's Sales & Service
% Jerry Bruce
Box 1755
Gallup, NM 87301

Bruce's TV
% Bruce Van Deventer
1 South Main
Elburn, IL 60119

Buckhanan Supply, Inc.
% Ed Bunn
P.O. Box 390
Vansant, VA 24656

CNI Satellite Systems, Inc.
Michael A. Ostwind, Vice President
P.O. Box 621
Leesburg, VA 22075

Cable Sat, Inc.
% Mr. Wayne Studniarz
114 Ebbesen Drive
De Kalb, IL 60115

Carbon County TV & Radio Shack
% Don Van Horn
701 West Spruce
Rawlings, WY 82301

Carlson TV and Electronics
% Earl Carlson
1523 Belknap Street
Superior, WI 54880

Carpenter Electric
% Val Carpenter
853 Redrock Road
St. George, UT 84770

Casablanca Video
% Jason Sherman, President
635 S. Broadway
Boulder, CO 80303

Chamness TV & Appliances
% Wallace M. Gayle
3532 South Broadway
Tyler, TX 75701

Channel One
Mr. Fred Hopengarten
79 Massasoit St.
Waltham, MA 02154

Christy Sales, Inc.
% Harlan "Bud" Christy
106 Birdland
Newton, IA 50208

Chuck's TV and Microwave Center
% Mr. Charles B. Hall, Sr.
290 Watervliet Shaker Road
Watervliet, NY 12189

Clark's TV & Communications, Inc.
% R. Marlene Clark
2140 Planet Avenue
Salina, KS 67401

Coin Sports Dist., Inc. d/b/a/SATVID
8834 E. 350 Hwy.
Raytown, MO 64133

Colorado Satellite Systems
% Ernie Hornbaker
13333 S. Blaney
Peyton, CO 80831

Colorworld TV Rental, Inc.
% Patricia Flowers
3301 San Mateo, N.E.
Albuquerque, NM 87110

Computers n' Stuff
% Jerry Coleman
612 S. Lake St.
Mundlein, IL 60060

Comstock TV
% Val Lane
1907 Central Avenue
Kearny, NE 68847

Consumer Satellite Systems, Inc.
Jon Powell, President
Mike Schroeder, Vice President
6202 LaPas Trail
Indianapolis, IN 46268

Cook's True Value Home Center
% Don Cook
525 Main Street
Osawatomie, KS 66064

Cooksey TV
% Kenney Cooksey
413 Main
Quinter, KS 67752

Country Cable, Inc.
% John Buchanan
329 Pike Lake
Duluth, MN 55811

Cove TV & Electronics, Inc.
% D. Shelton, President
925 Cherry Street
Panama City, FL 32401

Cross Curtis Mathes Show Room, Inc.
% Bob Cross
1944 Cy Ave.
Casper, WY 82601

Crouch TV Sales & Service
% Jim Crouch
1813 Grand Avenue
Fort Smith, AR 72901

Cryer's Electronics
% Woodrow Cryer
301 N. 3rd Street
Leesville, LA 71446

Curtis Mathes Home Entertainment Center
% Charles E. Lewis
6950 West Division Street
St. Cloud, MN 56301

Curtis TV
% Curtis Andersen
2510 Glenn Avenue
Glenwood Springs, CO 81601

Custom Sounds
% Kevin Fotorny
835 S. Sam Houston
Huntsville, TX 77340

D & D Electronics
% Dave Drobnitch
705 Elm
Pueblo, CO 81004

D & L Specialties
% H. Dave Schmidtke
214 Third Street, S.W.
Fairbault, MN 55021

D-R Hardware & Electronics
% Ron Endress
Box 237
Pearl City, IL 61062

DICOMM
Mr. Patrick E. O'Farrell
P.O. Box 536
Selah, WA 98942

Dave & Mike's TV & Appliances
% Gary Busenbark
3728 Wabash Ave
Terre Haute, IN 47803

Davis TV & Stereo
% John Davis
911 12th Street
Cody, WY 82414

Davis TV & Video
% Norm Davis
1426 W. Eisenhower
Loveland, CO 80537

de Waal's Video Services, Inc.
% Terry de Waal
449 McCarty
San Antonio, TX 78216

Deer Hunters Supply
P.O. Box 101
Hondo, TX 78861

Deitz Electronics Service
% James Deitz
918 West Central
Carthage, MO 64836

Del City Music
℅ Bob Wood
2901 Epperly
Del City, OR 73115

Desert TV & Appliance
Mr. Bud Wheeler
3661 Maryland Parkway
Maryland Square #10
Las Vegas, NV 89109

Don's Electronic Service
℅ Don Brown
P.O. Box 208
Ten Sleep, WY 82442

Donn Peaster TV
℅ Don Peaster
420 W. 4th
Claremore, OK 74017

Dunker's Radio & TV
℅ Harold Dunker
423 State Street
Atwood, KS 67730

E.K. Pendarvis
Rt. 2, Box 130
Denham Springs, LA 70726

EMV Company
℅ Larry Vaughn
915 W. 8th
Yuma, CO 80759

ETL Corp.
℅ Louis K. Linn
P.O. Box 9525
Asheville, NC 28815

Earth Stations Manning and Sumter
℅ Junius M. Johnson
P.O. Box 423
Manning, SC 29102

Earth Stations of Columbia, Inc.
℅ Norman Goldberg
1416 Bluff Road
Columbia, SC 29201

Eastern Microtech, Inc.
℅ Jean L. Roten, Vice President
7892 Cryden Way
Forestville, MD 20747

Eastern Satellite TV Systems, Inc.
℅ Sherwood Craig
77 Exchange Street
Bangor, ME 04401

Electronic Analysis & Repair
℅ Gene Koonce
3510 10th Street
Greeley, CO 80631

Electronic Media Services, Inc.
℅ Robert Lavretta, President
3614 Turner Heights Drive
Decatur, GA 30033

Electronics-By-Heck
℅ D. W. Heck
1504 Rodgers Drive
Graham, TX 76046

Energy Systems, Ltd.
℅ J. D. Fore
2306 Charles Avenue
Dunbar, WV 25064

England Dan's
℅ Gary Flower
440 N. 5th
Kankakee, IL 60901

Entertainment Warehouse, Inc.
℅ Mike Rhodes
125 N. Church
Palestine, TX 75801

Far Out
℅ Fred Ulrich
913 State Street
Belle Fourche, SD 57717

Fevold Electronics
℅ Clifford Fevold
103 3rd Street N.W.
Roseau, MN 56751

Fiber-Classics Sat. Corp.
℅ Mr. Robert N. Entrup
1640 No. Townsend
Montrose, CO 81401

Finger Lakes Communication Co., Inc.
℅ William F. Webster
189 Clark Street
Auburn, NY 13021

Flathead Electric Service
℅ Frank Mutch
11 3rd Ave. W.
Polson, MT 59860

Franklin Electronics
℅ Barry Franklin, President
600 N. Main, Box 621
Corsicana, TX 75110

Furniture Fashions, Inc.
% J. L. Cook
Box 29
Rayville, LA 71269

G/C Electronics
% Steven Labrue
114 N. Main
Fairview, OK 73737

Gene's TV
% Gene Weeks
705 East 8th
Coffeyville, KS 67337

George Alarm Co., Inc.
% Donald George
2307 Willemoore
Springfield, IL 62704

German Corner TV Sales & Service
% Lawrence E. Poyner
11425 E. 116th Street
North Collinsville, OK 74021

Gifford Radio & TV
% Morris Gifford
P.O. Box 453
Stephenville, TX 76401

Glentronics, Inc.
% Herb Glenn, Jr.
P.O. Box 6-L
St. Croix, VI 00820

Goble LP Gas & Appliance
% Charles Goble
Rt. 42 North, Box 157
Casey, IL 62420

Graves HVAC
% Doy R. Graves
5732 Noble Avenue
Van Nuys, CA 91411

Great Egg Earth Stations
% Mr. Olen B. Soifer
1608 Black Horse Pike
Cardiff, NJ 08232

Great Plains TV Satellite
% John Welch, President
2407 S.W. New York
Lawton, OK 73501

Greeley Gas & Electric, Inc.
% M. K. Harris
Box 10
Greeley, NE 68842

Green's Appliance, Inc.
% Vernon L. Green
1917 W. Austin
Port Lavaca, TX 77979

Grove Communications
% Arvin S. Grove
Box 1967
Wilson, ND 58801

H & B Corp., Satellite Division
% Steve Kuplen
Route One
Mulberry, KS 66756

H & H Electrical Systems, Inc.
% William S. Heinz
130 E. Main Street
New Albany, IN 47150

Hagens Furniture & TV
% Clarence Hagen
1214 S. Gilbert
Iowa City, Iowa 52240

Hank Turek
15921 S. Latrobe
Oak Forest, IL 60452

Hank's Radio, TV & Appliance, Inc.
% Dale Reichert
102 N. Dewey
North Platte, NE 69101

Hawkins Decorative Center
% R. B. Hawkins
808 Loop 59 North
Atlanta, TX 75551

Hayes TV, Inc.
% John A. Hayes
1843 E. Cherry
Springfield, MO 65802

Hephner TV & Electronics, Inc.
% J. E. Hephner
737 South Washington
Wichita, KS 67211

Hi-Pix Electronics
% Henry G. Hyde
3427 South 42nd Street
Omaha, NE 68102

Hiawatha Electronics, Inc.
% Ms. Leslie Hittner, President
P.O. Box 442—619 Huff Street
Winona, MN 55987

Hinkle-Young Satellite
P.O. Box 647 N. Hwy. 73
Falls City, NE 68355

Home TV Sales & Service
Mr. Keith Weed
3036 No. Emerson
Fowling Rt., Box 7A
Alliance, NE 69301

Home View Microwave
% Jon Reinhart
R.R. #1, Clay Road
Wapakoneta, OH 45895

House of Television
% Raleigh House
511 S. Main
Hillsboro, IL 62049

House's Good House Keeping Shoppe
% John S. House
1525 Williamsburg Shopping Center
Paris, TX 75460

Hub Supply
% Bill Knapp
P.O. Box 600
Durango, CO 81301

Huberts II, Inc.
% Paul G. Mintong, Pres.
2619 So. Caraway
Jonesboro, AR 72401

Indesat, Inc.
% Ronald Phillips
211 N. Delaware
Butler, MO 64730

Instructional Media Associates
% Dan W. Weggeland
7763 Buckboard Trail
Palo Cedro, CA 96073

Int'l TeleCom of Asheville
% Edmund Horgan
P.O. Box 1189
Asheville, NC 28801

International Satellite Sales, Inc.
917 S. 9th Street
Springfield, IL 62704

Invictus Telecommunications, Inc.
% Gregory V. Peck
807-F Loveland Madrira Road
Cincinnati, OH 45140

J & J Leasing
% John J. Thomas
1406 Market Street
Reading, OH 45215

J. Mc Distributors
% Jay McDonald
P.O. Box 30293
Lafayette, LA 70503

JV Electronics
Box D208, 41 Canal Street
Landing, NJ 07850

Jensen's TV
% Mark Jensen
P.O. Box 385
Elgin, NE 68636

Jerry Hall's
% Jerry Hall
R.R. 4, Box 157
Fredonia, KS 66736

Jerry's Radio Communications
% Jerry L. Dorr
809 Northeast 9th
Wagoner, OK 74467

Jersey Jim Towers TV
% Rick Towers
512 U.S. Hwy. 19, S.
Clearwater, FL 33515

Jimmy Aud Electronics
% Jimmy Aud
1001 South Wheeler
Jasper, TX 75951

John Iverson Company
% Bob Brown
3300 N. Broadway
Minot, ND 58701

Johnny's TV
242 North Main
Stillwater, MN 55082

Johnson TV & Sound
% Gilbert Johnson
9422 N. May
Oklahoma City, OK 73120

Joy Appliance & Video Center
% Bill Joy
109 Hesse Street
Buffalo, WY 82834

KKH Enterprises
% Sam Kamees
912 East 16th Street
Cheyenne, WY 82001

Karl's Music & TV Appliance
% Elmer Karl
605 Main
Gregory, SD 57533

Kauffman TV
% Mr. Dave Kauffman
507 W. Hwy. 7
Garden City, MO 64747

Kenny's TV
Mr. Kenneth Simmons
231 St. Louis Road
Collinsville, IL 62234

King's Antenna Service
% King D. Oberlin
812 W. Maumee
Angola, IN 46703

Kline's TV & Appliance
% Paul Kline
2730 N. Michigan
Plymouth, IN 46563

Lakeside Television
% Viver Granzella, Jr.
Box 190
Whestland, MO 65779

Lang's Danielson & Brost
% Don Lang
2705 6th Avenue, S.E.
Aberdeen, SD 57401

Laurie Video, Inc.
Mr. Jack Thompson
Box 136 B
Lake Plaza
Laurie, MO 65038

Lavon's TV Sales & Service
423 W. Bdwy.
Hobbs, NM 88240

Lewis TV Service
% G. R. Lewis
109 N. Main
Eureka, KS 67045

Lindly TV & Appliance
% Harold Lindly
721 Main Street
Winfield, KS 67156

Lock's Electric
% John Lock
Box 176
Sharon Springs, KS 67758

Logan's Antenna Service
% Gary Logan
106 N. 8th Street
Norfold, NE 68701

McMillan Television Stereo
% Larry J. McMillan
244 E. Idaho Street
Kalispell, MT 59901

Logsdan Magnavox
% Ron Logsden
1820 Stevenson
Springfield, IL 62703

Longview Home Center
% James A. Williams
1613 Loop 281 West
Longview, TX 75608

Lorenson Plumbing & Appliance
% Alfred Lorenson
105 S. Main
Monroe City, MO 63456

Mack's TV
Mr. Mack Szymanski
507 East Commerce
Fairfield, TX 75840

Main Electronics Company
% Gordon Main
5558 S. Pennsylvania Avenue
Lansing, MI 48910

Manee Consulting and Research, Inc.
A T Division, % Alan Manee
P.O. Box 1107
Santa Barbara, CA 93102

Master Antennas, Inc.
% Ralph E. Cannon
1012 N. Ocean Blvd.
Pompano Beach, FL 33306

Master Craft of Owatonna, Inc.
% Fred Hirsch
123 W. Broadway, Box 386
Owatonna, MN 55060

McCann Electronics
% Gerry McCann
100 Division Street
Metairie, LA 70001

Mel's TV
% Mel Helmann
2807 13th Street
Columbus, NE 68960

Metsat
% Mr. William L. Wheaton
P.O. Box 1420
Bailey's Crossroads, VA 22041

Michael's Fotoshop
% Dave Benedetti
114 North Commercial
Trinidad, CO 81082

Micro Star
% James Dunden
300 Mountain View Drive
Rock Springs, WY 82901

Micro-Link Technology
P.O. Box 9254
St. Louis, MO 63117

Midwest Communications
% Tony Hood
2803 Rangeline
Joplin, MO 64801

Midwest Communications
1036 Hampshire
Quincy, IL 62301

Mitchell Two-Way Radio
% Chuck Mavszycki, Sales Manager
210 North Main
Mitchell, SD 57301

Montana Television-Appliance
Box 3597
1525 S. Russell
Missoula, MT 59807

Moore TV, Inc.
% W. C. Moore, President
629 Grant Avenue
Junction City, KS 66441

Mortec Industries, Inc.
% Rick Cox
515 Industrial Road
Brush, CO 80723

Murray Electric TV
% Bill Murray
205 West Maple
Lancaster, WI 53813

National Satellite Communications
Corporation
% Sally DiDonato
Plaza 7
Latham, NY 12110

Neistadt, Inc.
% Earl Neistadt, President
20251 Highway One, P.O. Box 5
Manchester, CA 95459

Norman Antenna, Inc.
% Jim Norman, Pres.
4505 11th Avenue, S.W.
Rochester, MN 55901

Norwood-Lowell TV
% C. D. Andreasen
Rt. 1, Box 503
Lowell, AR 72745

Nunnery Electronics
% Larry Nunnery
112 West 1st
LaJunta, CO 81050

Odom Antennas, Inc.
% Larry Duke
West Mississippi & Pecan
Beebe, AR 72012

Odom TV Sales and Services
% Jack Odom
102 Cartersville Road
Petal, MS 39465

Orsag's Inc.
% Alvin Orsag
201 W. Austin Street
Giddings, TX 78942

Pacific Trading Company
% Albert Meriwald
17023 S.E. McLaughlin Blvd.
Milwaukie, OR 97222

Pacific West Space Communications, Inc.
% John B. Fonteno, Jr.
203 Auto Drive
Compton, CA 90221

Pfeiffer Feed & Seed, Inc.
% Gary Pfeiffer
Box 216
Holbrook, NE 68948

Phonoscope, Inc.
4125 Hollister
Houston, TX 77080

Pioneer Satellite Systems, Inc.
% James Carrick
4615 Crossroads Park Drive
Liverpool, NY 13088

Potomac Satellite Systems
% Peter C. Foley, Pres.
1021 S. Barton Street, Suite 128
Arlington, VA 22204

Precision Satellite Systems
% Darryl Van Kirk
I-70 & Hwy 83, Rt. 2 Box 117A
Oakley, KS 67748

Primetek Corporation
% Allan Block
2219 Edgewood Avenue So.
Minneapolis, MN 55426

Private Communications
% Stan Esegon
1980 S. Quebec
Denver, CO 80231

Quality Appliance Center
% James Pasckke
315 S. Pearl Street
New London, WI 54961

Quarles Electric Corp.
% Mr. Bill Quarles
P.O. Box 1367
Greenwood, SC 29646

Radio Shack #7178
% Robert E. Montgomery, Jr.
800 West Ormulsee
Muskogee, OR 74401

Rall's TV Sales & Service, Inc.
% William Rall
RR #1 Box 314M
Logansport, IN 46947

Ranchland Electronics
% Mr. E. Wade Wobig
West Hwy. 20
Valentine, NE 69201

Rapid Associates
% Dan M. Keaton
536 Laurel Lake Drive
N. Augusta, SC 29841

Raynel's, Inc.
% Kenneth Embree
5802 Summitview
Yakima, WA 98908

Reflections
% Denny McBrien
210 W. C Street
McCook, NE 69001

Reynolds Sales
% Edward Reynolds
P.O. Box 165
Red Springs, NC 28377

Rocky Mountain Satellite Systems
% Jeanne Zarback
4141 Sinton Road
Colorado Springs, CO 80907

SAT PRO
% Mr. Ricky Klein
P.O. Box 1042
Sedalia, MO 65301

SAT PRO Communications, Inc.
% Lew Robinson
P.O. Box 1017
Janesville, WI 53547

SAT-COM Marketing, Inc.
% Jim Mackerelle, President
Rt. 1, Box 268A
Kelleyville, OK 74039

Sat-View Systems, Inc.
Rt. 1, Box 67A
Berryville, AR 72616

Satellite Communications Corp.
% Joe Tongish
Box 8304
Topeka, KS 66609

Satellite Concepts, Inc.
% Thomas Karr Houston
2142 Brighton Ct.
Winston-Salem, NC 27103

Satellite Downlink
% Robert G. Shelton
410 N. Pine Street
Patchogue, NY 11772

Satellite Home Entertainment, Inc.
% Ronald J. Burton, Pres.
1185 Sagamore Parkway
Lafayette, IN 47906

Satellite Reception Systems
% Walter J. Everett
145 N. Columbus Road
Athens, OH 45701

Satellite Reception Technology
Mr. Robert J. Subr
2965 Alpen Glow Way—Box 1718
Steamboat Springs, CO 80477

Satellite Scanners, Inc.
% Kim A. Lord
1819 S. Central, #42
Kent, WA 98032

Satellite Supply, Inc.
% Robert Dominix
2075 N.E. 154th Street
N. Miami Beach, FL 33162

Satellite TV Systems
% Mr. Dana Harrington
3406 State Road 38 E.
Lafayette, IN 47905

Satellite Technology, Inc.
% Mary Jo Rosecan
2302 Preston Trails Cover
Austin, TX 78747

Satellite Television, Inc.
P.O. Box 817
Pocatello, ID 83201

Satellite Television System
a Subsidiary of Kayburn, Inc.
% Stephen Kaloroplos, Pres.
Robers Plaza, 123 By-Pass
Clemson, SC 29631

Satellite Video Distributors, Inc.
% Bill Wylds, President
P.O. Box 5145
McAllen, TX 78501

Satellite Video Supply
% Dale Kernen
307 Broad Street
Conway, IA 50834

Satellite Video Systems
% Jay G. Smith
P.O. Box 2042
Westminster, MD 21157

Satellite Video Systems, Inc.
Mr. C.R. Langsett
1285 Manheim Place
Lancaster, PA 17601

Schmidt TV
% Raymond Schmidt
R.R. 2, Box 9, Hwy 218 South
Austin, MN 55912

Scientific Earth Stations
% Andy Vargo
103 Benton Spur Road
Bossier City, LA 71111

Shafer Radio and TV, Inc.
% Murrel Morehart, Mgr.
617 E. Hartford
Ponca City, OK 74601

Shafer TV
% Larry Miller
2111 Oklahoma Avenue
Woodward, OK 73801

Shaw TV & Appliance
% Sherman Freeman
1621 Austin Avenue
Brownwood, TX 76801

Sky-Scan of Kentucky
Mr. Bob Banks
124 Lincoln Drive
Frankfort, KY 40601

Skyline Satellite Systems
% Craig Larson
RT. 1, Box 187-A
Brainerd, MN 56401

Smith Music & TV, Inc.
% Robert W. Hornbeck
109 N. Bullard
Silver City, NM 88061

Snider's TV
% Chris Snider
312 W. Broadway
Steeleville, IL 62588

Solid State Communications, Inc.
% Mr. Don Ruff
303 E. 29th
Topeka, KS 66605

Sooner TV
% Jerry Hari
328 E. Downing
Tahleguah, OK 74464

Sound Room
% Jent Hubbard
802 N. Federal
Riverton, WY 82501

Sperry TV
% John C. Sperry
1115 N. 47th Street
Lincoln, NE 68503

Stan R. Esecson
2193 S. Dayton
Denver, CO 80231

Stan's Electronics
% Stan Meyers
910 E. Court
Beatrice, NE 68310

Star Systems
% Ronald Walker
Box 1991
Arlene, TX 75751

Star-Com Satellite Systems
% Gary R. Moore, V.P.
1009 Gregg Street
Big Springs, TX 79720

Starpath II
% Roy Lee Helm
Box 212
Jamestown, KY 42629

Startech, Inc.
% Mr. Richard C. Goodwin
P.O. Box 9737
Roanoke, VA 24020

Steger TV Radio Corp.
Rick De La Cardis
45 W. Sauk Trail
South Chicago Heights, IL 60411

Stellarview Satellite Systems, Inc.
% Gary S. Knox, President
431 Hwy 17 S.
P.O. Box 4718
Surfside Beach, SC 29577

Steve Esslinger
1120 2nd Street
Clay Center, KS 67432

Stevenson's TV & Appliance
% Leon Stephenson
2740 West Main
Bozeman, MT 59715

Stocksmans Supply
% Ken Minnie
P.O. Box 343
Roundup, MT 59072

Stucky Bros.
% R. James Benninghoff
5601 Coldwater Road
Ft. Wayne, IN 46802

Sullivan Enterprises
% Mr. Thomas C. Sullivan
5-1268-D
Swanton, OH 43558

Summit Audio-Vision
% Larry C. Smith
Box 900 Suite 169
Silverthorne, CO 80498

Sunlife
% Ed Stahl
Box 517
Tabernash, CO 80478

Switlik's Eagle Electronics
% James Switlik
130 E. Cherry
Nevada, MO 64772

TV Appliance Center
% Gene Erlardson
409 W. Main
Lewistown, MT 59457

TV Headquarters
% Robert Hammett
400 N Hackley
Muncie, IN 47303

TV Land
Clifton J. Anderson
2461 I-40 West
Amarillo, TX 79109

Tabler Furniture, Inc.
% Charles Tabler
401 Broadway
Larned, KS 67550

Team Electronics
% Gerald A. Snyder
1101 Omaha Street
Rapid City, SD 57701

Ted McDonald TV
% Ted McDonald
P.O. Box 66
Gay's Mills, WI 54631

Tele-Sat, Inc.
Luther L. Hancock, President
P.O. Box 500
Batson, TX 77519

Television Satellite Systems, Inc.
% Barbara Witczak
2122 24th Place, N.E.
Washington, D.C. 20018

Telstar Enterprises
% Mr. Edmond S. Winslow
1733 Esplanade
Chico, CA 95926

The Satellite Link, Inc.
Mr. Frank Abruzzo
303 SW 76 Terrace
N. Lauderdale, FL 33068

The Satellite and Sound Connection, Inc.
% Avi A. Kay, President
166 West 25th Street
Hialeah, FL 33010

The Video Center
% David L. Shadley
3012 Main
Parsons, KS 67357

The Video Set
% Larry Williams
2361 North Academy Blvd.
Colorado Springs, CO 80909

Tom Padgitt, Inc.
% Wilton A. Lanning, Jr.
5054 Franklin Avenue
Waco, TX 76710

Town & Country Electronics
% Ed Gendes
135 S. Main
Walnut, IL 63176

Town & Country TV
% Chester Davis
202 N. Timberland
Lufkin, TX 75901

Translator TV, Inc.
% Dick Statham
1160 Woodstock Drive
Estes Park, CO 80517

Transvision Corp.
2100 Redwood Hwy.
Greenbrae, CA 94904

Tri-Lakes Electronics, Inc.
% Robert S. Anderson, President
106 2nd Street
Monument, CO 80132

Tri-State Aviation
% Gerald S. Beck
P.O. Box 820
Wahpeton, ND 58075

Tri-X Engineering, Inc.
P.O. Box 70
Kent, WA 98031

Tulsat Corporation
% Mr. Allan McCoy
1839 N 105 E Avenue
Tulsa, OK 74116

Turner Satellite Systems
% Charles C. Turner
RR 1
Neoga, IL 62447

US Stereo & Video
% Tim Casey
2430 U.S. 6&50 Mesa Mall
Grand Junction, CO 81501

Universal Home Satellite
% Dave Maron
250 26th Street, #202
Santa Monica, CA 90402

VIA SAT Comm
5201 Bridge Street
Ft. Worth, TX 76103

Vern's TV & Radio
% Vern Sallee
354 E. Avenue
Limon, CO 80828

Video Kingdom
% Lynn Weaver, President
2418 N. Webb Rd., Suite E
Grand Island, NE 68801

Video Technology, Inc.
% John J. Kripps
3716-18 Lorna Road
River Oaks Village
Hoover, AL 35216

Video Warehouse
P.O. Box 218
Woodland Park, CO 80863

Video World
% Terry Phelps
1524 Mappa St.
Eau Claire, WI 54701

Videosat, Inc.
% Vince Kelly
P.O. Box 449
Prescott, AZ 86302

Vidtech Communications, Inc.
% Eugene Park
107 N. 59th Street
Seattle, WA 98103

Visalli Satellite Systems, Inc.
RD 5, Tuckahoe Road
Vineland, NJ 08360

World Sat—Division of Lawhorn, Inc.
% Gordon H. Lawhorn
6325 Chamberlayne Road
Mechanicsville, VA 23111

Wadena TV Center
% Mr. James Lundquist
311 Jefferson Street North
Wadena, MN 56482

Walton Satellite TV
% Larry & Ralph Walton
905 E. Cumberland Street
Lebanon, PA 17042

Wax TV & Sound Center, Inc.
% William Wax
1716 Kings Hwy
Shreveport, LA 71103

West Winds Construction
% Gerald Garnick
1041 S. 5th Street
Douglas, WY 82633

West-Hills Electronics
% Fred W. Mack II
5992 Steubenville Pike
McKees Rocks, PA 15136

Westar Satellite and Communications
% Timothy Regnitz
N56 W6523 Center Street
Cedarburg, WI 53012

Weststar Enterprises, Inc.
% C.H. Spangler
6836 San Pedro, Suite 120
San Antonio, TX 78233

Wetherell Manufacturing Company
407 W. Grace
Cleghorn, IA 51014

Wheeler Communications
% Jim Wheeler
Route 116
Flanagan, IL 61740

White Refrigeration Co.
% Joe Summens
9th St. & 9th Ave.
South Great Falls, MT 59405

Whites TV
% Clarence Whites
P.O. Box 903
Rolls, MO 65401

Wyoming Satellite Television
% Dan Pince
Route 65, Box 927
Pavillion, WY 82523

APPENDIX **D**

Diameter vs Location: How to Calculate What Size Dish to Buy

This Appendix provides the answer to the question of how large a diameter antenna is required to successfully pick up sparkle-free television pictures from a given domestic or international satellite. Antenna gain formulas are provided along with several convenient charts and tables which eliminate any drudgery involved in calculations. Finally, a set of satellite footprint maps are included to provide approximate EIRP signal strengths of satellite transponders throughout their effective coverage area. Satellites age over time, and their power outputs decrease. Thus, it is usually prudent to factor in a 1 to 2 dB cushion in determining the minimum antenna size required for a given location.

HOW LARGE A DISH DO I NEED?

Todays modern generation of C-band satellites and newer, more efficient receiver and parabolic antenna designs have made the old 15-foot (diameter) industry standard dish almost obsolete.

In many parts of the country, excellent home reception of the popular tv satellites can be obtained with an 8-foot dish, and on some transponders with specially tuned high performance systems, six-foot and even four-foot dishes have been shown to work well. The realistic minimum-size antenna which will deliver good pictures throughout the country is 11 feet. In the central areas of the United States, 8-foot dishes are more than adequate. (See Fig. D–1.) These dish sizes assume that a 120 LNA and TVRO receiver with 8-dB threshold extension are used to complete the system package. It also assumes that the dish has an efficiency of at least 55%—a common figure today.

As the Clarke belt becomes more crowded, the orbital arc spacing between satellites will be reduced from the original 4° to 3° to as little as 2° by the late 1980s. When this ultimately happens, an 8-foot dish will

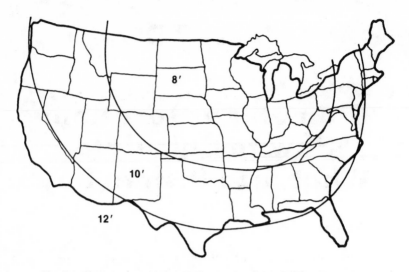

Fig. D-1. Footprint map of the US showing the minimum TVRO antenna size for good tv reception.

probably be too small to pick up just one satellite; the adjacent birds will cause "crosstalk" interference to occur. An 11-foot dish is a much safer choice to buy.

CALCULATING SYSTEM GAIN

The ability of a TVRO system to pick up a noise-free picture from a 5- to 12-watt transmitter located 22,300 miles away is a function of a number of elements: Satellite EIRP output as received on the ground, dish efficiency, LNA and antenna noise figures, antenna size, and receiver "sensitivity" and internal noise. The natural "black body" background noise radiated upwards by the earth itself and man-made terrestrial microwave interference also affect overall system gain and signal-to-noise ratios. Thus, to receive a clean satellite television picture, sufficient gain must be present in each of the system elements to overcome background and thermal radiation noise, and each system element must also have the lowest "noise figure" possible. This is why LNAs are rated in terms of their internal noise-generating figures, rather than in terms of raw gain.

The signal-to-noise (S/N) ratio of a TVRO system is a function of the received carrier-to-noise ratio (CNR) and it is this figure which is most frequently used in the industry. The formula for CNR is:

CNR = EIRP − Path Loss in Space + G/T − 10 Log (Bandwidth) +
      Boltzman Constant

where,
   EIRP is the satellite EIRP in dBW,
   Path loss in space equals 196.5 dB,
   G/T is the figure of merit or Gain/Noise in Kelvins of the antenna,
   10 log bandwidth = 10 log 36 MHz = 75.6,
   Boltzman's constant equals 228.6.

Thus, the equation becomes

$$CNR = EIRP - 196.5 + (G/T) - 75.6 - 228.6$$

The figure of merit (G/T) is calculated by the equation:

$$G/T = \frac{\text{antenna gain in dB}}{10 \log (\text{antenna noise} + \text{LNA noise})}$$

where, the antenna gain in dB is calculated by:

$$\text{Antenna Gain} = 10 \log \left[\left(\frac{4\pi(A)}{f_{wl}} \right) \times E \right]$$

where,
   A is the area of aperture of the antenna (πR^2)
   f_{wl} is the wavelength of the frequency,
   E is the illumination efficiency usually 55%).

   The footprint maps of the most popular satellites now in orbit are given in Figs. D−2 through D−9. Fig. D−2 is the footprint map of the EIRP power output as received at various points (measured in dBW) for Satcom 1 (now co-located with Satcom 2 at 119° W), the "Grandaddy of the cable television birds." The radiation output patterns vary slightly from transponder to transponder. Fig. D−3 is the footprint map for Satcom 3 (131° W), the "cable bird." The Satcom 4 (83° W) footprint map is given in Fig. D−4, and Westar 3 (91° W) in Fig. D−5. The horizontal transponder footprint maps for Westar 4 (99° W) and 5 (123° W) are given in Figs. D−6 and D−7. The horizontal transponders shown for these satellites share power with the Hawaii spot beam; therefore, add approximately 2 dB for the vertical transponder power levels. Fig. D−8 gives the footprint map for Comstar 4 (127° W). The Canadian Anik D satellite (104° W) foorprint map is given in Fig. D−9.
   The projected footprints expected to be obtained for the Galaxy 1 and 2 satellites are given in Figs. D−10 through D−12. The projected footprints for the horizontal transponders of Galaxy 1 (135° W) are given in Fig. D−10. These horizontal transponders will share power with the Hawaii spot beam (power level in Hawaii will be between 22 and 26

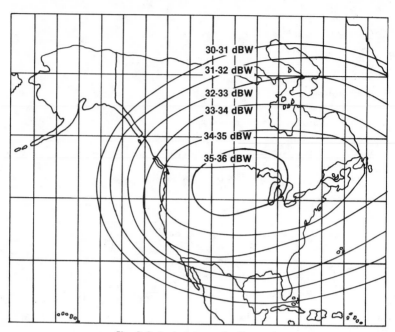

Fig. D-2. Footprint map for Satcom 1.

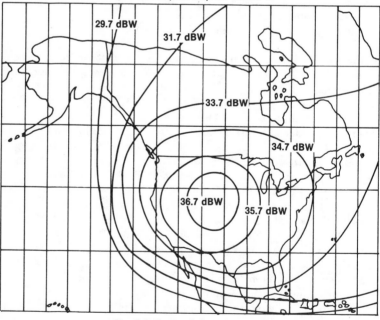

Fig. D-3. Satcom 3 footprint map.

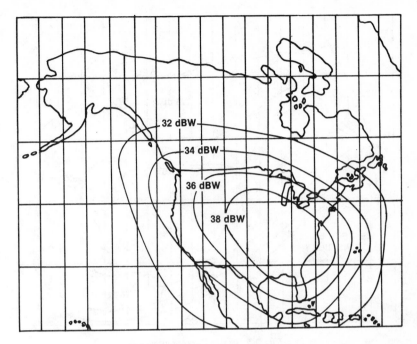

Fig. D-4. Footprint map for Satcom 4.

dBW). The footprint map for the horizontal transponders of Galaxy 2 (74° W) is given in Fig. D—11, while the vertical transponders for this satellite are mapped in Fig. D—12. To use the maps, assume that an 8-foot dish in St. Louis, Missouri is used to pick up a Satcom 3 satellite transponder radiating 37 dBW. The calculations are:

$$CNR = 37 \text{ dBW} - 196.5 \text{ dB} + 15.7 \text{ dB} - 75.6 + 228.6$$
$$= 9.2 \text{ dB}$$

assuming a 15.7 dB figure of merit (G/T) for the 8-foot antenna.

CNR is a measure of the received signal power level to the noise of the overall TVRO system. For an fm receiver to "capture" a video signal, without resorting to "sparkle-suppression" techniques (which also reduce the picture resolution), a minimum CNR of 7 dB is required, with an 8-dB CNR preferred. Thus, this 8-foot dish antenna has more than adequate CNR for presenting a noise-free picture to the home television viewer. Table D—1 presents a summary of antenna gain and G/T figures for popular antenna sizes.

TVRO receivers are now being built with both 8-dB and 7-dB CNR figures. It is expected that careful "tuning" of all the system elements

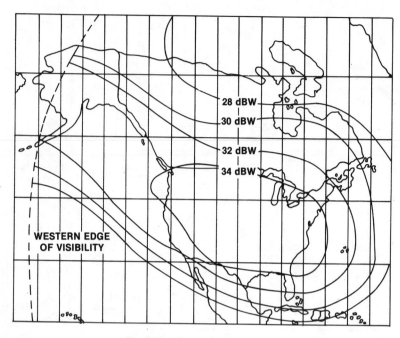

Fig. D-5. Westar 3 footprint map.

will result in a reduction to about 6 feet of the dish size as the downward limit for reception of the C-Band satellite transmissions. This assumes that "deep dish" carefully shaped parabolic antennas with 80% to 85% efficiency, 80 K LNAs, and narrow i-f, noise-processing receivers

Table D−1. Parabolic Antenna Gain*

Dish Size	Dish Gain (in dB)	G/T
6 foot/1.9 meters	35.5	13.1
6 foot/2.1 meters	36.5	14.4
8 foot/2.4 meters	37.5	15.7
9 foot/2.7 meters	38.5	16.7
10 foot/3.0 meters	39.5	17.8
11 foot/3.4 meters	40.0	18.6
12 foot/3.8 meters	41.0	19.4
13 foot/4.2 meters	41.5	20.1
15 foot/4.6 meters	43.0	21.4
16 foot/5.0 meters	44.5	21.9
20 foot/6.0 meters	45.5	24.0

*Assumes 55% efficient dish and a 120 K LNA is used.

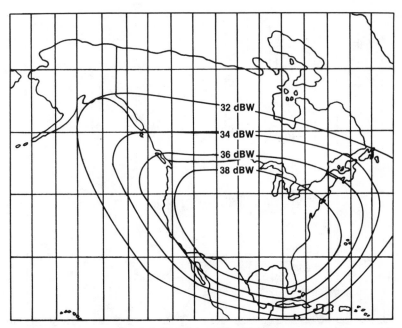

Fig. D-6. Footprint map for Westar 4.

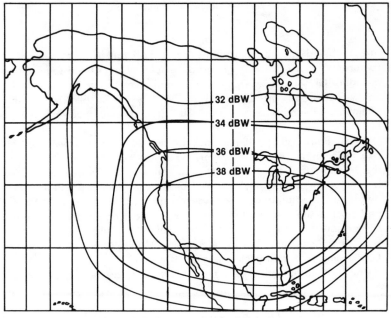

Fig. D-7. Footprint map for Westar 5.

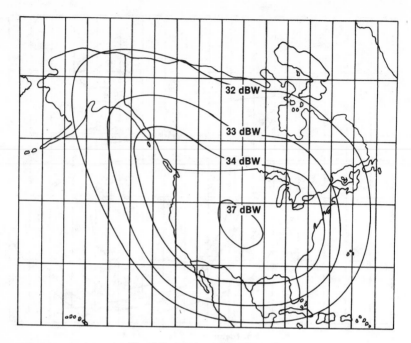

Fig. D-8. Comstar 4 footprint map.

are combined into an integrated systems package. A number of manufacturers are now working to introduce such systems, and several firms
including Gillaspie & Associates, Inc. and Automation Techniques, Inc.
now sell such "optimized" small-bore TVRO systems. Table D–2 presents
minimum antenna sizes for given EIRP outputs and receiver combinations.

The C-Band TVRO antenna may be limited to a slightly larger dish
size for another reason, however. When the FCC originally authorized
the launch of C-Band domestic communications satellites, it contemplated an orbital spacing of 4° from satellite-to-satellite in the Clarke
belt. The Canadian government uses 5° spacing. Recently, both the FCC
and WARC (the World Administrative Radio Conference) authorized
tighter satellite spacing. The international WARC rulings call for 2° between each satellite, allowing a doubling of the number of communications satellites in geosynchronous orbit.

The formula to compute the beamwidth of a parabolic TVRO antenna
is:

$$\text{Beamwidth (3 dB point)} = \left(\frac{52.3}{\sqrt{E}}\right) \left(\frac{f_{wl}}{\text{ant diameter}}\right)$$

358

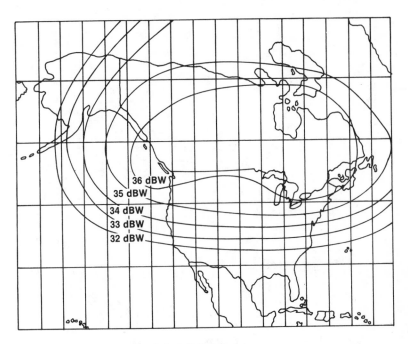

Fig. D-9. Anik D footprint map.

Table D-2. Satellite EIRP Output vs Antenna Size

EIRP (dBW)	Minimum Antenna Size	
	8 dB CNR Receiver	7 dB CNR Receiver
38	7 feet	6 feet
37	8 feet	7 feet
36	9 feet	8 feet
35	10 feet	9 feet
34	11 feet	10 feet
33	12 feet	11 feet
32	13 feet	12 feet
31	15 feet	13 feet
30	17 feet	15 feet
29	20 feet	18 feet
28	24 feet	20 feet
27	26 feet	22 feet
26	30 feet	24 feet
25	33 feet	27 feet

Assuming a 120 K LNA is used, with antenna efficiency of 55% and 30° dish elevation. Receiver provides full 30-MHz bandwidth. If a 100 K LNA, 65% efficiency antenna with a 20-MHz bandwidth is used, the overall dish diameter can be reduced by about 10–15%.

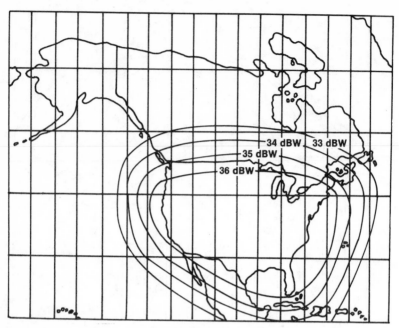

Fig. D-10. Projected footprint map for Galaxy 1.

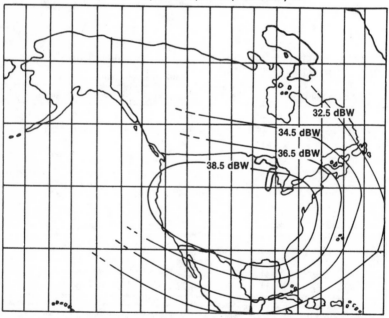

Fig. D-11. Projected footprint map for the horizontal transponders of Galaxy 2.

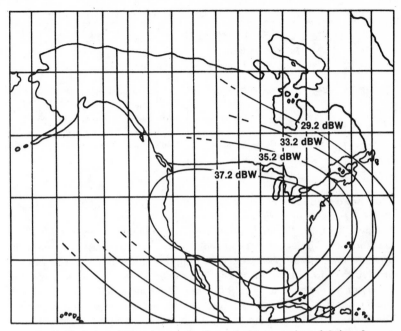

Fig. D-12. Projected footprint map for the vertical transponders of Galaxy 2.

where,

 E is the antenna efficiency,

 f_{wl} is the wavelength (.237 feet)

For an 8 foot antenna operating at 70% efficiency, the 3 dB beamwidth = 1.85°, which will likely be adequate for reception of a satellite which is 2° adjacent to other satellites on either side of it. A 6-foot dish has a beamwidth of 2.47°, however, and at spacing under 2.5 degrees significant "crosstalk" interference from adjacent satellites may occur. Thus, it seems likely that until a new generation of TVRO antennas with specially engineered sidelobe attenuators arrive on the scene, the 7- to 8-foot parabolic antenna will be the smallest size to be sold in a world of 2° satellite spacing.

SUBJECTIVE QUALITY OF TV PICTURES

The North American C-Band communications satellites have transponder channel bandwidths of 36-MHz, separated by 4-MHz guard channels. The NTSC 4.2 MHz video signal at baseband is typically modulated at the T-R uplink with a 10.75 MHz fm deviation. The i-f bandwidth of a TVRO receiver needed to successfully pass all picture information is 2 ×

Table D-3. CNR and S/N Ratios vs Viewer Acceptability

CNR	S/N	Rating by Viewers	Perceived Quality
18	55	99% = Excellent	Network industry standard.
16	53	99% = Excellent	TV station quality.
14	51	95% = Excellent	CATV system quality TVRO.
12	49	90% = Excellent	CATV quality. Home near headend of CATV system.
10	47	80% = Excellent	CATV quality. Home at furthermost point on cable from CATV headend.
8	45	70% = Excellent 99% = Fine	Good home TVRO system.
7	43	50% = Fine	Acceptable Home TVRO system.

(Baseband frequency + Peak Deviation) = 2 (4.2 + 10.75) = 2 (14.95) = 29.9 MHz. Therefore, most commercial TVRO receivers have an i-f bandwidth of 30 MHz or greater. However, the larger the bandwidth of a signal, the higher the fm threshold must be to successfully receive it. Thus, some home TVRO receiver manufacturers have purposely restricted their i-f bandwidth to 24, 22, or even 18 MHz. This increases the ability of the receiver to accept a lower CNR (and smaller dish size) with the tradeoff of reducing picture "sharpness" (some of the high frequency signal information which contributes to crisp edges between picture elements is lost). In fact, some of the satellite programmers intentionally restrict their peak deviations at uplink to as little as 10 MHz which can make their picture appear "hotter" when received from the bird. The difference can be dramatic. A receiver with a noise figure of 14 dB will have a −104.44 dBW threshold (input signal level) when its i-f bandwidth is 36 MHz. The same receiver will only require −107.96 dBW at a 16-MHz i-f bandwidth. This is an improvement of almost 3 dB, equivalent to a power increase of 100%.

Viewing a tv picture is very subjective, and acceptable CNRs can vary greatly from person to person. Over the years, however, studies have been conducted by the Television Allocations and Study Organization (TASO) to quantify the subjective picture quality acceptability versus CNR and S/N ratios. Table D-3 presents the findings of these studies extrapolated for the TVRO user. The goal is to achieve the highest CNR possible. With a CNR of 13 dB or higher, superb tv reception should be expected. However, for the vast majority of viewers who purchase state-of-the-art TVRO equipment, a CNR of better than 9 dB will be obtained with the popular 11-foot antenna throughout the United States. Since TVRO systems and satellite transponders usually deteriorate with age, if it is possible to use a slightly larger antenna, the effort will be worth it. An extra dB or two of gain will never be wasted.

Locating The Birds in Space: How to Aim Your Dish

In this Appendix, the basic geometry of orbital mechanics is presented. How to locate a satellite parked in the geosynchronous Clarke orbit is actually quite a simple procedure. Anyone who vaguely remembers his or her high school geometry—along with a dash of algebra—will find the following material a lot of fun.

After the basic concepts have been covered, an Apple® II* computer program, written in a very standard BASIC is given. This program will automatically compute the proper azimuth, elevation, and polar mount declination angles for over 100 communications satellites either now in orbit, or scheduled to be launched over the next few years. Finally, a short satellite tracking program, written for the Texas Instruments TI 58/59 series of programmable calculators, is also provided along with sample printouts. Information on how to obtain a customized satellite tracking/location printout for any place in the world is included at the end of this Appendix.

INTRODUCTION TO ORBITAL MECHANICS

Since the time of the ancient Egyptian kingdoms, sailors have used the stars to plot their course across the seas. By the middle ages, elaborate maps of the heavens had been produced to provide the navigator with celestial sightings to be used when the ship was beyond the sight of land. These sky maps varied from location to location, thus suggesting to the early scientific minds that the sun and stars were part of a big sphere rotating around the earth.

By the time of Napoleon Bonaparte, the proper relation of the planets to the sun had become known and astronomers had systematically divided the spherical earth into units of angular measurement. The British Navy established two systems of measurement: latitude and longitude.

---

*Apple is a registered trademark of Apple Computer Inc.

Using the equator as a starting point, defined as zero degrees latitude, the earth was divided into north and south angular measurements called degrees. The latitude increased as one traveled northward, until at the North Pole, the latitude was 90 degrees. As one traveled southward, the latitude decreased until reaching −90 degrees at the South Pole.

Likewise, using the Royal Observatory at Greenwich as a starting point, the earth was further divided into 360 degrees of longitude. Traveling westward around the globe from the Greenwich Meridian (zero degrees longitude), the west longitude increased until reaching 360 degrees back at the starting point. Traveling eastward in the opposite direction from Greenwich the east longitude would increase. This system of angular measurement was permanently codified at an international conference in Washington, DC, held at the turn of the century, and is in universal use today.

Since the radius of the earth is approximately 3,957 miles, its circumference at the equator is $2\pi R$, or about 24,860 miles.

Therefore, each degree of latitude or longitude (at the equator) represents a distance of approximately 69.057 miles (24,860 divided by 360 degrees). See Fig. E−1.

For a planet with the mass of the earth, Newton's Law of Gravity (the attraction of two bodies toward each other) indicates that any object placed in an orbit approximately 22,300 miles above the surface of the

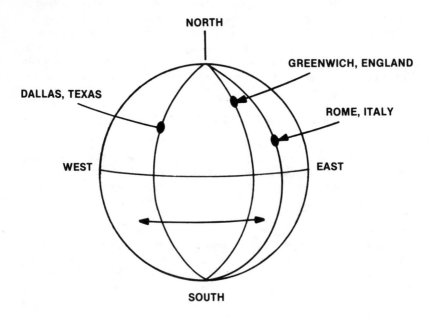

Fig. E-1. Longitude and latitude of the Earth.

earth (or 26,257 miles from the center of the earth) will rotate around the earth precisely once every 24 hours. If this body is positioned in an orbit which cuts a plane through the equator of the earth, then the object will travel in its orbit around the earth at the same speed the earth turns below it. Thus, the object will appear to remain fixed or "geostationary" with respect to an observer located on the surface of the earth. The majority of the satellites are parked in this orbit—known as the Clarke orbit—directly above the equator (latitude of the satellite = zero).

As the satellites are, in fact, flying through space at almost 7,000 miles per hour to complete their daily orbit of 163,628 miles, they occasionally hit minute particles of space dust. These impacts, along with the irregular influence of the gravity pull from mountains, causes the satellites to wobble a bit on their orbits.

Thus, "stationkeeping" techniques must be employed to keep the satellites correctly positioned at their designated longitude and precisely over the equator. Two types of mechanical positioning techniques are used, depending upon the specific manufacturer of the satellite. In the first instance, the satellite can be "spun stabilized"; that is, the body of the satellite, by necessity a cylindrical shape, can be set to spinning around its axis. This principle, used by gyroscopes and spinning tops, will minimize any tendency for the satellite to wander off its axis.

The second technique consists of periodically firing compressed gas jets, located at key points, on the satellite's body. These tiny jets release just enough propellant to counteract the slight movements of the satellite. Since the propellant has been pumped into the satellite's tanks at the launch site, a limited supply is carried—generally enough to stabilize the correctly orbited satellite for seven to ten years. After that time, the satellite is decommissioned, its transmitters are switched off and it begins to wander aimlessly around the heavens. To minimize the dangers of collision with other orbiting vehicles, sometimes a satellite is nudged into a different orbit which will accelerate its orbital decay back into the atmosphere—and a fiery death.

A specific location on the surface of the earth is measured in terms of its latitude and longitude in reference to the equator and Greenwich, England. A satellite TVRO antenna, however, is pointed at a heavenly body, and the dish's pointing angles, known as *azimuth* and *elevation* are computed differently.

The *azimuth* of an antenna is simply its angle of left-to-right rotation. It is its clockwise rotation when looking downward from above the antenna. By definition, when the antenna is pointed true (not magnetic) north, its azimuth is zero degrees. As the antenna is rotated, its azimuth increases. Due east is 90 degrees, south is 180 degrees, and west is 270 degrees. The *elevation* of an antenna is its angle of downward or upward tilt. When aimed straight up, the elevation is 90 degrees. The elevation angle is at zero degrees when the antenna is pointed at the horizon.

Two basic earth station antenna mounts (Fig. E–2) have been developed to permit the easy positioning of a satellite dish to point at a desired satellite in geosynchronous orbit. The Azimuth/Elevation (Az/El) mount (Fig. E–2A) is, perhaps, the easiest to align. To change the antenna position, two axes of rotation are varied. The polar mount (Fig. E–2B) replaces these two axes with a single hour angle axis through which the antenna is rotated. By properly aligning the polar mount's declination axis with the arc of geostationary satellites, the polar mount can be made to accurately track, with some approximation, all the satellites in the Clarke orbit. Thus, the polar mount antenna is tending to be the more popular, and lends itself to single-motor mechanical positioning.

While the Az/El mount is inherently more stable, the polar mount antenna must be rotated around the polar axis. Thus, for those satellites that are more than 20 degrees from due south (for a given TVRO location), the structural loading of the antenna is unbalanced. In addition, the polar mount is truly polar only directly under the satellite at the

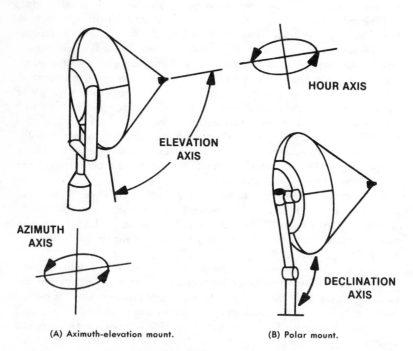

(A) Aximuth-elevation mount.   (B) Polar mount.

Fig. E-2. Popular TVRO mounts.

equator. At any other latitude, a declination adjustment must be made to approximate the correct antenna positioning.

To correctly position the polar mount antenna, the declination angle must be computed. See Fig. E-3. The formula for the declination angle average is:

$$\text{Declination Angle} = 90 \text{ degrees} - \tan^{-1}\left[\frac{3957 \text{ Sin Lat}}{22,300+3957 \ (1 \text{ Cos Lat})}\right]$$

where,

Lat is the latitude in degrees of the TVRO antenna,
3957 is the radius of the earth in miles,

Using this formula, for example, the declination angle for a polar mount antenna located at 40 degrees, 45 minutes North (New York City), is about 83.65 degrees. In normal TVRO antenna practice, however, the declination angle is usually related to the equatorial plane, which is at right angles to the earth's polar axis. Thus, the "real world" declination angle of our example is said to be 83.65 −90.00 = −6.35 degrees.

Almost all of the polar mount antennas manufactured today are, in reality, approximations to the true polar mount, and use a series of

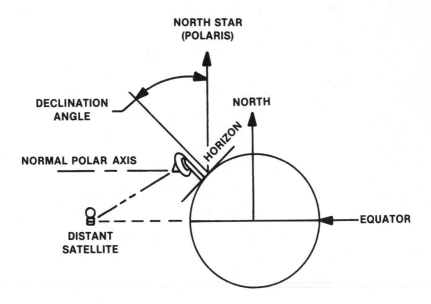

Fig. E-3. The declination angle using a polar mount offsets north.

shims, adjustable bars, etc., to minimize the tracking error. Fig. E-4 shows the reason why.

The polar mount antenna, once it has been properly aligned with true north, and with its declination angle correctly set, will cut out an arc in space which closely (but not exactly) matches the arc of the satellites in orbit. When located on the equator, a polar mount need only be pointed straight up. The arc of satellites would then be seen by simply rotating the antenna from left to right (east-to-west) about its hour-angle axis. Most TVRO dishes are located in northern latitudes ranging from +25 to +50 degrees north. Here, the antenna no longer cuts out an arc in space which matches the geosynchronous satellite belt. This error increases the farther away from the equator one travels. Thus, a polar mount will require some slight modification by the manufacturer to "warp" its declination angle upward and downward as the antenna is rotated through its hour axis.

For this reason, a dual-motor Az/E1 motorized antenna mount driven by a microprocessor controller is the ultimate choice of the TVRO enthusiast. This configuration, in many cases, can also enable the satellite television viewer to watch the Russian domestic Molyna-series satellites which are positioned in elongated polar orbits over the northern skies instead of geostationary orbits. These satellites require a steerable TVRO dish because they must be "tracked" as they move through the heavens.

Given the known longitude of a desired satellite in the Clarke orbit (for

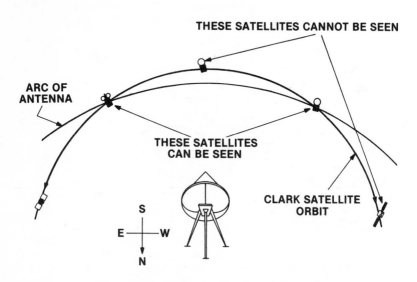

Fig. E-4. The polar mount arc vs the satellite orbit.

example the RCA Satcom 3 24-transponder bird located at 131 degrees west longitude) and the latitude and longitude of the TVRO antenna on the surface of the earth, a simple set of geometric equations will provide the desired azimuth and elevation angle settings. The first step, however, must be to properly align the antenna with respect to true north.

There are a variety of methods available to locate true north from any position in the northern hemisphere. Perhaps the easiest consists of using a magnetic compass. Since magnetic north varies greatly from true north, depending upon both longitude and latitude, mariners have, for years, used magnetic compass correction tables to compensate for the error. Such a table for North America is presented in Table E−1. To use this table, add or subtract the appropriate degrees from the actual compass reading depending upon westward correction. For example, in San Francisco, California, a magnetic compass will read to the east by

Table E−1. Magnetic Compass Corrections for True North

Alabama	2E	New Hampshire	16W
Alaska	26E	New Jersey	11W
Arizona	14E	New Mexico	13E
Arkansas	6E	New York	10W
California	17E	North Carolina	5W
Colorado	14E	North Dakota	11E
Connecticut	13W	Ohio	3W
Delaware	10W	Oklahoma	9E
Washington, DC	8W	Oregon	20E
Florida	2E	Pennsylvania	8W
Georgia	0	Rhode Island	15W
Hawaii	11E	South Carolina	2W
Idaho	19E	South Dakota	11E
Illinois	2E	Tennessee	1E
Indiana	0	Texas	10E
Iowa	6E	Utah	15E
Kansas	9E	Vermont	15W
Kentucky	1E	Virginia	6W
Louisiana	6E	Washington	22E
Maine	20W	West Virginia	5W
Maryland	8W	Wisconsin	2E
Massachusetts	15W	Wyoming	13E
Michigan	3W		
Minnesota	6E	Alberta	22E
Mississippi	5E	British Columbia	23E
Missouri	6E	Manitoba	10E
Montana	18E	Ontario	8W
Nebraska	11E	Saskatchewan	17E
Nevada	17E	Quebec	17W

17 degrees. True north is 17 degrees further west. Fig. E—5 presents a map of the United States with approximate magnetic correction lines drawn in. Using either the table or map will guarantee a true north error of under 1 degree, which should be more than adequate in initially aligning the antenna to find the birds. For locations in other countries, call the local airport control tower or consult a flight map for the proper magnetic compass correction figure.

Finding Polaris (the North Pole Star) is another method frequently used to locate true north. Polaris is the bright star at the top of the Little Dipper's handle. The Little Dipper (Ursa Minor) is a pattern of seven stars which is very easy to find, the pointer stars which form the right-hand wall of the Big Dipper lead straight to Polaris. Fig. E—6 is a map of the North Star in relation to the Big and Little Dippers. To use the map, rotate it so the particular season is at the bottom, and then constellations will appear approximately as shown if viewed between 25 and 50 degrees north latitude. Once Polaris has been located, a line pointing from the observer's position on the earth northward to where the North Star would be if it fell to earth will provide true north. In the southern hemisphere the pattern of stars known as the Southern Cross will provide a similar indication of true south. Determining true south provides the same information for use in correctly aligning the antenna with the true North-South polar axis of the earth.

Another technique used to find true north (or south) makes use of the sun. Since a shadow at the time of local noon points to due North (in the northern hemisphere), this procedure requires simply knowing when real local noon occurs. (Because the earth has been divided into standardized time zones, real local noon will rarely be the same as noon time indicated on a clock.) First, note the times of the local sunrise and sunset from the daily newspaper. For example if local sunrise occurs at 6:03 am, and local sunset occurs at 7:55 pm, then local noon time is halfway between these times or:

$$\text{Local Noon} = \frac{12:00 - 6:03 + 7:55 + 6:03}{2} = 12:59 \text{ pm}$$

The true north/south direction for this location will be found as the shadow cast by a vertical pole (or a person). Both the Polaris-sighting and sun-shadow methods have their limitations, of course. They are difficult at best to employ when it is raining.

After aligning the antenna with true north to establish the correct location of zero degree azimuth, the declination angle can then be used to tilt the polar mount antenna on its polar axis by the necessary amount.

When properly aligned, the polar mount antenna will pick up satellite after satellite as it is rotated from east to west (left to right) about its

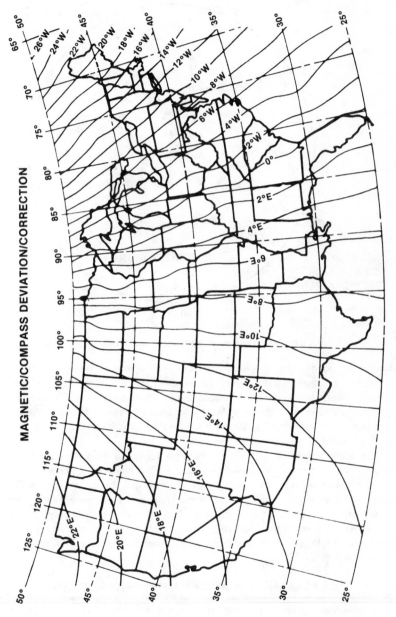

Fig. E-5. Magnetic map showing the compass corrections for the United States.
(Courtesy KLM Electronics, Inc.)

371

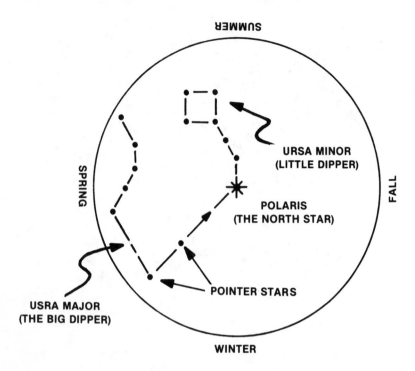

Fig. E-6. Sky map of the North Star (Polaris).

hour axis. To determine the correct elevation angle of an Az/El mount antenna, a simple inclinometer can be used. Commercial inclinometers used in site surveying work are sold by many dealers including the national chains like Montgomery Ward and Sears Roebuck & Co. A cut-out inclinometer, a piece of string and a weight will work just as well. Fig. E–7 provides a do-it-yourself version.

The inclinometer of Fig. E–7A can be removed from the book (or copied) and attached to a stiff cardboard. This scale is attached to the antenna with the line from the target symbol (at the point) to the 90° point on the scale positioned in exact horizontal alignment with the plane of the antenna. Then, attach a piece of string to the center of the target symbol and attach a weight to the other end of the string. The elevation angle is read at the point where the string passes the scale as shown in Fig. E–7B.

CALCULATING THE ANTENNA POINTING ANGLES

A half-dozen popular methods are available for determining what the correct TVRO azimuth and elevation angles should be for a given satellite. A commercially available plotting table can be used, or a hand

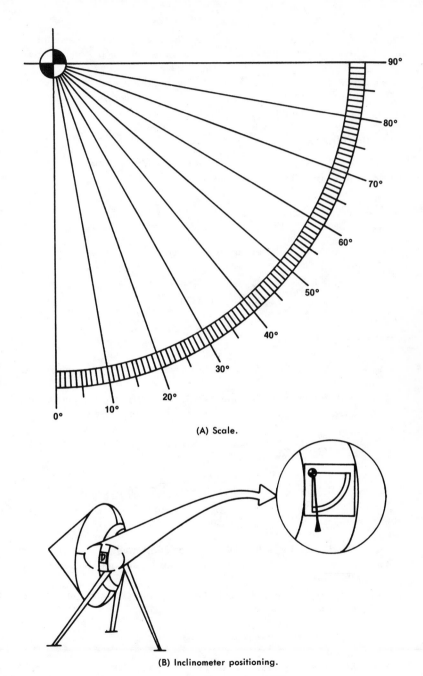

(A) Scale.

(B) Inclinometer positioning.

Fig. E-7. A "do-it-yourself" inclinometer.

calculation can quickly be performed. More sophisticated techniques may include using the power of a programmable calculator or a personal computer tracking program. This Appendix provides four ways for the reader to find the correct Az/El angles: hand calculation, look angle chart, Apple® II BASIC program, and a Texas Instruments programmable calculator routine.

Hand Calculation

Given the TVRO antenna latitude and longitude and the longitude of the satellite in the Clarke Orbit, the equations are:

$$\text{Azimuth} = 180° + \tan^{-1} \left(\frac{\tan B}{\sin A} \right)$$

for the northern hemisphere. For the southern hemisphere, the equation is:

$$\text{Azimuth} = \tan^{-1} \left(\frac{\tan B}{\sin A} \right)$$

where,

$\tan^{-1}$ is the arc tangent,

Angle A is the latitude of the TVRO antenna (north is positive and south negative,

Angle B is the longitude of the TVRO antenna (east is positive and west is negative) *less* the longitude of the desired satellite.

The range or distance between the TVRO antenna and the satellite in space is computed from the Laws of Cosines:

$$\text{Range} = \sqrt{R^2 + (R+D)^2 - 2R(R+D)\cos C}$$

where,

R is the radius of the earth (6367 km or 3957 miles)

D is the distance from the satellite to the earth's surface (35,800 km or 22,245 miles)

C is the "central angle" between the TVRO antenna and the satellite given by equation below:

$$C = \cos^{-1} (\cos A \times \cos B)$$

Given the range, the elevation angle may be computed from the following formula:

374

$$\text{Elevation} = \cos^{-1}\left[\frac{\text{Range} + R - (R-D)}{2 \times \text{Range} \times R}\right] - 90 \text{ degrees}$$

For example, a TVRO antenna located at San Francisco, California will be pointed at the Satcom 3 bird. The satellite is positioned at 131° west longitude. San Francisco is at 37° 45′ North Latitude, 122° 26′ West Longitude.

A = 37° 45′
B = 122° 26′ − (−131 degrees) = 8° 34′

$$\text{Azimuth} = 180° + \tan^{-1}[\tan(8° \, 34′) / \sin(37° \, 45′)]$$
$$= 193.7°$$

Taking the distance in miles,

R = 3957
D = 22,245
$$C = \cos^{-1}[\cos(37° \, 45′) \times \cos(8° \, 34′)]$$
$$= \cos^{-1}(0.7907 \times .9888)$$
$$= .6732$$

$$\text{Range} = $$
$$\sqrt{(3957)^2 + (3957 + 22{,}245)^2 - 2 \times 3957 \times (3957 + 22{,}245) \times \cos C}$$

$$= 23{,}240 \text{ miles}$$

$$\text{Elevation} = \cos^{-1}\left[\frac{(23{,}240^2) + (3957)^2 - (3957 + 22{,}245)^2}{2 \times (23{,}240) \times (3957)}\right] \, 90°$$

$$= 45.3°$$

For a planet the size of the earth, if the computed value of central angle C exceeds 81.3 degrees, then the satellite is below the horizon of the TVRO antenna, and cannot be seen. Likewise, if the value of angle B is less than 81.3 degrees, the satellite is also invisible to the TVRO antenna. In this case, the difference in longitude between the satellite and the TVRO antenna location is so great that the satellite will again be over the horizon.

Look Angle Chart

A far easier way to determine the correct azimuth and elevation pointing angles for satellites located in the northern hemisphere is to make use of a look angle chart which graphs the approximate angles for any given satellite location. Fig. E−8 presents such a chart for use in the

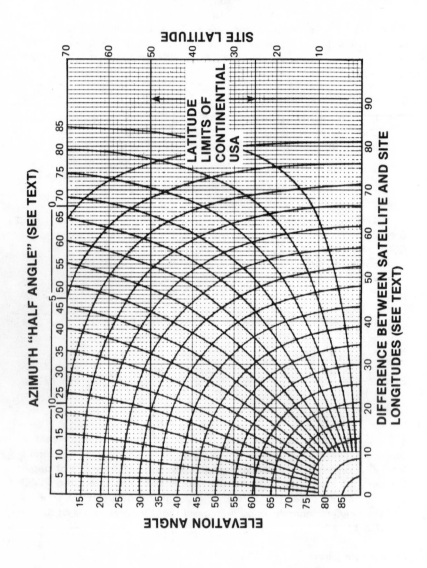

Fig. E-8. Satellite look angle chart.
(Courtesy Scientific-Atlanta, Inc.)

northern hemisphere. To use the chart, first determine the longitude and latitude of the desired TVRO site, and the west longitude of the desired satellite. Plot the TVRO latitude on the right-hand scale of the chart. Next, subtract the TVRO longitude from the satellite longitude, and plot this on the bottom scale of the chart. At the intersection of the two on the chart, read off the desired elevation and azimuth "half-angle" from the left-hand and top scales of the chart. The azimuth "half-angle" is used to determine the true azimuth by *adding* this figure to 180° *if* the satellite longitude less the TVRO longitude is positive. If the satellite longitude less the TVRO longitude is negative, *subtract* the azimuth "half-angle" from 180°.

APPLE® II BASIC Program

Many readers now own their own personal computers, and the following BASIC program will automatically calculate all of the parameters of interest for over 100 domestic, international, and future satellites in geostationary orbit. The Satellite Tracking Program II computes the azimuth, elevation, the declination angles along with the range and the distance-to-subpoint (that point on the equator that the satellite would hit if it fell to earth). The program also includes a built-in data base of 35 cities around the globe, along with their latitudes and longitudes. The program was written in a simple subset of Applesoft BASIC so that it can be easily converted from the Apple to run on another machine. With minor modification, notably to the printer commands found in lines 260, 350, 1204, and 4028, the program will run on the Radio Shack TRS-80®* and IBM personal computers.

In operation, the program asks the user to select among a number of options. The output will be displayed on the printer (located in slot 1—change line 391 if a different slot is used) or crt under user control. When "output option" is presented, the user can elect to obtain a listing of the current data base of satellites, their orbital positions, and whether they carry C-Band or Ku Band transponders by typing "LIST." Alternatively, the user can direct the computer to calculate the pointing angles for all the satellites in the data base ("ALL"), or just those US and Canadian domestic satellites of primary interest ("US"). Finally, by typing "KYBD," the computer will ask the user to provide a particular west longitude orbital position for which the antenna pointing angles will be computed.

Other options allow for the computation of distances in miles or kilometers, and for the automatic retrieval of the latitude and longitude for à city from the built-in data base of cities. If a location is specified which is not in the computer's memory, the program will then query the

---

*TRS-80 is a registered trademark of Tandy Corp.

user to enter the latitude and longitude of that location which may be found in any atlas.

Fig. E—9 presents the listing of the Satellite Tracking Program II. The list of satellites and their orbital positions currently in the satellite data base is included as Fig. E—10. (A * in the C-Band or Ku Band indicates the satellite operates in this band; a ● indicates it does not.) Note that this list of satellites and their locations may also be useful to the reader who may wish to compute the pointing angles by hand or programmable calculator. Fig. E—11 presents some typical printouts. Finally, Fig. E—12 is the list of cities and their latitudes and longitudes currently maintained in the program.

Additional cities can easily be added by placing them in data statements between lines 5214 and 5999. Be sure to change the city number count in line 5020 to indicate the total number of cities in the new list.

Likewise, additional satellites may be added by placing them in data statements between lines 6274 and 9999. Again, the number of North American domestic satellites and total satellite count must also be updated in line 6020.

```
50   REM : ------------------------------------------
51   REM :
53   REM : CERTAIN LINES IN THIS PROGRAM HAVE BEEN
54   REM : MODIFIED, NOTABLY 260 & 350 FOR
55   REM : USE WITH 80-COLUMN BOARDS LOCATED
56   REM : IN SLOT #3.  TO USE THIS PROGRAM
57   REM : WITH A 40-COLUMN APPLE, DELETE
58   REM : ALL LINES WHICH HAVE A 'PR#3' IN THEM
59   REM :
94   REM : INITIALIZE PRINTER
95   PRINT "";"PR#1"
96   PRINT
97   PRINT "";"PR#3"
99   GOTO 350
100  REM : SATELLITE TRACKING PROGRAM II
101  REM : FOR GEOSTATIONARY SATELLITES
102  REM : IN THE 'CLARKE ORBIT'
103  REM :
104  REM : CALCULATES AZIMUTH & ELEVATION
105  REM : ANTENNA POINTING ANGLES FOR
106  REM : OVER 50 SATELLITES IN ORBITS;
107  REM : ALSO COMPUTES DECLINATION FOR
108  REM : POLAR MOUNT ANTENNAS.
109  REM : AZ/EL FOR A SPECIFIC WEST-
110  REM : LONGITUDE ORBITAL LOCATION
111  REM : ENTERED BY THE USER CAN ALSO
112  REM : BE COMPUTED.
113  REM :
```

Fig. E-9. Satellite Tracking
(Copyright © 1983 by The Satellite Center.

```
114    REM : A DATA BASE OF POPULAR WORLD
115    REM : CITIES (LONGITUDE & LATITUDE)
116    REM : AND A SECOND DATA BASE OF
117    REM : GLOBAL SATELLITES IS PROVIDED.
118    REM :
119    REM : COPYRIGHT  1983, THE SATELLITE CENTER
120    REM : BY ANTHONY T. EASTON, JULY, 1983
121    REM : ALL RIGHTS RESERVED
122    REM :
123    REM : THIS PROGRAM MAY BE USED ONLY
124    REM : BY THE BUYER OF THIS BOOK FOR
125    REM : PERSONAL USE ONLY. IT MAY NOT
126    REM : BE SOLD OR USED COMMERCIALLY
127    REM : WITHOUT SPECIFIC WRITTEN
128    REM : PERMISSION OF THE SATELLITE CENTER.
129    REM :
130    REM : -------------------------------
131    REM : A COMPLETE PRINTOUT OF ALL SATELLITES
132    REM : VISIBLE FROM YOUR OWN SITE IS AVAILABLE
133    REM : FOR $10 (NON-US ORDERS: $15 IN US FUNDS
134    REM : DRAWN ON A NEW YORK BANK) FROM:
135    REM :
136    REM : THE SATELLITE CENTER
137    REM : POST OFFICE BOX 330045
138    REM : SAN FRANCISCO, CA, 94133
139    REM : PHONE: (415) 673-7000
140    REM :--NO COD ORDERS PLEASE/VISA & MC OK--
141    REM :-------------------------------
142    REM :
143    REM : OTHER PRODUCTS AVAILABLE FROM
144    REM : THE SATELLITE CENTER INCLUDE:
145    REM : 1) THIS PROGRAM ON APPLE II
146    REM :    DOS 3.3 DISK W/INSTRUCTIONS
147    REM :    ON ITS USE.......$20 (NON-US:$30)
148    REM : 2) THE 'SAT FINDER TOOL'
149    REM :    A MULTI-FUNCTION COMPASS,
150    REM :    SITE FINDER & PROTRACTOR,
151    REM :    W/INSTRUCTIONS...$40 (NON-US:$50)
152    REM : 3) COMBO SITE FINDER & DISK
153    REM :    ON DISCOUNT......$50 (NON-US:$60)
154    REM :
155    REM : POSTAGE & HANDLING ARE INCLUDED!
156    REM :
157    REM : PLEASE ALLOW 6-8 WEEKS FOR SHIPPING.
158    REM :
159    REM :-------------------------------
160    REM : STRUCTURE OF THIS PROGRAM
161    REM : LINES 100-199    REMARKS
162    REM :      200-299    INITITIALIZATION & MISC. R
       OUTINES
```

Program II listing.
All rights reserved. Used with Permission.)

```
163   REM :         300-399    VARIABLES LIST & HNITIALIZ
      ATION
164   REM :         400-999    USER INPUT OPTION ROUTINE
165   REM :         1000-1199  ANTENNA COORDINATE ENTRY R
      OUTINE
166   REM :         1200-1999  PRINTING HEADER ROUTINE
167   REM :         2000-2099  READ SATELLITES FROM DATA
      BASE
168   REM :         2100-2199  CHECK IF SATELLITE IS BELO
      W HORIZON
169   REM :         2200-2999  MAIN CALCULATION & PRINT R
      OUTINE
170   REM :         4000-4999  SATELLITE DATA BASE LIST P
      RINTING ROUTINE
171   REM :         5000-5999  CITY LONGITUDE/LATITUDE DA
      TA BASE
172   REM :         6000-6999  SATELLITE DATA BASE
173   REM :
174   REM : THIS. PROGRAM HAS INTENTIONALLY BEEN WRITTEN

175   REM : IN A SIMPLE SUBSET OF APPLESOFT BASIC SO TH
      AT
176   REM : IT CAN BE EASILY CONVERTED TO ANOTHER MACHI
      NE'S
177   REM : BASIC, THUS CODE COMPACTION HAS NOT BEEN DO
      NE.
178   REM :
199   REM :
200   REM : INITIALIZATION & MISC. ROUTINES
201   REM : PRINT SUBROUTINE
220   PRINT
230   PRINT "-------------------------------------------"
240   RETURN
250   REM : RERUN-PROGRAM-OPTION ROUTINE
260   PRINT "";"PR#3"
265   PRINT "RUN AGAIN (Y OR N)";
270   INPUT Q$
280   IF Q$ = "Y" THEN 350
290   GOTO 9999
299   REM :
300   REM : VARIABLES LIST & INITIALIZATION
301   REM :
302   REM : THE FOLLOWING VARIABLES ARE USED IN THE PRO
      GRAM:
303   REM :    K/K1/K2 = COUNTERS
304   REM :    T$      = OUTPUT OPTION (US,LIST,ALL,KYB
      D)
305   REM :    D$      = CITY NAME REQUESTED (ALSO P$ I
      N D.B.READ ROUTINE)
306   REM :    E$      = STATE/COUNTRY NAME FROM CITY D
      ATA BASE
307   REM :    U$      = USER PRINTING OPTION (P OR C)
```

Fig. E-9-cont. Satellite

```
308  REM :     W$        = MILES/KM OPTION USER OPTION (M
         OR K)
309  REM :     Z1        = RADIUS OF EARTH
310  REM :     Z2        = DISTANCE FROM EARTH TO SATELLI
         TE BELT
311  REM :     M1/M2/M3= LATITUDE OF CITY IN DEGREES, M
         INS, SECS
312  REM :     M$        = LATITUDE DESCRIPTION (N OR S)
313  REM :     L1/L2/L3= LONGITUDE OF CITY IN DEGREES,
         MINS, SECS
314  REM :     L$        = LONGITUDE DESCRIPTION (E OR W)

315  REM :     J1        = NUMBER OF US/CANADIAN SATS IN
         SAT DATA BASE
316  REM :     J2        = TOTAL NUMBER OF SATELLITES IN
         SAT DATA BASE
317  REM :     N$        = SATELLITE NAME IN SAT DATA BAS
         E
318  REM :     N         = WEST LONGITUDE OF A SATELLITE
         IN SAT DATA BASE
319  REM :     D1        = NUMBER OF CITIES IN CITY DATA
         BASE
320  REM :     I/J       = LOOP COUNTERS
321  REM :     C1-C4     = CALCULATION SUB-COMPONENTS
322  REM :     L4/X/Y    = CALCULATION SUB-COMPONENTS
323  REM :     A         = AZIMUTH OF ANTENNA
324  REM :     R         = RANGE TO SATELLITE
325  REM :     E         = ELEVATION OF ANTENNA
326  REM :     D         = DECLINATION ANGLE OF A POLAR A
         NTENNA
349  REM :
350  PRINT "";"PR#3"
351  TEXT
352  HOME
353  RESTORE
360  LET K = 0
361  LET K1 = 0
362  LET K2 = 0
370  LET C$ = ""
380  LET E$ = ""
390  REM : PRINTER IS ASSUMED TO BE IN SLOT 1, S=SLOT
         #
391  LET S = 1
399  REM :
400  REM : INPUT USER OPTIONS & SAT/CITY LOCATION ROUT
         INE
401  REM : ----- ---- ------- - -------- ------,, ----
         ---
402  REM :
410  REM : 'ALL' PRINTS AZ/EL FOR ALL SATELLITES
411  REM : IN THE DATA BASE
412  REM :
```

Tracking Program II listing.

```
420  REM : 'US' PRINTS US & CANADIAN SATELLITES ONLY
421  REM :
430  REM : 'KYBD' ASKS USER FOR A SPECIFIC SATELLITE
431  REM : W. LONGITUDE ORBCTAL 'SLOT', IN DEGREES
432  REM :
440  REM : 'LIST' PRINTS A LIST OF ALL SATELLITES IN T
     HE
441  REM : DATA BASE, THEIR NAMES & W. LONGITUDES
445  REM :
446  GOSUB 220
448  PRINT
450  PRINT "SATELLITE TRACKING PROGRAM II"
455  PRINT "COMPUTES ANTENNA POINTING ANGLES"
460  PRINT
465  PRINT "COPYRIGHT  C 1983, THE SATELLITE CENTER"
500  GOSUB 220
502  PRINT "PRINTER OR CRT (P OR C)";
504  INPUT U$
506  PRINT
510  PRINT "OUTPUT OPTION (ALL,US,KYBD,LIST)";
520  INPUT T$
521  PRINT
524  IF T$ = "LIST" THEN 4000
525  PRINT "WHAT IS YOUR CITY";
526  INPUT D$
527  PRINT
530  IF T$ < > "KYBD" THEN 600
540  PRINT "ENTER W. LONGITUDE OF DESIRED SLOT";
550  INPUT N
560  LET N$ = "SLOT AT "
570  LET O$ = "N/A  "
600  PRINT
640  PRINT "MILES OR KILOMETERS (M OR K)";
650  INPUT W$
655  PRINT
660  IF W$ = "M" THEN 690
670  REM : RADIUS OF EARTH & DISTANCE TO SATELLITE
671  REM : BELT FROM EQUATOR IN KILOMETERS
680  LET Z1 = 6367
685  LET Z2 = 35800
687  LET Z3 = 111.136
688  PRINT "DISTANCES ARE IN KILOMETERS"
689  GOTO 700
690  REM : RADIUS OF EARTH & DISTANCE TO SATELLITE
691  REM : BELT FROM EQUATOR IN MILES
695  LET Z1 = 3957
697  LET Z2 = 22245
698  LET Z3 = 69.057
699  PRINT "DISTANCES ARE IN MILES"
700  PRINT
799  REM :
800  REM : CITY DATA BASE READ ROUTINE
```

Fig. E-9-cont. Satellite

```
801    REM :
802    REM : CHECK CITY NAME D$ FROM USER TO SEE IF IT'S
803    REM : ALREADY IN THE DATA BASE. IF NOT, QUERY USE
       R
804    REM : FOR ITS LATITUDE & LONGITUDE (L'S & M'S)
805    REM : FORMAT IS CITY, STATE/COUNTRY, LATITUDE, LO
       NGITUDE
806    REM :
810    RESTORE
815    REM : GET SIZE OF CITY DATA BASE LIST
820    READ D1
830    FOR I = 1 TO D1
840    READ P$,E$,M1,M2,M$,L1,L2,L$
845    IF D$ = P$ THEN 875
850    NEXT I
855    PRINT
860    PRINT D$;" NOT FOUND IN DATA BASE"
865    PRINT
866    LET E$ = ""
870    GOTO 1000
874    REM : ADVANCE POINTER TO START OF SATELLITE DATA
       BASE
875    LET C$ = ", "
876    RESTORE
880    READ D1
881    FOR I = 1 TO D1
882    READ P$,P$,I1,I1,P$,I1,I1,P$
883    NEXT I
890    GOTO 1050
899    REM :
900    REM : IF CITY NAME IS NOT IN DATA BASE, GET ITS C
       OORDINATES
901    REM : IN DEGREES (NOT RADIANS) FROM THE USER.
902    REM :
903    REM : NOTE THAT TRIG FUNCTIONS COMPUTE IN RADIANS
       , NOT DEGREES,
904    REM : THEREFORE CONVERSION MULTIPLIERS OF 57.2957
       8 AND .0174533
905    REM : ARE USED IN THE FORMULAS!
906    REM :
994    REM :
1000   REM : ANTENNA COORDINATE ENTRY ROUTINE
1001   REM : ------- ---------- ----- -------
1002   REM :
1005   PRINT "ENTER LATITUDE IN FORMAT:"
1006   PRINT "DEGREES, MINUTES, SECONDS, N OR S"
1010   INPUT M1,M2,M3,M$
1020   PRINT
1030   PRINT "ENTER LONGITUDE IN FORMAT:"
1031   PRINT "DEGREES, MINUTES, SECONDS, E OR W"
1040   INPUT L1,L2,L3,L$
```

Tracking Program II listing.

```
1050    LET L = L1 + (L2 + (L3 / 60)) / 60
1060    LET M = M1 + (M2 + (M3 / 60)) / 60
1070    IF L$ = "E" THEN 1090
1080    LET L =  - L
1090    IF M$ = "N" THEN 1100
1095    LET M =  - M
1099    REM :
1100    REM : DECLINATION ANGLE COMPUTATION ROUTINE
1101    REM :
1110    LET C5 = 3964 *  SIN (M * .0174533)
1120    LET C6 = 22300 + 3964 * (1 -  COS (M * .0174533)
        )
1130    LET D =  - 57.29578 *  ATN (C5 / C6)
1199    REM :
1200    REM : PRINTING HEADER ROUTINE
1201    REM : --------- ------ -------
1202    IF U$ = "C" THEN 1212
1203    PRINT "TURN ON PRINTER AND PUSH RETURN"
1204    PRINT "";"PR#";S
1205    GET Q$
1206    GOSUB 220
1207    PRINT "GEOSTATIONARY SATELLITE LOCATION TABLE"
1208    PRINT "-------------- --------- -------- -----"
1209    PRINT
1210    PRINT "COURTESY OF THE SATELLITE CENTER, SAN FRA
        NCISCO"
1211    PRINT " (PRODUCED BY SATELLITE TRACKING PROGRAM
        II)"
1212    PRINT
1213    PRINT
1214    PRINT "LOCATION: ";D$;C$;E$;"      ";M1;"-";M2;"
        ";M$;" LATITUDE   ";L1;"-";L2;" ";L$;" LONGITUDE"

1215    PRINT
1216    PRINT "DECLINATION= ";( INT (D * 100)) / 100;" D
        EGREES (POLAR MOUNT)"
1217    PRINT
1218    PRINT
1219    IF U$ = "C" THEN 2000
1240    PRINT
1250    PRINT "SATELLITE","DATE"; TAB( 3);"DEGREE","OWNE
        R","AZIMUTH","ELEVATION","DISTANCE","SUBPOINT"
1251    PRINT "----------","----"; TAB( 3);"------","----
        -","-------","---------","---------"
2000    REM : READ SATELLITES FROM DATA BASE
2001    REM : ---- ---------- ---- ---- ----
2003    REM :
2005    IF T$ = "KYBD" THEN 2025
2010    READ J1,J2
2011    IF T$ <  > "US" THEN 2014
2012    LET J = J1
2013    GOTO 2015
```

Fig. E-9-cont. Satellite

```
2014   LET J = J2
2015   FOR I = 1 TO J
2020   READ N$,O$,N,L$,C$,K$
2025   LET C1 = 0
2026   LET C2 = 0
2027   LET C3 = 0
2028   LET C4 = 0
2099   REM :
2100   REM : CHECK IF SATELLITE IS BELOW HORIZON
2101   REM : ----- -- ----------- -- ----- -------
2110   REM :
2115   LET K1 = 1
2120   LET L4 = L + N
2125   IF L4 < = 180 THEN 2135
2130   LET L4 = L4 - 360
2135   IF L4 > = - 180 THEN 2145
2140   LET L4 = L4 + 360
2145   IF L4 < 81.3 THEN 2180
2150   IF U$ = "P" THEN 2600
2155   LET X = L4 - 81.3
2160   PRINT N$;" AT ";N;" W. LONGITUDE"
2165   PRINT "IS BELOW THE LOCAL HORIZON BY"
2170   PRINT  INT (X + .5);" DEGREES. NOT VISIBLE."
2175   GOTO 2452
2180   LET Y = M
2185   LET X = L4
2200   REM : MAIN CALCULATION & PRINT ROUTINE
2201   REM : ---- ----------- - ----- -------
2202   REM 0
2203   REM : COMPUTE GREAT CIRCLE MEASUREMENTS
2204   REM :
2210   LET C1 =  COS (0.0174533 * X) *  COS (0.0174533 *
       Y)
2219   REM :
2220   REM : COMPUTE GREAT CIRCLE ANGLE BETWEEN SITE
2221   REM : AND POINT ON THE EQUATOR BELOW SATELLITE
2225   REM : INVERSE COSINE FORMULA FOR C2
2226   REM :
2230   LET C2 = 57.29578 * ( -  ATN (C1 /  SQR ( - C1 *
       C1 + 1)) + 1.5708)
2232   IF C2 < 81.3 THEN 2240
2234   IF U$ = "P" THEN 2600
2235   LET X = C2 - 81.3
2236   PRINT N$;" AT ";N;" W. LONGITUDE"
2237   PRINT "IS BELOW THE LOCAL HORIZON BY"
2238   PRINT  INT (X + .5);" DEGREES. "N$;" NOT VISIBLE
       ."
2239   GOTO 2452
2240   REM :
2241   REM : CALCULATE DISTANCE TO SUB-POINT
2242   REM :
2250   LET C4 = Z3 * C2
```

Tracking Program II listing.

385

```
2259  REM :
2260  REM : CALCULATE AZIMUTH OF ANTENNA TO SATELLITE
2261  REM :
2270  LET A = 180 + 57.29578 * ( ATN ( TAN (0.0174533 *
      X) /  SIN (0.0174533 * Y)))
2280  IF A > 0 THEN 2300
2290  LET A = A + 180
2299  REM :
2300  REM : CALCULATE RANGE FROM EARTH STATION
2301  REM : TO SATELLITE IN ORBIT
2302  REM :
2310  LET R =  SQR (Z1 ^ 2 + (Z1 + Z2) ^ 2 - 2 * Z1 *
      (Z1 + Z2) *  COS (0.0174533 * C2))
2319  REM :
2320  REM : CALCULATE ELEVATION FROM SITE TO SATELLITE

2321  REM :
2325  LET C3 = ((R ^ 2 + Z1 ^ 2 - (Z1 + Z2) ^ 2) / (2 *
      Z1 * R))
2330  LET E =  - 90 + 57.29578 * ( -  ATN (C3 /  SQR (
      - C3 * C3 + 1)) + 1.5708)
2410  IF U$ = "C" THEN 2450
2422  PRINT N$,L$; TAB( 5);N,O$,( INT (A * 10)) / 10,(
      INT (E * 10)) / 10, INT (R), INT (C4)
2440  GOTO 2600
2450  PRINT N$;" ";N;"   AZ=";( INT (A * 10)) / 10;" E
      =";( INT (E * 10)) / 10
2451  PRINT "RANGE ="; INT (R);" SUBPOINT ="; INT (C4)

2452  PRINT
2455  IF U$ = "P" THEN 2600
2460  LET K = K + 1
2470  IF K < 7 THEN 2600
2480  PRINT
2490  PRINT "MORE (Y OR N)";
2500  INPUT Q$
2510  IF Q$ = "N" THEN 2700
2520  LET K = 0
2530  PRINT
2600  IF K2 = 1 THEN 2700
2605  IF T$ = "KYBD" THEN 2700
2610  NEXT I
2620  IF K1 = 1 THEN 2700
2630  LET K2 = 1
2640  GOTO 2025
2700  GOSUB 220
2710  GOTO 260
2999  REM :
4000  REM : SATELLITE DATA BASE LIST PRINTING ROUTINE
4001  REM : ---------- ---- ---- ---- -------- -------
4002  REM :
```

Fig. E-9-cont. Satellite

```
4010   REM : READ NAMES & W. LONGITUDE OF SATELLITES
4015   RESTORE
4016   REM : MOVE POINTER DOWN D.B. PAST CITY LIST
4017   READ Dl
4018   FOR I = 1 TO Dl
4019   READ P$,E$,M1,M2,M$,L1,L2,L$
4020   NEXT I
4025   READ J1,J2
4026   IF U$ = "C" THEN 4040
4027   PRINT "TURN ON PRINTER AND PUSH RETURN"
4028   PRINT "";"PR#";S
4029   GET Q$
4030   GOSUB 220
4031   PRINT
4033   PRINT "THE SATELLITE CENTER, SAN FRANCISCO"
4034   PRINT
4035   PRINT "LISTING OF GEOSTATIONARY SATELLITES IN GL
       OBAL DATA BASE"
4036   PRINT "-------- -- -------------- ---------- -- --
       ---- ---- ----"
4037   PRINT
4040   PRINT "US/CAN SAT","DATE"; TAB( 3);"DEGREE","OWN
       ER","C-BAND","KU-BAND"
4050   PRINT "-----------","----"; TAB( 3);"------","---
       --","-------","--------"
4060   PRINT
4070   FOR I = 1 TO J1
4080   READ N$,O$,N,L$,C$,K$
4090   PRINT N$,L$; TAB( 5);N,O$,C$,K$
4110   NEXT I
4120   PRINT
4130   PRINT "WORLD SATS","DATE"; TAB( 3);"DEGREE","OWN
       ER","C-BAND","KU-BAND"
4140   PRINT "------------","----"; TAB( 3);"------","---
       --","-------","--------"
4145   PRINT
4150   FOR I = 1 TO J2 - J1
4160   READ N$,O$,N,L$,C$,K$
4170   PRINT N$,L$; TAB( 5);N,O$,C$,K$
4190   NEXT I
4210   GOSUB 220
4220   GOTO 250.
4998   REM : ***********************************
4999   REM :
5000   REM : CITY DATA BASE OF LATITUDES & LONGITUDES
5001   REM : ---- ---- ---- -- ---------- - ----------
5005   REM :
5006   REM : TO ADD MORE CITIES, SIMPLY PLACE THEM
5007   REM : BEFORE LINE 5999.
5008   REM : BE SURE TO CHANGE THE CITY # COUNT
5009   REM : IN LINE 5020 TO MATCH DATA COUNT.
5010   REM : NUMBER OF CITIES IN LIST
```

Tracking Program II listing.

```
5020   DATA   35
5030   REM : FORMAT IS* CITY, STATE/COUNTRY,
5031   REM : LATITUDE, LONGITUDE
5040   REM :
5101   DATA   ANCHORAGE,ALASKA,61,12,N,149,48,W
5102   DATA   ATLANTA,GEORGIA,33,45,N,84,24,W
5103   DATA   BOSTON,MASS,42,21,N,71,03,W
5104   DATA   CHICAGO,ILLINOIS,41,52,N,87,38,W
5105   DATA   DALLAS,TEXAS,32,47,N,96,48,W
5106   DATA   DENVER,COLORADO,39,45,N,104,59,W
5107   DATA   HONOLULU,HAWAII,21,20,N,158,0,W
5108   DATA   HOUSTON,TEXAS,29,45,N,95,22,W
5109   DATA   LAS VEGAS,NEVADA,36,10,N,115,09,W
5110   DATA   LOS ANGELES,CALIFORNIA,34,.03,N,118,14,W
5111   DATA   MIAMI,FLORIDA,25,47,N,80,12,W
5112   DATA   NEW YORK,NEW YORK,40,45,N,74,00,W
5113   DATA   PHILADELPHIA,PENNSYLVANIA,39,57,N,75,09,W
5114   DATA   SALT LAKE CITY,UTAH,40,45,N,111,53,W
5115   DATA   SAN FRANCISCO,CALIFORNIA,37,46,N,122,27,W
5116   DATA   SEATTLE,WASHINGTON,47,37,N,122,20,W
5117   DATA   WASHINGTON,D.C.,38,54,N,77,01,W
5118   DATA   SAN JUAN,PUERTO RICO,18,22,N,66,.07,W
5119   DATA   MONTREAL,CANADA,43,31,N,73,34,W
5120   DATA   TORONTO,CANADA,43,39,N,79,23,W
5121   DATA   VANCOUVER,CANADA,49,16,N,123,.07,W
5201   DATA   AMSTERDAM,NETHERLANDS,55,22,N,4,54,E
5202   DATA   BERLIN,GERMANY,52,30,N,13,20,E
5203   DATA   GENEVA,SWITZERLAND,42,12,N,6,09,E
5204   DATA   HONG KONG,UK,22,15,N,114,10,E
5205   DATA   LONDON,ENGLAND,51,30,N,0,10,W
5206   DATA   MADRID,SPAIN,40,24,N,3,41,W
5207   DATA   MANILA,PHILIPPINES,41,53,N,95,14,W
5208   DATA   MEXICO CITY,MEXICO,23,0,N,102,0,W
5209   DATA   MOSCOW,RUSSIA,55,45,N,37,35,E
5210   DATA   PARIS,FRANCE,48,52,N,2,20,E
5211   DATA   ROME,ITALY,41,54,N,12,29,E
5212   DATA   SINGAPORE,CITY STATE,1,22,N,103,48,E
5213   DATA   SYDNEY,AUSTRALIA,33,52,S,151,13,E
5214   DATA   TOYKO,JAPAN,35,42,N,139,46,W
5999   REM :
6000   REM : SATELLITE DATA BASE
6001   REM : ---------- ---- ----
6005   REM :
6006   REM : TO ADD MORE SATELLITES, SIMPLY PLACE
6007   REM : THEM BEFORE LINE 9999
6008   REM : BE SURE TO CHANGE THE SATELLITE #
6009   REM : COUNT IN LINE 6020.
6010   REM : NUMBER OF US/CANADIAN SATELLITES,
6011   REM : TOTAL # OF SATELLITES IN LIST
6020   DATA   36,110
6025   REM
```

Fig. E-9-cont. Satellite

```
6030    REM : FORMAT IS SAT, OWNER, W. LONGITUDE, LAUNCH
        DATE, C & KU-BAND
6040    REM :
6098    REM : NORTH AMERICAN SATELLITES
6099    REM : ————— ————————— —————————
6100    DATA   SATCOM 2R,RCA,72,ON,*,.
6101    DATA   WESTAR 2,WU,79,ON,*,.
6102    DATA   SATCOM 4,RCA,83,ON,*,.
6103    DATA   COMSTAR 3,ATT,87,ON,*,.
6104    DATA   SBS 3,SBS,94,ON,.,*
6105    DATA   COMSTAR 1/2,ATT,95,ON,*,.
6106    DATA   SBS 2,SBS,97,ON,.,*
6107    DATA   WESTAR 4,WU,99,ON,*,.
6108    DATA   SBS 1,SBS,100,ON,.,*
6109    DATA   ANIK D,TELSAT,104.5,ON,*,.
6110    DATA   ANIK B1,TELSAT,109,ON,*,*
6111    DATA   ANIK A3,TELSAT,114,ON,*,.
6112    DATA   SATCOM 2,RCA,119,ON,*,.
6113    DATA   WESTAR 5,WU,123.5,ON,*,.
6114    DATA   COMSTAR 4,ATT,127,ON,*,.
6115    DATA   SATCOM 3R,RCA,131,ON,*,.
6116    DATA   GALAXY 1,HUGHES,134.5,ON,*,.
6117    DATA   SATCOM 1R,RCA,139,ON,*,.
6118    DATA   AURORA 1,ALASCOM,143,ON,*,.
6119    DATA   SPACENET 2,SPC,70,84,*,*
6120    DATA   GALAXY 2,HUGHES,74,83,*,.
6121    DATA   ADV-W 2,WU,79,84,*,*
6122    DATA   TELSTAR 2,ATT,87,84,*,.
6123    DATA   WESTAR 6,WU,91,83,*,.
6124    DATA   ADV-W 1,WU,91,84,*,*
6125    DATA   TELSTAR 1,ATT,95,83,*,.
6126    DATA   GSTAR 1,GTE,103,84,.,*
6127    DATA   GSTAR 2,GTE,106,84,.,*
6128    DATA   ANIK C3,TELSAT,109,84,.,*
6129    DATA   ANIK D3,TELSAT,109,84,*,.
6130    DATA   ANIK D2,TELSAT,114,83,*,.
6131    DATA   ANIK C2,TELSAT,116,83,.,*
6132    DATA   GALAXY 3,HUGHES,119,84,*,.
6133    DATA   SPACENET 1,SPC,119,84,*,*
6134    DATA   AMSAT 1,AMSAT,122,84,*,*
6135    DATA   TELSTAR 3,ATT,127,85,*,.
6197    REM :
6198    REM : INTERNATIONAL SATELLITES
6199    REM : —————————————— ——————————
6201    DATA   INT IV,INTL,.1,ON,*,.
6202    DATA   INT IV,INTL,1,ON,*,.
6203    DATA   TELCOM 1B,FRANCE,7,84,*,*
6204    DATA   TELCOM 1A,FRANCE,10,83,*,*
6205    DATA   SYMPHONIE 1/2,FR/GERMANY,11.5,ON,*,.
6206    DATA   GORIZONT 2, USSR,14,ON,*,.
6207    DATA   MARISAT-1,MARINE,16,ON,*,.
6208    DATA   INT IVA,INTL,18.5,ON,*,.
```

Tracking Program II listing.

```
6209  DATA    INT V,INTL,18.5,83,*,*
6210  DATA    TV-SAT,GERMANY,19,ON,.,*
6211  DATA    TDF-1,FRANCE,19,86,.,*
6212  DATA    LUXSAT,LUXEMBOURG,19,87,.,*
6213  DATA    L-SAT,EUROPE,19,87,.,*
6214  DATA    NATO-3B,INTL,20,ON,*,.
6215  DATA    INT IVA,INTL,21.5,ON,*,.
6216  DATA    INT V,INTL,22,83,*,*
6217  DATA    INT V,INTL,24.5,ON,*,*
6218  DATA    INT V,INTL,27.5,ON,*,*
6219  DATA    INT V,INTL,31,83,*,*
6220  DATA    HALLEY 1,UK,31,86,.,*
6221  DATA    INT IVA,INTL,34.5,ON,*,.
6222  DATA    INT V,INTL,35,ON,*,*
6223  DATA    TDRS-2,USA-NASA,41,83,.,*
6224  DATA    INT IV,INTL,53,ON,*,.
6225  DATA    BRASILSAT 2,BRAZIL,60,86,*,.
6226  DATA    BRASILSAT 3,BRAZIL,65,87,*,.
6227  DATA    BRASILSAT 1,BRAZIL,70,85,*,.
6228  DATA    SATCOL 2,COLUMBIA,70,84,*,.
6229  DATA    SATCOL 1,COLUMBIA,75.5,84,*,.
6230  DATA    MEXSAT,MEXICO,85,85,*,*
6231  DATA    MEXSAT 2,MEXICO,102,85,*,*
6232  DATA    TDRS 1,USA-NASA,171,83,.,*
6233  DATA    INT IV,INTL,181,ON,*,.
6234  DATA    INT V,INTL,181,83,*,*
6235  DATA    MARECS 3,MARINE,183,84,*,.
6236  DATA    MARISAT 2,MARINE,184,ON,*,.
6237  DATA    INT IV,INTL,186,ON,*,.
6238  DATA    AUSSAT 2,AUSTRALIA,196,86,.,*
6239  DATA    AUSSAT 3,AUSTRALIA,200,87,.,*
6240  DATA    AUSSAT 1,AUSTRALIA,204,85,.,*
6241  DATA    CS 1,JAPAN,225,ON,*,*
6242  DATA    CS 2,JAPAN,225,84,*,*
6243  DATA    ETS 2,JAPAN,230,ON,.,*
6244  DATA    CS 2a,JAPAN,230,83,*,*
6245  DATA    ETS 3,JAPAN,238,ON,.,*
6246  DATA    PALAPA B3,INDONESIA,242,85,*,.
6247  DATA    PALAPA B2,INDONESIA,247,84,*,.
6248  DATA    BS 2A,JAPAN,249,84,.,*
6249  DATA    BSE 2,JAPAN,250,83,.,*
6250  DATA    BS2B,JAPAN,251,85,.,*
6251  DATA    PALAPA B1,INDONESIA,252,83,*,.
6252  DATA    APPLE 1,INDIA ,255,ON,*,.
6253  DATA    INSAT 1B,INDIA,266,83,*,.
6254  DATA    GORIZONT 4,USSR,270,ON,*,.
6255  DATA    RADUGA ,USSR,275,ON,*,.
6256  DATA    PALAPA A2,INDONESIA,279,ON,*,.
6257  DATA    KOSMOS 1366,USSR,280,ON,.,*
6258  DATA    PALAPA A1,INDONESIA,283,ON,*,.
6259  DATA    INSAT 1A,INDIA,286,ON,*,.
```

Fig. E-9-cont. Satellite Tracking Program II listing.

```
6260    DATA    MARISAT 3,MARINE,288,ON,*,.
6261    DATA    MARECS 2,MARINE,295,ON,*,.
6262    DATA    INT IVA,INTL,297,ON,*,.
6263    DATA    INT V,INTL,297,ON,*,*
6264    DATA    INT IVA,INTL,300,ON,*,.
6265    DATA    INT V,INTL,300,ON,*,*
6266    DATA    GORIZONT 5,USSR,307,ON,*,.
6267    DATA    GORIZONT 3,USSR,309,ON,*,.
6268    DATA    RADUGA 9,USSR,325,ON,*,.
6269    DATA    ARABSAT 2, ARAB,334,85,*,.
6270    DATA    ARABSAT 1, ARAB,341,84,*,.
6271    DATA    ECS 2,EUROPE,347,83,.,*
6272    DATA    ECS 1, EUROPE,350,83,.,*
6273    DATA    OTS 2, EUROPE,355,ON,.,*
6274    DATA    TELE X,NORDIC,355,87,.,*
9999    END
```

Fig. E-9-cont. Satellite Tracking Program II listing.

THE SATELLITE CENTER, SAN FRANCISCO

LISTING OF GEOSTATIONARY SATELLITES IN GLOBAL DATA BASE

US/CAN SAT	DATE	DEGREE	OWNER	C-BAND	KU-BAND
SATCOM 2R	ON	72	RCA	*	.
WESTAR 2	ON	79	WU	*	.
SATCOM 4	ON	83	RCA	*	.
COMSTAR 3	ON	87	ATT	*	.
SBS 3	ON	94	SBS	.	*
COMSTAR 1/2	ON	95	ATT	*	.
SBS 2	ON	97	SBS	.	*
WESTAR 4	ON	99	WU	*	.
SBS 1	ON	100	SBS	.	*
ANIK D	ON	104.5	TELSAT	*	.
ANIK B1	ON	109	TELSAT	*	*
ANIK A3	ON	114	TELSAT	*	.
SATCOM 2	ON	119	RCA	*	.
WESTAR 5	ON	123.5	WU	*	.
COMSTAR 4	ON	127	ATT	*	.
SATCOM 3R	ON	131	RCA	*	.
GALAXY 1	ON	134.5	HUGHES	*	.
SATCOM 1R	ON	139	RCA	*	.
AURORA 1	ON	143	ALASCOM	*	.
SPACENET 2	84	70	SPC	*	*
GALAXY 2	83	74	HUGHES	*	.
ADV-W 2	84	79	WU	*	*
TELSTAR 2	84	87	ATT	*	.
WESTAR 6	83	91	WU	*	.
ADV-W 1	84	91	WU	*	*
TELSTAR 1	83	95	ATT	*	.
GSTAR 1	84	103	GTE	.	*
GSTAR 2	84	106	GTE	.	*
ANIK C3	84	109	TELSAT	.	*
ANIK D3	84	109	TELSAT	*	.
ANIK D2	83	114	TELSAT	*	.
ANIK C2	83	116	TELSAT	.	*
GALAXY 3	84	119	HUGHES	*	.
SPACENET 1	84	119	SPC	*	*
AMSAT 1	84	122	AMSAT	*	*
TELSTAR 3	85	127	ATT	*	.

Fig. E-10. Listing of geostationary satellites in the global data base.
(Courtesy The Satellite Center)

WORLD SATS	DATE	DEGREE	OWNER	C-BAND	KU-BAND
INT IV	ON	.1	INTL	★	•
INT IV	ON	1	INTL	★	•
TELCOM 1B	84	7	FRANCE	★	★
TELCOM 1A	83	10	FRANCE	★	★
SYMPHONIE 1/2	ON	11.5	FR/GERMANY	★	•
GORIZONT 2	ON	14	USSR	★	•
MARISAT-1	ON	16	MARINE	★	•
INT IVA	ON	18.5	INTL	★	•
INT V	83	18.5	INTL	★	★
TV-SAT	ON	19	GERMANY	•	★
TDF-1	86	19	FRANCE	•	★
LUXSAT	87	19	LUXEMBOURG	•	★
L-SAT	87	19	EUROPE	•	★
NATO-3B	ON	20	INTL	★	•
INT IVA	ON	21.5	INTL	★	•
INT V	83	22	INTL	★	★
INT V	ON	24.5	INTL	★	★
INT V	ON	27.5	INTL	★	★
INT V	83	31	INTL	★	★
HALLEY 1	86	31	UK	•	★
INT IVA	ON	34.5	INTL	★	•
INT V	ON	35	INTL	★	★
TDRS-2	83	41	USA-NASA	•	★
INT IV	ON	53	INTL	★	•
BRASILSAT 2	86	60	BRAZIL	★	•
BRASILSAT 3	87	65	BRAZIL	★	•
BRASILSAT 1	85	70	BRAZIL	★	•
SATCOL 2	84	70	COLUMBIA	★	•
SATCOL 1	84	75.5	COLUMBIA	★	•
MEXSAT	85	85	MEXICO	★	★
MEXSAT 2	85	102	MEXICO	★	★
TDRS 1	83	171	USA-NASA	•	★
INT IV	ON	181	INTL	★	•
INT V	83	181	INTL	★	★
MARECS 3	84	183	MARINE	★	•
MARISAT 2	ON	184	MARINE	★	•
INT IV	ON	186	INTL	★	•
AUSSAT 2	86	196	AUSTRALIA	•	★
AUSSAT 3	87	200	AUSTRALIA	•	★
AUSSAT 1	85	204	AUSTRALIA	•	★
CS 1	ON	225	JAPAN	★	★
CS 2	84	225	JAPAN	★	★
ETS 2	ON	230	JAPAN	•	★
CS 2a	83	230	JAPAN	★	★
ETS 3	ON	238	JAPAN	•	★
PALAPA B3	85	242	INDONESIA	★	•
PALAPA B2	84	247	INDONESIA	★	•
BS 2A	84	249	JAPAN	•	★
BSE 2	83	250	JAPAN	•	★
BS2B	85	251	JAPAN	•	★
PALAPA B1	83	252	INDONESIA	★	•
APPLE 1	ON	255	INDIA	★	•
INSAT 1B	83	266	INDIA	★	•
GORIZONT 4	ON	270	USSR	★	•
RADUGA	ON	275	USSR	★	•
PALAPA A2	ON	279	INDONESIA	★	•
KOSMOS 1366	ON	280	USSR	•	★
PALAPA A1	ON	283	INDONESIA	★	•
INSAT 1A	ON	286	INDIA	★	•
MARISAT 3	ON	288	MARINE	★	•
MARECS 2	ON	295	MARINE	★	•
INT IVA	ON	297	INTL	★	•
INT V	ON	297	INTL	★	★
INT IVA	ON	300	INTL	★	•
INT V	ON	300	INTL	★	★

Fig. E-10-cont. Listing of geostationary satellites in the global data base.

```
GORIZONT 5      ON    307     USSR            *              .
GORIZONT 3      ON    309     USSR            *              .
RADUGA 9        ON    325     USSR            *              .
ARABSAT 2       85    334     ARAB            *              .
ARABSAT 1       84    341     ARAB            *              .
ECS 2           83    347     EUROPE          .              *
ECS 1           83    350     EUROPE          .              *
OTS 2           ON    355     EUROPE          .              *
TELE X          87    355     NORDIC          .              *

-------------------------------------
```

Fig. E-10-cont. Listing of geostationary satellites in the global data base.

Texas Instruments TI 58/59 Calculator Program

One of the great advances of the electronic age is the invention of the scientific calculator with its built-in trig function. This type of calculator totally eliminates the need for trig tables when computing satellite pointing angles by hand. Yet another big step forward is the programmable calculator which can easily be used to compute the satellite pointing angles. The azimuth and elevation formulas have been incorporated into a short programmable calculator routine by James P. Kennedy, W3DBG,

```
RUN


-------------------------------------------


SATELLITE TRACKING PROGRAM II
COMPUTES ANTENNA POINTING ANGLES

COPYRIGHT  C 1983, THE SATELLITE CENTER

-------------------------------------------
PRINTER OR CRT (P OR C)?P

OUTPUT OPTION (ALL,US,KYBD,LIST)?US

WHAT IS YOUR CITY?NEW YORK


MILES OR KILOMETERS (M OR K)?M

DISTANCES ARE IN MILES

TURN ON PRINTER AND PUSH RETURN

-------------------------------------------
GEOSTATIONARY SATELLITE LOCATION TABLE
-------------- --------- -------- -----

COURTESY OF THE SATELLITE CENTER, SAN FRANCISCO
  (PRODUCED BY SATELLITE TRACKING PROGRAM II)


LOCATION: NEW YORK, NEW YORK      40-45 N LATITUDE   74-0 W LONGITUDE

DECLINATION= -6.35 DEGREES (POLAR MOUNT)
```

Fig. E-11. Typical printouts produced by the Satellite Tracking Program II.
(Courtesy The Satellite Center)

SATELLITE	DATE	DEGREE	OWNER	AZIMUTH	ELEVATION	DISTANCE	SUBPOINT
SATCOM 2R	ON	72	RCA	176.9	42.8	23349	530
WESTAR 2	ON	79	WU	187.6	42.6	23360	533
SATCOM 4	ON	83	RCA	193.6	41.9	23389	540
COMSTAR 3	ON	87	ATT	199.4	41	23433	551
SBS 3	ON	94	SBS	209.1	38.6	23549	579
COMSTAR 1/2	ON	95	ATT	210.4	38.1	23570	584
SBS 2	ON	97	SBS	213	37.3	23613	595
WESTAR 4	ON	99	WU	215.5	36.3	23660	606
SBS 1	ON	100	SBS	216.7	35.8	23685	612
ANIK D	ON	104.5	TELSAT	222	33.5	23808	640
ANIK B1	ON	109	TELSAT	227	30.9	23948	671
ANIK A3	ON	114	TELSAT	232.1	27.7	24121	708
SATCOM 2	ON	119	RCA	236.8	24.4	24313	748
WESTAR 5	ON	123.5	WU	240.8	21.3	24498	786
COMSTAR 4	ON	127	ATT	243.8	18.9	24650	817
SATCOM 3R	ON	131	RCA	247	16	24832	853
GALAXY 1	ON	134.5	HUGHES	249.7	13.4	24996	885
SATCOM 1R	ON	139	RCA	253	10.1	25215	927
AURORA 1	ON	143	ALASCOM	255.9	7.1	25414	965
SPACENET 2	84	70	SPC	173.8	42.7	23355	531
GALAXY 2	84	74	HUGHES	180	42.8	23347	529
ADV-W 2	84	79	WU	187.6	42.6	23360	533
TELSTAR 2	83	87	ATT	199.4	41	23433	551
WESTAR 6	83	91	WU	205	39.7	23494	566
ADV-W 1	84	91	WU	205	39.7	23494	566
TELSTAR 1	83	95	ATT	210.4	38.1	23570	584
GSTAR 1	84	103	GTE	220.3	34.3	23765	630
GSTAR 2	84	106	GTE	223.7	32.6	23853	650
ANIK C3	84	109	TELSAT	227	30.9	23948	671
ANIK D3	84	109	TELSAT	227	30.9	23948	671
ANIK D2	83	114	TELSAT	232.1	27.7	24121	708
ANIK C2	83	116	TELSAT	234	26.4	24196	724
GALAXY 3	84	119	HUGHES	236.8	24.4	24313	748
SPACENET 1	84	119	SPC	236.8	24.4	24313	748
AMSAT 1	84	122	AMSAT	239.5	22.4	24435	774
TELSTAR 3	85	127	ATT	243.8	18.9	24650	817

RUN AGAIN (Y OR N)?

Fig. E-11-cont. Typical printouts produced

RUN

SATELLITE TRACKING PROGRAM II
COMPUTES ANTENNA POINTING ANGLES

COPYRIGHT C 1983, THE SATELLITE CENTER

PRINTER OR CRT (P OR C)?C

OUTPUT OPTION (ALL,US,KYBD,LIST)?US

WHAT IS YOUR CITY?MIAMI

MILES OR KILOMETERS (M OR K)?K

DISTANCES ARE IN KILOMETERS

LOCATION: MIAMI, FLORIDA   25-47 N LATITUDE 80-12 W LONGITUDE

DECLINATION= -4.35 DEGREES (POLAR MOUNT)

SATCOM 2R 72   AZ=161.6 E=58.5
RANGE =36606 SUBPOINT =2997

WESTAR 2 79   AZ=177.2 E=59.8
RANGE =36540 SUBPOINT =2868

SATCOM 4 83   AZ=186.4 E=59.7
RANGE =36546 SUBPOINT =2881

COMSTAR 3 87   AZ=195.3 E=58.9
RANGE =36585 SUBPOINT =2956

SBS 3 94   AZ=209.4 E=56.1
RANGE =36729 SUBPOINT =3225

COMSTAR 1/2 95   AZ=211.2 E=55.6
RANGE =36757 SUBPOINT =3275

SBS 2 97   AZ=214.7 E=54.5
RANGE =36820 SUBPOINT =3384

MORE (Y OR N)?Y

WESTAR 4 99   AZ=218 E=53.2
RANGE =36890 SUBPOINT =3503

SBS 1 100   AZ=219.6 E=52.6
RANGE =36928 SUBPOINT =3566

ANIK D 104.5   AZ=226 E=49.5
RANGE =37120 SUBPOINT =3872

ANIK B1 109   AZ=231.6 E=46
RANGE =37348 SUBPOINT =4212

ANIK A3 114   AZ=236.9 E=41.9
RANGE =37640 SUBPOINT =4618

by the Satellite Tracking Program II.

```
SATCOM 2 119     AZ=241.5 E=37.7
RANGE =37970 SUBPOINT =5049

WESTAR 5 123.5     AZ=245.2 E=33.7
RANGE =38297 SUBPOINT =5451

MORE (Y OR N)?N

------------------------------------------
RUN AGAIN (Y OR N)?N

]
```

Fig. E-11-cont. Typical printouts produced by the Satellite Tracking Program II.

```
CITY DATA BASE CURRENTLY ON-LINE
---- ---- ---- --------- -------
```

ANCHORAGE	ALASKA	61-12	N	149-48	W
ATLANTA	GEORGIA	33-45	N	84-24	W
BOSTON	MASS	42-21	N	71-3	W
CHICAGO	ILLINOIS	41-52	N	87-38	W
DALLAS	TEXAS	32-47	N	96-48	W
DENVER	COLORADO	39-45	N	104-59	W
HONOLULU	HAWAII	21-20	N	158-0	W
HOUSTON	TEXAS	29-45	N	95-22	W
LAS VEGAS	NEVADA	36-10	N	115-9	W
LOS ANGELES	CALIFORNIA	34-3	N	118-14	W
MIAMI	FLORIDA	25-47	N	80-12	W
NEW YORK	NEW YORK	40-45	N	74-0	W
PHILADELPHIA	PENNSYLVANIA	39-57	N	75-9	W
SALT LAKE CITY	UTAH	40-45	N	111-53	W
SAN FRANCISCO	CALIFORNIA	37-46	N	122-27	W
SEATTLE	WASHINGTON	47-37	N	122-20	W
WASHINGTON	D.C.	38-54	N	77-1	W
SAN JUAN	PUERTO RICO	18-22	N	66-7	W
MONTREAL	CANADA	43-31	N	73-34	W
TORONTO	CANADA	43-39	N	79-23	W
VANCOUVER	CANADA	49-16	N	123-7	W
AMSTERDAM	NETHERLANDS	55-22	N	4-54	E
BERLIN	GERMANY	52-30	N	13-20	E
GENEVA	SWITZERLAND	42-12	N	6-9	E
HONG KONG	UK	22-15	N	114-10	E
LONDON	ENGLAND	51-30	N	0-10	W
MADRID	SPAIN	40-24	N	3-41	W
MANILA	PHILIPPINES	41-53	N	95-14	W
MEXICO CITY	MEXICO	23-0	N	102-0	W
MOSCOW	RUSSIA	55-45	N	37-35	E
PARIS	FRANCE	48-52	N	2-20	E
ROME	ITALY	41-54	N	12-29	E
SINGAPORE	CITY STATE	1-22	N	103-48	E
SYDNEY	AUSTRALIA	33-52	S	151-13	E
TOYKO	JAPAN	35-42	N	139-46	W

Fig. E-12. Listing of cities and their coordinates.
(Courtesy The Satellite Center)

of Gaithersburg, Maryland, an ex-Apollo project manager and old friend of this author. He originally wrote the program for his own use and we thank him for sharing it with us here. There are actually two versions, Fig. E—13 presents a printing calculator version, while Fig. E—14 lists the nonprinting calculator routine.

To use either program, simply load the steps one-by-one into the calculator memory according to the instructions that come with the calculator. Depending upon the type of Texas Instruments calculator used, the program may be either temporarily loaded into a volatile memory, or permanently saved on a small magnetic memory strip for easy retrieval. Since this program makes computing satellite pointing angles a snap for the field installer, satellite TVRO dealers reading this book might wish to purchase their own calculators. Radio Shack also sells these calculators under their own Realistic® brand name for well under $50, with some models available for less than $25.

The programs can easily be converted to run on other programmable calculators that have scientific routines (trig functions) built in. Hewlett Packard line is a notable example.

Both programs function in a similar manner. The printing-calculator version utilizes the TI printing cradle option to output the Az/El computed angles to the small printer.

The second version functions with a stand-alone TI calculator which will display the Az/El angles on its built-in screen.

The latitude of the TVRO antenna is entered in location 01 of the calculator memory in the format: degrees, minutes, and seconds. North latitude is a positive number, south latitude is negative. Thus an earth station in San Francisco at 37° 45' 25'' North is entered as 37.4525. The longitude is likewise input into memory location 03 with similar format. West longitude is positive, east longitude is negative. For the San Franciscan TVRO whose longitude is 122° 26' 75'' West, the input is 122.2675.

The west longitude of a desired orbital slot for which the Az/El pointing angles will be computed is next entered into memory location 06 in decimal format. For example, the Satcom 3 satellite at 131.2° W is entered as 131.2.

After loading the data, the user need only depress the RUN/ START key. The calculator will then compute the azimuth and elevation angles (rounded to the nearest degree) and will display them in sequence (azimuth followed by elevation).

To compute the Az/El angles for a new satellite, the new west longitude of the bird can be loaded into location 06 and the RUN/ START key depressed again.

HOW TO OBTAIN YOUR SATELLITE PRINTOUT

A number of excellent articles have been written on the subject of tracking earth-orbit satellites.

The May 1978 issue of *Ham Radio* published an article (p. 67) by H.

```
000  43  RCL        051  42  STO        101  75  -
001  01   01        052  04   04        102  93  .
002  88  DMS        053  43  RCL        103  01  1
003  42  STO        054  00   00        104  05  5
004  00   00        055  30  TAN        105  01  1
005  43  RCL        056  55   ÷         106  03  3
006  03   03        057  43  RCL        107  54  )
007  88  DMS        058  04   04        108  95  =
008  75   -         059  30  TAN        109  24  CE
009  43  RCL        060  95   =         110  22  INV
010  06   06        061  24  CE         111  30  TAN
011  95   =         062  94  +/-        112  42  STO
012  42  STO        063  22  INV        113  08  08
013  02   02        064  39  COS        114  00  0
014  00   0         065  42  STO        115  32  X:T
015  32  X:T        066  05   05        116  43  RCL
016  43  RCL        067  61  GTO        117  02  02
017  02   02        068  00   00        118  77  GE
018  67  EQ         069  73   73        119  01  01
019  00   00        070  00   0         120  29  29
020  41   41        071  42  STO        121  43  RCL
021  53   (         072  05   05        122  07  07
022  53   (         073  53   (         123  58  FIX
023  43  RCL        074  03   3         124  00  00
024  00   00        075  06   6         125  99  PRT
025  39  COS        076  00   0         126  61  GTO
026  54   )         077  75   -         127  01  01
027  65   ×         078  43  RCL        128  34  34
028  53   (         079  05   05        129  43  RCL
029  43  RCL        080  54   )         130  05  05
030  02   02        081  42  STO        131  58  FIX
031  39  COS        082  07   07        132  00  00
032  54   )         083  43  RCL        133  99  PRT
033  54   )         084  04   04        134  43  RCL
034  22  INV        085  39  COS        135  08  08
035  39  COS        086  42  STO        136  99  PRT
036  42  STO        087  09   09        137  22  INV
037  04   04        088  43  RCL        138  58  FIX
038  61  GTO        089  09   09        139  98  ADV
039  00   00        090  33  x²         140  91  R/S
040  53   53        091  94  +/-        141  81  RST
041  00   0         092  85   +
042  32  X:T        093  01   1
043  43  RCL        094  95   =
044  01   01        095  34  ⌐X
045  22  INV        096  35  1/X
046  77  GE         097  65   ×
047  00   00        098  53   (
048  70   70        099  43  RCL
049  43  RCL        100  09   09
050  00   00
```

Fig. E-13. Printing calculator program to compute Az/El angles.

Paul Shuch on using an HP-25 programmable calculator to find Az/El angles. This program outputs not only for geostationary satellites, but also for satellites positioned in polar and other moving orbits such as the Russian domestic Molyna series of birds.

In the March 1978 issue of QST magazine (p. 23), Bill Johnston wrote

000	43	RCL		051	42	STO		101	75	-
001	01	01		052	04	04		102	93	.
002	88	DMS		053	43	RCL		103	01	1
003	42	STO		054	00	00		104	05	5
004	00	00		055	30	TAN		105	01	1
005	43	RCL		056	55	÷		106	03	3
006	03	03		057	43	RCL		107	54	)
007	88	DMS		058	04	04		108	95	=
008	75	-		059	30	TAN		109	24	CE
009	43	RCL		060	95	=		110	22	INV
010	06	06		061	24	CE		111	30	TAN
011	95	=		062	94	+/-		112	42	STO
012	42	STO		063	22	INV		113	08	08
013	02	02		064	39	COS		114	00	0
014	00	0		065	42	STO		115	32	X:T
015	32	X:T		066	05	05		116	43	RCL
016	43	RCL		067	61	GTO		117	02	02
017	02	02		068	00	00		118	77	GE
018	67	EQ		069	73	73		119	01	01
019	00	00		070	00	0		120	29	29
020	41	41		071	42	STO		121	43	RCL
021	53	(		072	05	05		122	07	07
022	53	(		073	53	(		123	58	FIX
023	43	RCL		074	03	3		124	00	00
024	00	00		075	06	6		125	91	R/S
025	39	COS		076	00	0		126	61	GTO
026	54	)		077	75	-		127	01	01
027	65	×		078	43	RCL		128	34	34
028	53	(		079	05	05		129	43	RCL
029	43	RCL		080	54	)		130	05	05
030	02	02		081	42	STO		131	58	FIX
031	39	COS		082	07	07		132	00	00
032	54	)		083	43	RCL		133	91	R/S
033	54	)		084	04	04		134	43	RCL
034	22	INV		085	39	COS		135	08	08
035	39	COS		086	42	STO		136	91	R/S
036	42	STO		087	09	09		137	22	INV
037	04	04		088	43	RCL		138	58	FIX
038	61	GTO		089	09	09		139	68	NOP
039	00	00		090	33	X²		140	91	R/S
040	53	53		091	94	+/-		141	81	RST
041	00	0		092	85	+				
042	32	X:T		093	01	1				
043	43	RCL		094	95	=				
044	01	01		095	34	ГX				
045	22	INV		096	35	1/X				
046	77	GE		097	65	×				
047	00	00		098	53	(				
048	70	70		099	43	RCL				
049	43	RCL		100	09	09				
050	00	00								

Fig. E-14. Nonprinting calculator program to compute Az/El angles.

"Locating Geosynchronous Satellites," which provides basic geometry and further hand-calculation examples.

The microcomputer magazine *BYTE* published a computer program in its article "The Geosat Program," in its January 1982 issue (p. 420). This program was designed to operate on the Apple II computer, and displays its output on the CRT screen.

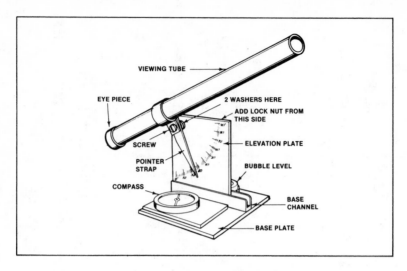

Fig. E-15. A popular "site finder" tool.
(Courtesy KLM Electronics, Inc.)

For those readers who would like to use a programmable calculator to eliminate the drudgery of hand calculations of Az/El angles, a Texas Instruments calculator package, complete with preloaded and tested satellite tracking program, and detailed instructions on its use, along with all the necessary satellite data is available from The Satellite Center at P.O. Box 330045, San Francisco, California, 94133. As calculator prices are continually dropping, inquire about the current price. Ask for the "TI Satellite Tracking System."

The Apple II Satellite Tracking Program is being continuously updated by this author, and the latest version is available on 5 ¼" minifloppy DOS 3.3 Apple II disk, in unprotected format, along with complete documentation, for $20 (includes postage and handling), also from The Satellite Center. (The cost is $30 for orders from Canada and other non-US locations and must be paid in US funds on a US bank.)

Perhaps the quickest way to get started is to obtain a complete printout of all the satellites (both current and future, C-Band and Ku Band) and their pointing angles from any given city. This can also be ordered from The Satellite Center for $10 (US) and $20 (nonUS). When ordering, please provide the latitude and longitude of the location desired along with its name.

Finally, a combination inclinometer, magnetic compass, and site-finding tool (Fig. E−15), complete with instructions on its use in determining precise elevation and azimuth angles for a given location, is available from The Satellite Center for $40 ($60 overseas and Canada). Re-

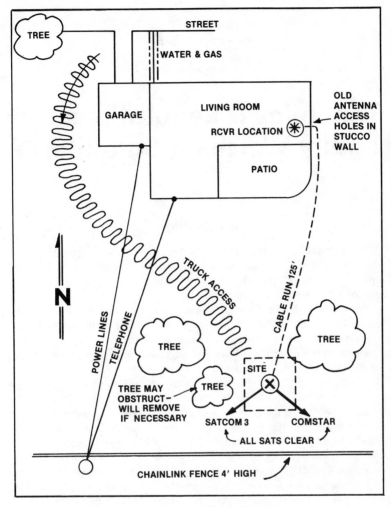

Fig. E-16. A rough sketch of the site.
(Courtesy KLM Electronics, Inc.)

quest the "Site-Finder" Tool. The tool comes with a $25 discount coupon good toward the purchase of a number of satellite television TVRO systems.

Using a tool like that of Fig. E–15 makes it a snap to check the property for clearance of trees, buildings, etc. which might block the view of the satellite arc. Used in conjunction with a rough sketch of the site (Fig. E–16), the user should have little trouble putting in his own backyard TVRO earth station.

The Rules & Regulations for Legally Watching Satellite TV

INTRODUCTION

This Appendix discusses the basic legal framework governing the Federal registration of a cable-tv system by the Federal Communications Commission, as well as the rules and regulations of the Copyright Tribunal and the local state agencies.

Although an SMATV or private cable system may not need to be licensed by the state or local government—as long as its cables do not cross city or county property—these private cable tv companies will come under the jurisdiction of the FCC, and will have to register with the Commission to legally do business in many instances.

The FCC registration process is straightforward, and will benefit the operator by providing official "certification" of the system. This may be useful in establishing the professional business credentials for the SMATV.

Registration with the FCC will also bring the SMATV company under the jurisdiction of the US Copyright Tribunal. This organization is responsible for collecting the fees from the CATV companies for the use of (retransmission of) broadcast tv station copyrighted material. In most cases, these copyright fees are negligible. By complying with the FCC and copyright rules (and paying the small annual fees) the SMATV operator immediately assumes the status of a full-fledged, government-recognized CATV system.

This also works to the advantage of the SMATV operator when dealing with the program suppliers.

Although some satellite-based programmers (notably HBO and several of the other movie suppliers) will not deal directly with an individual "nonfranchised" SMATV operator, the Federal Government regulation of the SMATV operator works toward creating a more equitable negotiating position on the part of the SMATV system. Simply because the SMATV operator does not hold an exclusive monopoly franchise from some local government agency over a particular territory should not prevent the major movie programmers from selling their products to the private cable companies. The hesitation to do so is based upon purely economic concerns—the HBOs appear to be afraid of angering their big cable tv MSO customers. This is now being argued as illegal in a number of states. Look for major breakthroughs for the private cable operator during the next several years as competition appears, finally, in the CATV business at the consumer level.

The entrepreneur who is contemplating establishing an SMATV operation should carefully read this Appendix as well as consider joining SPACE as an SMATV member. The National Satellite Cable Association is another organization which represents the interests of the private cable operators throughout the United States.

In addition, if a major operation (1000 or more units) is contemplated, the company should consider retaining a reputable Washington, DC law firm which specializes in FCC work. A list of such firms is available from the FCC Bar Association and may also be found in the annual *Broadcasting Yearbook* (published by *Broadcasting Magazine*, Washington, DC). See the other Appendices for complete address information on these organizations.

Finally, The Satellite Center, P.O. Box 330045, San Francisco, CA, 94133 (415-673-7000) publishes a detailed private CATV report for entrepreneurs and organizations who wish to explore the potential of this exciting and lucrative business. The report costs $100 and comes complete with all the necessary rules, regulations, forms, and documents necessary to start an SMATV system and register it with the government. Essential reading.

CABLE TELEVISION AND THE FCC

For regulatory purposes, the FCC rules distinguish between the terms "cable television system" and "system community unit." By definition, a cable television system is described as a "non-broadcast facility consisting of a set of transmission paths and associated signal generation, reception, and control equipment, under common ownership and control, that distributes or is designed to distribute to subscribers the signals of one or more television broadcast stations . . ." This definition does not include such facilities which serve fewer than 50 subscribers. This means that an SMATV or private cable operator, who has up to 49 paying subscribers, is completely exempt from FCC rules. In addition, those SMATV facilities which serve subscribers in one or more multiple-unit dwellings under common ownership or management [usually through the master antenna television (MATV) system] are also exempted under the FCC definition, regardless of whether they serve 50 or more subscribers.

The term "system community unit" means a cable tv system, or portion of a cable system, operating within a separate and distinct community or municipal entity, including unincorporated areas and separate unincorporated areas within them. This definition of system is used in relation to registration, signal carriage, and network program nonduplication protection requirements.

Registration

As a precondition to commencing operation or adding any television broadcast signals to existing operations, a cable system operator must separately register each system community unit with the FCC. If a cable television facility serves fewer than 50 subscribers but is part of a larger system which, taken as a whole, serves 50 or more subscribers, the smaller facility is considered a community unit and is required to register. Effective October 20, 1978, the Commission eliminated the certification of compliance application process for cable systems and instituted this new procedure. To register, a cable television operator must send the following information to the Secretary of the Commission:

1. The legal name of the operator, entity Identification or Social Security number, and whether the operator is an individual, private association, partnership or corporation; if the operator is a partnership, the legal name of the partner responsible for communications with the commission.
2. The assumed name (if any) used for doing business in the community.
3. The mail address, including zip code, and the telephone number to which all communications are to be directed.
4. The date the system provided service to 50 subscribers.
5. The name of the community or area served and the county in which it is located.
6. The television broadcast signals to be carried which previously have not been certified or registered.
7. For a cable system (or an employment unit) with five or more full-time employees, a statement of the proposed community unit's equal opportunity program, unless such program has previously been filed for the community unit or is not required to be filed based on an anticipated number of fewer than five full-time employees during January, February and March of the year following commencement of operation; an explanation must be submitted if no program statement is filed.

Registration statements must be signed by an authorized representative of the cable television company.

The Commission issues a public notice setting forth the details of each registration statement as it is received.

No further filings, other than required annual reports, are necessary unless new signals are added.

Requirement for Small Systems

Cable television systems serving fewer than 1000 subscribers are exempt from most of the FCC rules. Generally, only the following requirements apply to smaller cable systems:

1. Comply with the registration requirements described above.
2. Comply with requests from local television stations for carriage on the cable systems.
3. Comply with the Commission's technical standards for cable television systems, including the frequency use requirements (except that annual proof of performance tests are not required).
4. Correct and/or furnish information in response to the following forms sent to the cable operator annually by the Commission:

—Form 325: "Annual Report of Cable Television System" (Schedules 1 and 2 only)
—Form 326: "Cable Television Annual Financial Report"
—Form 395A: "Annual Employment Report" (including the annual report of complaints) (Exempt for mini-CATV systems with fewer than 5 employees)

Franchising and Local Regulation

Acknowledging that Federal licensing would be an unmanageable burden and that the industry was uniquely suited to "a deliberately structured dualism," the Commission adopted a regulatory plan allowing local or state authorities to select a franchisee and to regulate in any areas that the FCC does not preempt. Typically, local governments have adopted laws and/or regulations on franchising, basic subscriber rates, theft of service, taxation, and pole attachment. In most states, the private CATV company which restricts its operations to private property is *exempt* from franchising laws and local regulations. A general business license may, of course, be required. The Commission preempts local regulation of signal carriage, pay cable, and technical quality. However, in the areas of technical standards enforcement the FCC may authorize local arrangements in some cases.

In 1972, the Commission adopted minimum standards for franchises issued by local governments. These standards related to the process of selecting a franchisee, franchise duration, and fees, the establishment of contruction timetables, and procedures for regulating rates and handling subscriber complaints. In 1976, the Commission deleted the franchise standard regarding local rate regulation procedures and, effective November 15, 1977, amended the remaining franchise standards (Docket 21002). The standards, which had been mandatory, were maintained on a voluntary or guideline basis, with the exception of the franchise fee limit.

The Commission's franchise standards retain the following nonmandatory recommendations or provisions that local governments might include in the franchise or as part of the franchising process:

1. The franchising authority should approve a franchisee's qualifications only after a full public proceeding affording due process.
2. Neither the initial franchise period nor the renewal period should exceed 15 years, and any renewal should be granted only after a public proceeding affording due process.
3. The franchise should accomplish significant construction within one year after registering with the Commission and make service available to a substantial portion of the franchise area each following year, as determined by the franchise authority.
4. A franchise policy requiring less than complete wiring of the franchise area should be adopted only after a full public proceeding, preceded by specific notice of such policy.
5. The franchise should specify that the franchisee and the franchisor have adopted local procedures for investigating and resolving complaints.

The standards also suggest that the franchise require the franchisee to maintain a local business office or agent for handling complaints and that it specify that procedures have been adopted for investigating and resolving complaints and that each new subscriber will be given notice of these procedures. The standards also recommend that the franchise designate, by title, the franchising authority office or official responsible for the continuing administration of the franchise and implementation of complaint procedures.

The only mandatory franchise standard in the FCC rules limits a franchise fee, the amount the system operator pays annually to the local franchising authority, to no more than 3 percent of the franchisee gross revenues per year from all cable services in the community. The Commission may approve a fee in the range of 3 to 5 percent if the franchisee shows that it will not interfere with Federal regulatory goals and if the franchising authority justifies the entire fee as appropriate in the light of the planned local regulatory program.

In April 1979, the Commission denied reconsideration of its decision to relax the franchising standards and began a further rulemaking in Docket 21002 to consider the desirability of continuing to have a Federal limit on franchise fees. To date, the proceeding remains pending.

Rates for Service

The FCC does not regulate rates for cable television service. Local governments may decide whether or not to regulate most rates; however, the FCC preempts local regulation of pay cable rates.

Most systems have a rate for basic subscriber service, one or more charges for pay cable services, and charges for installation and additional hook ups. An increasing number of systems have several tiers of programming, each with separate charges. The use of computerized billing methods is facilitating more complex rate structures.

According to the most recent FCC annual financial data, the average monthly subscriber rate for 1980 was $7.69. Among the States, the average low rate was $6.37 in Vermont and the average high rate was $24.76 in Alaska. For pay cable service, the average monthly rate was $9.13, with a high rate of $13.36 in Alaska and low of $6.00 in the Virgin Islands. The average installation fee was $17.66.

Signal Carriage

The 1972 carriage rules set up standards of television signal carriage which vary with market size. The determining factor is whether a community is located wholly or partially within a 35-mile radius—called the specified zone—of a commercial tv station licensed to a television market as defined in the Commission's cable television rules.

Certain television broadcast signals are required to be carried by cable systems, if the stations request it. Systems serving communities in major or smaller television markets must carry, on request, signals from the following broadcast sources:

1. All television stations licensed to communities within 35 miles of the cable system community.
2. Noncommercial educational television stations within whose Grade B contour the system community is located.
3. Commercial tv translator stations with 100 watts or higher power serving the community of the system and noncommercial translators with 5 watts or higher power serving the community.
4. Stations whose signals are significantly viewed in the community.
5. All Grade-B signals from other small markets (applies to systems serving communities in smaller markets only).

Where a cable system community is located entirely or partially within both a major and a smaller market, the carriage provisions for the major market apply.

In communities located outside of all markets, cable systems must carry signals from the following sources:

1. All stations within whose Grade-B signal contour the system community is located.
2. Commercial translator stations with 100 watts or higher power and noncommercial translators with 5 watts or higher power serving the community.
3. Stations whose signals are significantly viewed in the community.
4. All educational stations licensed to communities within 35 miles of the system community.

The cable-tv rules adopted in 1972 contained restrictions on the importation of distant signals into the local cable television community, with the exception of those systems serving fewer than 1000 subscribers per headend. In general the rules permitted carriage of a full network station of each major national television network and from one to three independent stations, depending on the size of the market in which a system was located. Systems located outside of all television markets were not restricted as to a certain number or type of distant signal which they could add.

In June 1977, the Commission began an inquiry into the economic relationship between television broadcasting and cable television (Docket 21284). In April 1979, on the basis of its economic inquiry report which concluded that elimination of the distant signal carriage restrictions would benefit consumers without harming the ability of broadcast stations to meet their public interest responsibilities, the Commission issued a notice of proposed rulemaking to delete these rules. In July 1980, the Commission adopted the recommendations in Report and Order in Dockets 21284 and 20988, 789 FCC 2d 652 (1980). The rule changes became effective July 7, 1981, following affirmation of the Commission decision by the US Supreme Court of Appeals for the Second Circuit in Malrite TV of New York, Inc. et al. v. FCC and USA, 652 F. 2d 1140 (2nd Cir. 1981), cert denied, 50 U.S. L.W. 3547 (1982).

Manner of Carriage

The programs of television broadcast signals must be carried in full, without alteration or deletion of their content. In cases where all parties agree, however, commercial inserts or changes may be permitted if the Commission grants a waiver of its rules.

Educational Stations

All educational stations within 35 miles of a cable system community located outside of all television markets and all educational stations that place a Grade-B signal contour over the community of a system in a major or smaller market must be carried by the cable system. Also, cable systems serving communities in states where noncommercial educational stations are run by a state agency may carry any such stations. Any additional educational stations may be carried if there are no convincing objections.

Radio Programming

The Commission places no restrictions on either the carriage of radio stations or cable system origination of radio programming.

Sports Programming

Cable television community units serving 1000 or more subscribers may not carry local sports events broadcast by distant television stations when the events are not broadcast on local television stations. The purpose of this rule is to maintain the integrity of local television sports blackouts. The holder of broadcast rights is responsible for notification of programming deletions at least no later than Monday preceding the calendar week during which deletions are desired. Programming from any other television station may be substituted for the deleted program.

STV Stations

The signal carriage rules were not intended to and do not require a cable operator to carry the subscription, or scrambled programming portion of a subscription television station (STV). If the station qualifies for mandatory carriage, however, the cable operator would be obliged to carry, on request, the unscrambled portion of the signal.

Pay Cable

Cable systems frequently offer pay cable services, on a per-program or per-channel basis, to supplement the basic retransmission service. Included are feature films, sports events of regional or national interest, and entertainment programs specifically produced for the cable audience.

Two recent Federal Circuit Court decisions involved the FCC jurisdiction to regulate pay cable. In Home Box Office Inc. v. FCC, 567 F. 2d 9, cert. denied, 434 U.S. 829 (D.C. Cir. 1977), the court overturned the FCC pay cable program content rules because of failure to demonstrate a genuine problem which required regulation. The FCC does not now regulate pay cable program content.

In Brookhaven Cable TV Inc. v. Kelly, 573 F. 2d 765 (2d Cir. 1978), cert. denied, 99 S. Ct. 1991 (1979), the court affirmed a lower court finding that FCC preemption of pay cable rates precluded the New York State Commission on Cable Television from regulating such rates. The court found that FCC preemption in this area was clearly asserted and within its jurisdiction.

The latest data indicate that approximately 13 million pay cable subscribers are served by more than 3000 cable systems. Program distribution is provided by means of cassettes, common carrier microwave, common carrier multipoint distribution service (MDS), and satellites. Since beginning operation in late 1975, satellite delivery of pay cable programming has proved popular and economical.

The cable industry has been the largest single user of satellite receive-only earth stations. As of March 1, 1978, approximately 1350 of a total of 1525 earth station applications received were from cable companies. Most owners of earth stations continue to register, although the Commission deleted the mandatory registration requirement as of October 18, 1979. As of December 14, 1981, 5535 of a total of 6540 domestic earth station applicants were cable television operators.

Local Record Keeping Requirements

All operators of cable systems serving 1000 or more subscribers must have a current copy of the Commission's cable television rules. In addition, the operators must maintain locally for public inspection a copy of the franchise, application(s) to the FCC including any petitions for special relief or show cause, FCC annual reports and related documents, and any application for transfer of control of a Cable Television Relay Station that may serve the system.

Furthermore, all operators must maintain in the public inspection file certain records for specific periods of time. Records required on equal employment opportunity and annual performance tests must be retained for five years. Records of origination of cablecasts by candidates for public office and lists identifying sponsors of originated programming must be maintained for two years. Records required on network program nonduplication private agreements must be maintained for the run of the contract. These documents must be maintained on file either at the system's local office or another accessible place in the community. Operators of systems serving fewer than 1000 subscribers are not required to provide nonduplication protection or to conduct performance tests.

All systems of all sizes must retain for three years all records of subscribers served during the last month of each quarter of operation. This record does not have to be included in the public inspection file, but it must be made available upon request by an authorized Commission representative.

Equal Employment Opportunity

Nondiscrimination rules were adopted shortly following the 1972 Report and Order and have remained largely unchanged, although in 1978 the Commission restructured the regulations to imporve their readability and adopted some minor revisions (Docket 20829). All operators of cable systems and cable relay stations are required to afford equal opportunity in employment to all qualified persons and are prohibited from discrimination in employment because of race, color, religion, national origin, or sex.

The rules require all cable operators and CARS licensees to establish and maintain programs designed to assure equal opportunity for females, Blacks, Hispanics, American Indians, or Alaskan Natives, and Asians or Pacific Islanders in recruitment, selection, training, placement, promotion, pay, working conditions, demotion, layoff, and termination.

In addition, if systems have five or more full-time employees, they must keep on file with the Commission an up-to-date statement of their equal opportunity program and file any changes to existing programs on or before May 31 of each year. They also must submit an annual employment report (FCC Form 395A). An employment unit, as defined in the FCC rules, may consist of one cable system or of two or more systems which are under common ownership or control and are interrelated in their local management and operation. Employment units having fewer than five employees are required to submit the 395A Form but not the employment statistics portion of that report. An annual report of EEO complaints, required of all employment units, is permitted to be filed in conjunction with the annual employment report.

Reports and Forms

The Commission sends computer printed forms on the following subjects and asks for verification of their accuracy:

1. Schedules 1 and 2 of Form 325 (annual report of cable systems);
2. Form 326 (annual financial report). Schedule 1 of which is preprinted.
3. Form 395A (annual employment report).

Whenever a change occurs in a system's mail address, operator legal name, or operational status, e.g. it obtains 50 or 1000 subscribers, the operator must notify the Commission of the change within 30 days. The operator must furnish the following information:

1. Operator legal name and type (individual, private association, partnership, corporation), including EI (Employment Identity) number or the Social Security number (if the company is an individual or partnership)
2. Assumed name, if any.
3. Mail address.
4. Nature of operational status change.
5. The names of the system communities affected and their FCC identifiers (e.g., CA0001), which are assigned routinely when the system registers with the Commission.

411

Various FCC information reports are available to the public based on data collected from cable companies. Photocopies of these reports of general statistical information, tv station distribution, mailing addresses, and other data may be obtained through the Downtown Copy Center, 1114 21st St., N.W., Washington, DC, 20037 (telephone 202/452-1422) for a nominal charge per page. The center also provides research services to assist in locating materials (telephone 202/233-9765). Some reports are available on microfiche. The National Technical Information Service (NTIS), 5285 Port Royal Road, Springfield, VA, 22161, is another source of FCC data, available on magnetic tape and microfiche. An NTIS accession number, needed to order a particular computer file, may be obtained by calling 703/487-4763; orders may be placed by dialing 703/487-4650.

Cable System Ownership

The Commission prohibits cable system ownership by telephone companies within their local exchange areas, by television stations within the same local service area (Grade-B contour) and by national television networks anywhere in the country.

In October 1980, the Commission requested a staff study of the broad range of ownership issues. The economic study, conducted by the Commissions Office of Plans and Policy and released for comment in November 1981, concluded that cable television is a highly flexible, high-capacity technology capable of providing a wide range of services and that its market is local and workably competitive. It endorsed a policy of free entry into cable and concluded that consumer preference and entrepreneurial incentives could best determine an industry structure which meets consumer needs. The study recommended deleting all of the broadcast-cable and network-cable cross-ownership rules and retaining the current general ban on telephone cable cross-ownership. It also recommended against imposing multiple ownership limitations or a separations policy (i.e. separation of ownership and operation of facilities from control of transmission or program content).

Technical Performance Requirements

To assure the delivery of satisfactory television signals to cable subscribers, the Commission requires cable operators to meet certain technical performance requirements. These requirements, however, do not apply to nonbroadcast services (such as access channels), cable tv receivers, "ghosting" and cable carriage of am and fm programming. In general, the rules require every cable operator to conduct annual performance tests and to use these results to determine whether the system is performing satisfactorily. Systems serving fewer than 1000 subscribers are subject to technical standards but are not required to take annual measurements. It should be noted, however, that any system, regardless of its number of subscribers, which operates in the frequency bands 108—136 and 225—400 MHz must comply with the signal leakage standards of the rules, including annual performance measurements.

The FCC preempts state or local regulation of technical performance requirements to prevent the establishment of nonuniform requirements that might hinder system interconnectability and impede the development and marketing of new cable services and equipment. However, the Commission welcomes local assistance in the resolution of subscriber complaints involving technical standards. Waivers may be granted in the area of additional standards only upon a showing of local capability to enforce them.

As a result of the final report of the Cable Technical Advisory Committee (CTAC), the Commission modified its technical standards (Docket 20765) to allow systems to make technical performance measurements based on headends rather than communities.

In the same proceeding, the FCC relaxed several standards for signal levels, clarified frequency standards for converters, and relaxed the frequency tolerance for cable carriage of translator signals.

Major current technical standards include standards for signal leakage, frequency channeling plans, and standards for cable compatible receivers. Research in progress on the nature of signal leakage must precede frequency channelling decisions, which in turn must precede standardization of cable compatible receivers. The problem of signal leakage is being addressed in Docket 21006. A Final Report of the Advisory Committee on Cable Signal Leakage, released January 24, 1980, is available through the National Technical Information Service [(703) 487-4650], publication number PB80-119605. This study resulted in a further rulemaking in Docket 21006 proposing new rules to control interference to aeronautical radio services due to signal leakage from cable television systems using aeronautical radio frequencies.

Pole Attachments

Most franchised CATV systems distribute tv signals through coaxial cable strung on existing poles owned by telephone or electric utility companies. Cable operators also may use their own poles, place their cable underground, or use transmission facilities or rights-of-way owned or controlled by a utility. Some may use combinations of these arrangements.

Sometimes conflicts arise between cable television systems and utility companies over pole attachment issues, particularly the rates for use of utility facilities. The Communications Act Amendments of 1978 (P.L. 95—234, approved February 21, 1978) require the FCC to regulate the rates, terms, and conditions for cable tv pole attachments to ensure they are just and reasonable unless a state regulates such factors. The law requires such a state to certify to the Commission that it regulates the rates, terms, and conditions for pole attachments, and that it has the authority to consider the interests of both cable tv subscribers and utility consumers.

The FCC has developed regulations to deal with pole attachment complaints (CC Docket 78–144) in states which have not certified their authority to the FCC. Alaska, California, Connecticut, Hawaii, Illinois, Kentucky, Louisiana, Massachusetts, Michigan, Nebraska, Nevada, New Jersey, New York, Ohio, Oregon, Pennsylvania, Puerto Rico, Tennessee, Utah, Vermont, Washington, and Wisconsin have certified to the Commission that pole attachment regulation is within their jurisdiction (Public Notice, February 2, 1982).

How to Obtain the Rules

Every cable operator serving 1000 or more subscribers is required by the FCC to have an up-to-date copy of the Cable Television Rules and Regulations (47 CFR Part 76) and to keep track of Commission actions that might alter them. These rules are available in loose-leaf form on a subscription basis through the Superintendent of Documents, US Government Printing Office, Washington, D.C. 20402. Changes in the rules are forwarded to subscribers within a few months after their adoption by the Commission, without additional charges. To order, request Volume XI of the FCC Rules and Regulations, August 1976 edition (subscription price $17.00).

Another means of keeping informed of rule changes is through the *Federal Register*, published daily by the Office of the Federal Register and distributed by the US Government Printing Office. A yearly subscription to the Federal Register is $75 (Catalog No. GS 4.107). The Office of the Federal Register also publishes an annual cumulation of the cable tv rules as Code of Federal Regulations, Title 47, Parts 70–79. This publication is revised as of October 1st each year; the current edition (1980) is available in paperback for $8.50 (Catalog No. GS 4.108:47 Parts 70–79). Mail orders should be sent to the Government Printing Office at the above address; phone orders are accepted. Payment may be made by check or charge; prices given include postage. For additional information contact the Office of Federal Register at (202) 532-5240.

Cable Television Bureau

The Cable Television Bureau of the FCC is responsible for implementing the Commission's cable television regulatory program. It is composed of the following five operating divisions:

1. *Compliance* —enforces the rules and handles subscriber complaints.
2. *Special Relief and Microwave* —handles petitions for special relief or waiver, and licenses CARS stations.
3. *Policy Review and Development* —conducts rulemaking proceedings and advises the Commission, cable operators, local governments, and the public on cable television rules and policy matters.
4. *Research* —conducts analytical studies largely of an economic, financial, or statistical nature and performs a broad range of advisory functions; includes equal employment opportunity (EEO) unit.
5. *Records and Systems Management* —examines the records filed by cable operators and produces reports for Commission and public use.

Members of the Bureau's staff welcome inquiries from the public. By contacting the appropriate division, a person can obtain assistance in participating in the Commission's processes. See Table F—1 for a list of key Cable Television Bureau offices and telephone numbers.

Table F—1. FCC Cable Television Bureau

Bureau Organization	Telephone (Area Code 202)
Office of Bureau Chief	632-6480
Policy Review & Development Division	632-6468
Research Division (including EEO Unit)	632-9797
Records & Systems Management Div. (including Public Reference Room)	632-7076
Compliance Division	632-7480
Complaints & Information	632-9703
Enforcement Branch	254-3420
Special Relief & Microwave Div.	632-8882
Special Relief Branch	632-8882
Microwave Branch	254-3420

The Cable Television Bureau is located at 2025 M St. N.W., 6th floor. The mailing address is: Cable Television Bureau, Federal Communications Commission, Washington, DC 20554.

COPYRIGHT RULES

The Act for General Revision of the Copyright Law (Public Law 94—533, Title 17 U.S.C.), effective January 1, 1978, established for the first time an obligation for cable operators to pay copyright fees for the retransmission of broadcast signals. Regulations implementing the law may be obtained from the Copyright Office which administers the law. The office is part of the Library of Congress.

According to the provisions of the law, cable operators must pay a copyright royalty fee for which they receive a compulsory license to retransmit radio and television broadcast signals. The fee for each cable system is based on the system's gross revenues from the carriage of broadcast signals and the number of "distant signal equivalents," a term identifying nonnetwork programming from distant television stations carried by the system.

The law requires a cable operator to file, semiannually, a statement of accounts, including information about system revenue and signal carriage as well as the royalty fee payment. The amount of the fee is the sum of 0.675 percent of the gross revenues for the privilege of carrying distant signal equivalents or fixed rates for the carriage of each signal, ranging from 0.675 percent of gross revenues for the first distant signal to 0.2 percent for the fifth and each additional distant signal equivalent. Systems whose total gross revenues are less than $160,000 for the accounting period pay a reduced fee; systems grossing less than $41,500 are required to pay a flat fee of $15.

The law also established a Copyright Royalty Tribunal, composed of five commissioners, to distribute the royalty fees and resolve disputes among copyright owners and to review the fee schedule in 1980 and every five years subsequently. The Copyright Tribunal also is responsible for revising the rates whenever the FCC amends its carriage rules in a manner that would allow carriage of additional programming from distant stations. Any revision of rates should reflect changes impacting on the copyright owner's property value.

The question of whether copyright fees may be added onto subscriber rates is entirely within the purview of local regulatory authorities.

Several bills proposing amendments to the existing copyright law were introduced in the US House of Representatives in the 97th Congress. The Subcommittee on Courts, Civil Liberties and the Administration of Justice held hearings in December 1981 on H.R. 3560 concerning possible revisions of the law as it pertains to cable carriage of television broadcast signals. Congress continues to consider these issues, and other amendments are possible before a final resolution is adopted. For further details on the status of pending legislation, contact the Subcommittee at (202) 225-3926.

For further information regarding copyright regulations, contact the Licensing Division, Copyright Office, Library of Congress, Washington, DC 20557 [(202) 287-8130].

References for Further Study

In this Appendix lists of magazines and periodicals, contacts and associations, and magazines and books on satellite television are given for your further reference.

MAGAZINES AND PERIODICALS

Channel Guide
Post Office Box 2027
Englewood, CO 80150

    Weekly, 1 year subscription = $48.00

Computers and Electronics (formerly Popular Electronics)
Ziff-Davis Publishing Co.
One Park Avenue
New York, NY 10016
(212) 725-3500

    Monthly, 1 year subscription = $16.00

Coops Satellite Digest
Satellite Television Technology, Inc.
Post Office Box 100858
Fort Lauderdale, FL 33310
(305) 771-0505

    Monthly, 1 year subscription = $50.00

Orbit Magazine
Post Office Box 1048
Hailey, ID 83333
(208) 788-4938

    Monthly, 1 year subscription = $48.00

Popular Communications
76 North Broadway
Hicksville, NY 11801
(516) 681-2922

    Monthly, 1 year subscription = $12.00

QST (Ham Radio)
International Radio Relay League, Inc.
225 Main Street
Newington, CT 00611

    Monthly, 1 year subscription = $12.00

Radio Electronics
Gernback Publications, Inc.
200 Park Avenue, South
New York, NY 10003
(212) 777-6400

    Monthly, 1 year subscription = $14.97

Satellite Channel Chart
Post Office Box 434
Pleasanton, CA 94566
(415) 846-7380

    Bi-monthly, 1 year subscription = $19.00

Satellite TV
P.O. Box 2384
Shelby, NC 28150
(704) 482-9673

    Monthly, 1 year subscription = $30.00

Satellite TV News
41/47 Derby Road
Heanor, Derbyshire
England DE7 7QH

    Monthly, 1 year subscription = 21.00
    UK 31.00 Airmail

Satellite TV Week
Post Office Box 308
Fortuna, CA 95540
(707) 725-2476

Weekly, 1 year subscription = $65.00

Satvision
SPACE
1920 N. Street, NW, Suite 510
Washington, DC 20036

Monthly, 1 year subscription = $30.00

Video Magazine
Reese Publishing Company, Inc.
235 Park Avenue, South
New York, NY 10003

Monthly, 1 year subscription = $15.00

Video Action
Video Action Incorporated
21 West Elm Street
Chicago, IL 60610

Monthly, 1 year subscription = $10.00

Video Equipment
Post Office Box 309
Fraser, Michigan 48026
(313) 776-0543

Bi-monthly, 1 year subscription =
$8.50

Videography Magazine
United Business Publications Incorporated
475 Park Avenue, South
New York, NY 10016

Monthly, 1 year subscription = $15.75

Video Play Magazine
C.S. Tepfer Publishing Company, Inc.
51 Sugarhollow Road
Danbury, CT 06810
(203) 743-2120

Bi-monthly, 1 year subscription =
$8.25

Video Review
CES Publishing
350 East 81st St.
New York, NY 10028
(212) 734-4440

Monthly, 1 year subscription = $18.00

Cable TV and SMATV Magazines

Cable Business (formerly TVC)
Cardiff Publishing
6430 S. Yosemite St.
Englewood, CO 80111

Bi-weekly, 1 year subscription =
$27.00

Cable News Newsletter
Phillips Publishing, Inc.
7315 Wisconsin Avenue
Washington, DC 20014

Weekly, 1 year subscription = $145.00

Cable Products News
Post Office Box 2772
Palm Springs, CA 92263
(714) 323-2000

Monthly, Free to qualified subscribers

Cable Vision
Titsch Publishing, Inc.
2500 Curtis Street
Post Office Box 5400-TA
Denver, CO 80205

Weekly except last week of December,
1 year subscription = $64.00

CED Magazine
Titsch Publishing, Inc.
2500 Curtis Street
Post Office Box 5400-TA
Denver, CO 80217

Monthly, 1 year subscription = $20.00

Community Antenna Television Journal
TPI, Inc.
4209 Northwest 23rd, Suite 106
Oklahoma City, OK 73107

Monthly, 1 year subscription = $18.00

Home Video Report Newsletter
Knowledge Industry Publications, Inc.
2 Corporate Park Drive
White Plains, NY 10604
(914) 694-8686

Weekly, 1 year subscription = $175.00

Multichannel News Newspaper
300 S. Jackson
Denver, CO 80209

Weekly, 1 year subscription = $18.50

Private Cable
Wiesner Publishing
2690 West Main
Littleton, CO 80120

Monthly, Free to qualified subscribers

SAT-Guide
Commtek Publishing Company
Post Office Box 1700
Hailey, ID 83333
(208) 788-4936

Monthly, 1 year subscription = $48.00

Dealer Magazines

Audio Video International
Dempa Publications, Inc.
380 Madison Avenue
New York, NY 10017

Monthly, 1 year subscription = $20.00

Video Product News
Post Office Box 2772
Palm Springs, CA 92263
(714) 323-2000

Monthly, Free to qualified subscribers

The Video Retailer
The National Video Clearing House, Inc.
100 LaFayette Drive
Syosset, NY 11791
(516) 364-3686

Monthly, Free to qualified subscribers

Video Retailing
Audio Mirror Publications, Inc.
60 East 42nd Street
New York, NY 10017
(212) 682-7320

Monthly, 1 year subscription = $15.00

Video Store
Hester Communications, Inc.
1700 East Dyer Road, Suite 250
Santa Ana, CA 92705
(714) 549-4834

Monthly, Free to qualified sbuscribers

Broadcasting Magazines

Broadcast Communications
Globecom Publishing Ltd.
4121 W. 83rd Street, Suite 265
Prairie Village, KS 66208

Monthly, Free to qualified subscribers

Broadcast Engineering
Intertec Publishing Co.
9221 Quivira Road
Post Office Box 12901
Overland Park, KS 66212

Monthly, Free to qualified customers

Broadcast Management/Engineering (BME)
Broadband Information Services, Inc.
295 Madison Avenue
New York, NY 10017
(212) 685-5320

Monthly, Free to qualified subscribers

Broadcasting
Broadcasting Publications, Inc.
1735 DeSales Street, NW
Washington, DC 20036

Weekly, 1 year subscription = $60.00

Broadcasting Systems & Operation
BSO Publications Ltd.
Post Office Box 141 High Street Wivenhoe
Colechester, CO7 9EA
England

> Monthly, 1 year subscription = 15
> British pounds per year (surface)
> 25 British pounds (airmail)

DBS News
Phillips Publishing, Inc.
7315 Wisconsin Ave.
Washington, DC 20014

> Monthly, 1 year subscription =
> $197.00

Satellite Communications
Cardiff Publishing Co.
6430 S. Yosemite St.
Englewood, CO 80111
(303) 694-1522

> Monthly, 1 year subscription = $25.00

Satellite News Newsletter
Phillips Publishing, Inc.
7315 Wisconsin Avenue
Washington, DC 20014

> Weekly, 1 year subscription = $327.00

Satellite Times
Radio World Newspaper
Post Office Box 1214
Falls Church, VA 22041

> Monthly; Free to qualified subscribers

Satellite Week
Television Digest, Inc.
1836 Jefferson Place, NW
Washington, DC 20036
(202) 872-9200

> Weekly, 1 year subscription = $345.00

Television/Radio Age
Television Editorial Corporation
1270 Avenue of the Americas
New York, NY 10020
(212) 757-8400

> Bi-weekly, 1 year subscription =
> $40.00

Video News Newsletter
Phillips Publishing, Inc.
7315 Wisconsin Avenue
Washington, DC 20014
(301) 986-0666

> Weekly, 1 year subscription = $197.00

Video Systems
Intertec Publishing Co.
9221 Quivira Road
Post Office Box 12901
Overland Park, KS 66212

> Monthly, $2.00 per issue

Video Week
Television Digest
1836 Jefferson Place, NW
Washington, DC 20036
(202) 872-9200

> Weekly, 1 year subscription = $345.00

CONTACTS AND ASSOCIATIONS

American Radio Relay League, Inc. (ARRL)
225 Main Street
Newington, CT 06111
(203) 666-1541

Cable Television Administration and
  Marketing Society (CTAM)
  Society (CTAM)
2033 M Street, N.W.
Washington, DC 20036
(202) 296-4219

Cable Television Information Center
2100 M Street, NW
Washington, DC 20037
(202) 872-8888

California Community Television Association
(CCTA)
3636 Castro Valley Boulevard, No. 10
Castro Valley, CA 94546
(415) 881-0211

Community Antenna Television Association
(CATA)
4209 N.W. 23rd, Suite 106
Oklahoma City, OK 73107
(405) 947-7664

Copyright Royalty Tribunal
Washington, DC 20003
(202) 653-5775

Electronic Industries Association (EIA)
2001 I Street, NW
Washington, DC 20006
(202) 457-4900

Cable Television Bureau or
  Public Information Office
Federal Trade Communications Commission
1919 M Street, NW
Washington, DC 20554
(202) 632-9703

National Association of Broadcasters (NAB)
1771 N Street, NW
Washington, DC 20006
(202) 293-3500

National Association of Satellite Dealers
  of America
7107 South 400 West #3
Midvale, UT 84047
(801) 566-5603

National Association of Television Program
  Executives (NATPE)
Box 5272
Lancaster, PA 17601
(717) 626-4424

National Cable Television Association
(NCTA)
1724 Massachusetts Avenue, NW
Washington, DC 20036
(202) 775-3350

National Federation of Local Cable
  Programmers
% Miami Valley Cable Council
3200 Far Hills Avenue, Room 109
Kettering, OH 45429
(513) 298-7890

National Institute for Low Power Television
11800 Sunrise Valley Drive
Reston, VA 22091
(703) 476-8896

Satellite Television Technology
Post Office Box G
Arcadia, OK 73007
(405) 396-2574

Society of Private and Commercial Earth
  Station Users (SPACE)
1920 N Street, NW, Suite 510
Washington, DC 20036
(202) 887-0605

Society of Cable Television Engineers, Inc.
1900 L Street, NW
Washington, DC 20036
(202) 223-0353

The Satellite Center
P.O. Box 330045
San Francisco, CA 94133
(415) 673-7000

The Public Service Satellite Consortium
1660 L Street, NW
Washington, DC 20036
(202) 331-1154

ARTICLES

"A Cautious Welcome for Hotel Teleconferencing," Video Systems, April, 1982, p. 28.

"A Consumer's Guide to Satellite Television," Mechanix Illustrated, August, 1983, p. 52.

"And Now, Direct From Space . . .," Video, January, 1983, p. 72.

"A Satellite-Dish Odyssey," Video Review, Oct. 1981, p. 68.

"All Feedhorns Are Not Alike," Videoplay Magazine, Oct./Nov. 1981, p. 26.

"Back to the Basics," Videoplay Magazine, October, 1982, p. 46.

"Backyard Satellite TV Receiver," Radio-Electronics, March 1980, p. 42, and April 1980, p. 47.

"Backyard Satellite TV Reception: Fact or Fantasy," *Radio-Electronics*, June 1980, p. 68.

"Build a Low-Cost, High-Performance Satellite TV Antenna," *Mechanix Illustrated*, August, 1983, p. 64.

"Build Your Own Satellite-TV Receiver," 3-Part series, *Radio-Electronics*, May-July, 1982.

"Building a Home Satellite Receiver That Really Works," *Video Review*, Aug. 1980, p. 40.

"Buyer Beware," *Videoplay Magazine*, April/May 1981, p. 64.

"DBS: Up the Down Link," *Video Store*, August 1982, p. 20

"Direct Satellite-To-Home Television," *Video Action*, Winter, 1982, p. 78.

"Direct to Dish Transmission and the Cable Empire," Home Electronics and Entertainment, February, 1983, p. 19.

"Directory of Home Satellite Equipment," *Videoplay Magazine*, April/May 1981, p. 72.

"Doing the Dishes," *Home Video*, December, 1981, p. 36.

"DXing Those TV Satellites," *Popular Electronics*, Oct. 1981, p. 49.

"Earth Station Gear: Present and Future," *Videoplay*, May, 1983, p. 32.

"Get the Signal: Shopping for an Earth Station," *Video*, Aug. 1981, p. 50.

"Guide to Programming Services Available via Satellite," *Videoplay Magazine*, October, 1982, p. 50.

"Guide to Satellite Dishes," *Video Review*, Jan. 1982, p. 58.

"Guide to Satellite Programming Services," *Videoplay*, May, 1983, p. 88.

"Home Earth Stations for Satellite Transmissions," *Popular Electronics*, July 1981, p. 24.

"Home Reception Using Backyard Satellite TV Receivers," *Radio-Electronics*, Jan. 1980, p. 55.

"Home Reception Via Satellite," *Radio-Electronics*, Aug. 1979, p. 47; Sept. 1979, p. 47; Oct. 1979, p. 81.

"Home Viewing Via the Satellite," *Engineering*, July 17, 1980, p. 25.

"How Safe is Satellite Radiation?" *Video Review*, February, 1983, p. 32.

"Install a Backyard Antenna to Tune-In Satellite TV," *Popular Science*, March 1980, p. 122.

"Living with Satellite TV," *Video*, Jan. 1982, p. 64.

"Low Cost Backyard Satellite TV Earth Station," *Radio-Electronics*, Feb. 1980, p. 47.

"Mounting the Dish," *Videoplay Magazines*, July, 1982, p. 75.

"Now: Receive Home TV Directly from Satellites," *Popular Mechanics*, Sept. 1980, p. 112.

"Planning Your Videoconference," *Videography*, May, 1981, p. 39.

"Satellite Dish Live: Ready for Prime Time," *Video Review*, March 1981, p. 50.

"Satellite Earth Station Systems and Components," *Videoplay Magazine*, Buyers Guide, Fall, 1982, p. 56.

"Satellite Gear Goes Stereo," *Video Review*, July, 1983, p. 32.

"Satellite TV Antenna: Part I," *Radio-Electronics*, August 1981, p. 45.

"Satellite TV Antenna: Part II," *Radio-Electronics*, September 1981, p. 59.

"Satellite TV Antenna: Part III," *Radio-Electronics,* October 1981, p. 48.

"Something for Everyone—Programming Service Available Via Satellite," *Videoplay Magazine,* April/May 1981, p. 58.

"So You Want to Build a Home Satellite Earth Station," *Videoplay,* Aug./Sept. 1981, p. 62.

"Testing a Satelcast Tactic," *Video Systems,* January, 1983, p. 68.

"The Complete Guide to Home Satellite Dishs," *Video Review,* January, 1982, p. 58.

"The DBS Frontier," *Video Systems,* March, 1983, p. 38.

"The 'Eight-Ball' Satellite Antenna Kit," *Videoplay Magazine,* Oct. 1980, p. 52.

"The $100 TVRO Receiver," 10-Part Series, *"73" Magazine,* December, 1981—November, 1982.

"The Scramble to Scramble Satellite TV," *Video Review,* March, 1983, p. 30.

"The View from Video Net," *Video Systems,* April, 1982, p. 14.

"Tracking the Birds," *Videoplay Magazine,* June/July 1981, p. 61.

"TVRO Receiver Buyer's Guide," Videoplay, March, 1983, p. 48.

"12 GHz DBS—The Next Generation," *Videoplay Magazine,* Buyers Guide, Fall, 1982, p. 90.

"Video Primer: Satellite Earth Station," *Home Video,* March, 1982, p. 72.

BOOKS

Bakan, Joseph D., and Chandler, David L., *The Independent Producer's Handbook of Satellite Communications.* New York: Shared Communications Systems, 1980.

Birkill, Stephen J., *International Satellite Television Reception Guidebook,* Arcadia, OK: Satellite Television Technology, 1982.

Bowick, Chris, and Kearney, Tim, *Introduction to Satellite TV,* Indianapolis: Howard W. Sams & Co., Inc., 1983.

Brown, Rick, *The Satellite Earth Station Zoning Book,* Washington, DC: SPACE, 1983.

Cannon, Don L., and Luecke, Gerald, *Understanding Communications Systems.* Dallas: TI Inc., 1980.

Chander, R., and Karnik, K., *Planning for Satellite Broadcasting: The Indian Instructional Television Experiment.* New York: UNESCO, 1977.

Cook, Rick and Vaughan, Frank, *All About Home Satellite Television,* Blue Ridge Summit, PA: TAB Books, 1983.

Cooper, Bob, Jr., *Coop's Satellite Business Opportunity Journal.* Arcadia, OK: Satellite Television Technology, 1980.

Cooper, Bob L. Jr., *Coop's Satellite Operations Manual.* Arcadia, OK: Satellite Television Technology, 1980.

Cooper, Bob, Jr., *Home Satellite TV Reception Handbook.* Arcadia, OK: Satellite Television Technology, 1980.

Cavens, Lloyd, *The Whole Earth Station Catalog.* Denver: Channel Guide Publications, 1982.

Easton, Anthony T., *The Home Satellite TV Book: How to Put the World in Your Backyard.* New York: Playboy Books/Seaview Books, 1982.

Ethier, Nelson, *The Nelson Parabolic TVRO Antenna Manual.* Arcadia, OK: Satellite Television Technology, 1980.

Gibson, Steve, *The Gibson Satellite Navigator Manual.* Arcadia, OK: Satellite Television Technology, 1980.

Goddard Memorial Symposium (14th), *Satellite Communications in the Next Decade: Proceedings.* American Astronaut, 1977.

Gould, R. G. and Lum, L. F., *Communications Satellite Systems: An Overview of the Technology.* New York: IEEE, 1976.

Howard, Herbert H. and Carroll, Sidney L., *SMATV: Strategic Opportunities in Private Cable* Washington, DC: National Association of Broadcasters, 1982.

Howard, M. Taylor, *The New Howard Terminal Manual.* Arcadia, OK: Satellite Television Technology, 1981.

Jansky, Donald M., *World Atlas of Satellites,* Dedham, MA: Artech House, 1983.

Jansky, Donald M. and Jeruchim, Michel C., *Communication Satellites in the Geostationary Orbit,* Dedham, MA: Artech House, 1983.

Kinsley, Michael E., *Outer Space and Inner Sanctums: Government, Business and Satellite Communications.* New York: Wiley, 1976.

Long, Mark and Keating, Jeffrey, *The World of Satellite Television,* Summertown, TN: The Book Publishing Company, 1983.

Martin, James, *Communication Satellite Systems.* New Jersey: Prentice-Hall, 1978.

Policies for Regulation of Direct Broadcast Satellites. Washington, D.C.: FCC Staff Reports, 1980.

Queeney, K. M., *Direct Broadcast Satellites and the United Nations.* London: Sijthoff and Noordbaff, 1978.

Reed, Stephen, *Satellite Television Handbook and Buyers Guide.* Maitland, Fla.: Reed Publications, 1980.

Salvati, M. J., *TV Antennas and Signal Distribution Systems.* Indianapolis: Howard W. Sams & Co., Inc., 1979.

Satellite News Editorial Staff, *The 1983 Satellite Directory.* Washington, DC, 1983.

Satellite Systems of the U.S. Domestic Communications Carriers, Gaithersburg, MD: Future Systems, Inc., 1983.

Schnapf, Abraham, *Communication Satellites: Overview and Options for Broadcasters,* Washington, DC: National Association of Broadcasters, 1982.

Schultheiss, Chris, *Satellite Aiming Guide,* Shelby, NC: Triple D, Inc., 1983.

Signitzer, Benno, *Regulations of Direct Broadcasting from Satellites: The U.N. Involvement.* New York: Praeger, 1976.

Snow, Marcellus, *International Commercial Satellite Communications: Economic and Political Issues of the First Decade of Intelsat.* New York: Praeger, 1976.

Taylor, John P., *Direct-to-Home Satellite Broadcasting.* New York: Television/Radio Age, 1980.

Taylor, John P., *What Broadcasters Should Know About Satellites*. New York: Television/ Radio Age, 1981.

Traister, Robert J., *Build A Personal Earth Station for Worldwide Satellite TV Reception*. Blue Ridge Summit, PA: Tab Books, 1982.

Van Trees, Harry, *Communications*. New York: Wiley-Interscience, 1980.

Washburn, Clyde, *Washburn High Performance Receiver Manual*. Arcadia, OK: Satellite Television Technology, 1980.

Index

THE SATELLITE CENTER

If you enjoyed reading this book and would like to obtain more information of the exciting field of satellite television, The Satellite Center in San Francisco can supply you with the answers. Founded by the author, Anthony T. Easton, The Satellite Center provides TVRO business opportunity reports, satellite TV programming newletters, computer satellite-finding programs, printouts and site finder tools, and other satellite television publications for both the home satellite TV enthusiast and the professional or business owner.

You can use the order form on the reverse side to request our latest catalog of publications and materials, or to order your own unique satellite printout of what "birds" you can see—and the latest up-to-date listing of what TV programming is on them.

SPECIAL OFFER: For readers of this book only, The Satellite Center offers you a special 10% discount on any of our products ordered using the order form on the reverse side! To get your 10% discount, simply *circle* the word "Satellite" in our name at the bottom of the order form, and send in the order with your check or money order. Happy viewing!

Product Description	Cost (US)	Cost (NON-US)	Order Code
1. Printout of Az/El and declination angles for any site in the world! Includes all satellites that can be seen from that location. (Please specify your latitude and longitude if you know them.)	$10	$15	P1
2. Apple II Computer Program (Satellite Tracking Program II) complete with latest global data base of satellites and cities! Automatically finds Az/El angles and declination for all satellites. Apple II disk format only. (Note: with minor modification, included with documentation, program will also run on TRS-80 and IBM machines.)	$20	$30	STPII
3. Molniya Russian Satellite Apple II Computer Program. Automatically finds Az/El angles and declination for Russian domestic satellites! Complete with instructions on how to modify your TVRO receiver to pick up the Russian signals. Now you can listen in on Moscow directly! Apple II disk format only (but can be modified to run on TRS-80 and IBM machines also).	$30	$40	RSTP
4. Site Finder Tool. A professional tool to quickly locate satellites in space! Combo inclinometer, compass, level, and site tool, complete with documentation.	$40	$50	SFT
5. Special Deluxe Satellite Program Package for Apple II. Includes Satellite Tracking Program II (item 2 above), Molniya Russian Satellite Tracking Program (item 3 above), and over a dozen other satellite programs including OSCAR (amateur radio) satellite locator, and more. Two disks.	$50	$70	SPP

6. The *Whole Earth Station Catalog*. Chock full of the latest information on TVRO equipment and manufacturers. Over 80 pages thick! $6 $11   ESC

7. The *Mini-CATV System* (How to Start and Operate). Report on the SMATV industry and running an SMATV operation. Complete with budgets, charts, drawings, findings, legal rulings, contracts. $95 $105   SMATV

8. One year subscription to *Channel Guide*, the weekly industry newspaper for the TVRO business! $48 $60   CG

------------------------------------CUT HERE------------------------------------

SATELLITE CENTER ORDER FORM

Product Order Code	Quantity Desired	Price	Total
_____	_____	_____	_____
_____	_____	_____	_____
_____	_____	_____	_____
_____	_____	_____	_____
_____	_____	_____	_____

Latest Catalog of Satellite
Center Publications   FREE   FREE

SALES TAX
Calif. residents add 6% .... $_____

TOTAL .... $_____

Type of payment (check one):   SORRY, NO COD's
☐ US check enclosed*
☐ VISA/MC (add 5% service charge)   Signature _____

Account number_____ Exp. date _____

*Canadian and overseas orders: because of excessive handling fees levied by our bank, we cannot accept corporate or personal checks from non-US banks. When ordering, please send either a postal money order in US dollars (available from your local post office), or a US cashier's check issued on a New York bank. Thanks for understanding!

Mail to: **The Satellite Center**
P.O. Box 33045
San Francisco, CA, 94133

Or call our automated order line at: (415) 673-7000

22055